*This book is dedicated to two outstanding scholars
in the physical activity field,
Dr. Charles B. Corbin and Dr. Steven N. Blair.*

*Dr. Corbin was a pioneer
in the area of "concepts-based" fitness education
and continues to be a major leader in school-, college-,
and community-based physical activity promotion.*

*Dr. Blair is a world-renowned epidemiologist
who has greatly advanced research
on the health benefits of physical activity during his career.*

*I have had the privilege of working with,
and learning from, both men
and know that I wouldn't be the same
(both personally or professionally) without their mentoring.*

*I also dedicate this book to my wife and four children
who have provided tremendous love and support over the years.*

Contents

Preface

Over the past 40 years, there has been growing interest in the study of the health benefits of regular physical activity. The 1996 publication of the Surgeon General's Report (SGR) on Physical Activity and Health brought further legitimacy to the study of physical activity and increased attention on physical inactivity as a major public health problem. Although the health benefits of physical activity are now clearly established, there is still much to be learned about the nature and mechanisms underlying these benefits. The high prevalence of inactivity in the population also confirms that major efforts are needed to promote the behavior.

A major research priority cited in the SGR and other research reviews is to improve the accuracy of available physical activity assessment techniques. Physical activity assessments serve as independent variables in studies examining relationships with health outcomes and behavioral or physiological correlates of physical activity. In this case, an accurate measure is important for accurately determining the amount of activity needed to improve health and clarify links with other outcomes or behaviors. Physical activity assessments serve as dependent variables in studies designed to promote physical activity. In these studies, the assessment is necessary to determine whether a particular behavioral intervention is successful in changing activity behavior. Accurate assessments are also needed to test and refine models used to predict physical behaviors and to develop and test measures of potential mediators. Finally, continued refinements in current surveillance techniques are needed to better understand population trends in physical activity and patterns across the lifespan. To continue to answer more complex questions in the physical activity field, the measurement techniques that are available to assess physical activity must be further refined.

The importance of physical activity assessments to the exercise science and physical activity epidemiology fields is evident in the increased attention focused on it in research journals and professional meetings. A conference was held at The Cooper Institute in the fall of 1999 to discuss practical issues associated with improving physical activity assessments. The intent was to bring together experts in physical activity assessment and experts in measurement techniques so that the collective thoughts from both groups would foster advances in this area. The proceedings from this conference were published in *Research Quarterly for Exercise and Sport* (RQES, 2000) and provide a valuable reference for researchers interested in the topic. In 1998, the International Life Sciences Institute commissioned a series of studies to improve current assessments of lifestyle physical activity. These studies, along with other papers on physical activity assessment, were compiled into a comprehensive supplement published in *Medicine and Science in Sports and Exercise* (International Life Sciences Institute, 2000). The supplement includes an updated listing for the "Compendium of Physical Activities" as well as other innovative studies on physical activity assessment techniques.

The recent attention given to physical activity assessments reveals its position as a unique and important area of research. It also serves as the impetus for this book. The Cooper Institute conference and proceedings in *RQES* point out the importance of applying sound measurement principles to the design and analyses of studies on physical activity assessment. The *MSSE* supplement provides a valuable collection of data-based studies that reflect the current state of the art in physical activity assessment research. A perceived gap in the current literature is a reference book that describes *how* physical activity instruments should be utilized for conducting public health research. Therefore, the goal of this book is to provide researchers with practical information about the most effective ways to collect, score, and interpret physical activity data for research purposes. There are four major sections that progress from general principles to more specific applications and concepts.

Part I includes a general introduction about physical activity assessments and an overview of the importance of physical activity assessments for health-related research. The introduction, by Dr. Greg Welk, provides an overview of the different realms of physical activity research and how they are interrelated. Research issues and priorities are discussed with an emphasis on how improvements in activity assessments may address some of these research challenges. In chapter 1, Dr. Darren Dale, Dr. Greg Welk, and Dr. Charles Matthews provide a review of different assessment techniques and general guidelines for collecting and interpreting physical activity data. These sections provide a valuable foundation for the subsequent sections of the book.

In part II, specific measurement issues are explored. Dr. Jim Morrow provides an overview of the importance of validity, reliability, and generalizability of physical activity instruments in chapter 2. Because of the current lack of definitive criterion measures of physical activity, it has proven difficult to develop measures to accurately estimate physical activity under field conditions. Dr. Matt Mahar and Dr. David Rowe address this issue in chapter 3 by providing detailed guidelines and procedures to assess construct validity in physical activity research. In chapter 4, Dr. Jerry Thomas and Dr. Katherine Thomas present a tutorial on the use of nonparametric techniques to analyze physical activity data. This area is critical because the distribution of most physical activity data violates the normality assumptions on which most parametric techniques are based. In chapter 5, Dr. Weimo Zhu provides a tutorial on ways to examine test equivalence in physical activity assessments. Numerous instruments are currently used in research studies, but without systematic analyses it is not possible to equate the scores on different assessments. In chapter 6, Dr. Diane Catallier describes the issues regarding statistical power for various research applications.

Part III describes the application of various techniques (self-report, activity monitors, heart rate monitors, pedometers, direct observation, and doubly labeled water) for assessing physical activity. Each of the techniques has specific advantages and disadvantages that must be considered during the selection of an instrument. The recent publication of the proceedings from the measurement conference also provides further information about each of the measures. This section emphasizes how these various instruments can be most effectively used in research settings. The lead authors for chapters 7 to 12 (Matthews, Welk, Janz, Bassett, McKenzie, and Starling) are resident experts on each of the different techniques. They provide valuable insights and comparisons about the application of each of these techniques for various research purposes.

Part IV describes some additional measurement applications for physical activity research. These techniques represent some new or innovative applications to better assess specific measurement needs. In chapter 13, Dr. Margarita Treuth describes the benefits of incorporating multiple measures to better estimate activity and energy expenditure. In chapter 14, Dr. Claudio Nigg describes the use of stages of change for evaluating activity patterns in behavioral interventions. In chapter 15, Dr. Adrian Bauman, Dr. Jim Sallis, and Dr. Neville Owen describe the use of environmental and policy indicators for physical activity research.

Physical Activity Assessments for Health-Related Research represents the collective work of dedicated professionals in exercise science and physical activity epidemiology and is intended to assist other professionals in applying physical activity assessment techniques to their research. This book is also appropriate for a graduate course in physical activity epidemiology or a special topics course on physical activity assessment. We hope that all who use this text find the information useful.

References

International Life Sciences Institute. (2000). Measurement of moderate physical activity: Advances in assessment techniques. *Medicine and Science in Sports and Exercise, 32* (Supplement).

RQES. (2000). Special issue. *Research Quarterly for Exercise and Sport, 71* (Supplement).

U.S. Department of Health and Human Services. (1996). *Physical activity and health: A report of the Surgeon General.* Atlanta, GA: U.S. Department of Health and Human Services, Centers for Disease Control and Prevention.

Introduction and Overview of Physical Activity Assessments

Introduction to Physical Activity Research

Gregory J. Welk, PhD
Iowa State University

The human body is designed for movement. Bones in the body are connected through a diverse array of joints that provide stability and efficient locomotion. Movement occurs through the coordinated contraction and synchronization of muscles and the mechanical transfer of this movement to the bones through a complicated series of levers. Complex physiological processes within the body provide the energy required to do this work and the homeostatic regulation to keep the other body functions operating effectively. Collectively, these functions allow the body to perform an array of complex movements.

Movement is an essential aspect of life for most animal species, but over the course of civilization, movement has become a less critical component of everyday life for humans. Advances in modern technology have made it possible to live, work, and even play without much effort or movement. Automobiles are the most notable influence, but other labor-saving devices like escalators, garage door openers, riding lawn mowers, and golf carts also reduce the amount of activity expended in a normal day.

While these developments have contributed to the ease and perhaps the quality of life, they have been the scourge of our public health system. It has become increasingly evident that the movement made possible by the body is also essential for its health and maintenance. A number of diseases are directly associated with physical inactivity and can be viewed as "hypokinetic" conditions (Corbin et al., 2001)—conditions at least partially attributable to a lack of movement. With respect to public health, Powell and Blair (1994) estimated that sedentary living habits account for a third of the deaths from coronary heart disease, colon cancer, and diabetes. McGinnis and Foege (1993) also reported that physical inactivity is second only to smoking as a leading cause of preventable mortality. These statistics are often cited to indicate the overall impact of physical inactivity on society. Although the absolute impact on public health can be debated, there can be no denying the overwhelming evidence documenting the critical importance of regular physical activity for optimal health.

A branch of science that specifically seeks to study and improve health in the population is the discipline of epidemiology. It has been defined as "the study of the distribution and determinants of health-related states and events in the population, and the application of this study to the control of health problems" (Last, 1983). In the early 1900s, the field of epidemiology was focused on tracking and controlling the spread of various infectious diseases that plagued society. Today, the epidemiology field is confronted with the challenge of reducing the incidence of chronic diseases caused primarily by unhealthy lifestyle behaviors. Work is conducted on a variety of lifestyle behaviors, but considerable attention is now devoted to the importance of regular physical activity for optimal health. The collective efforts in this realm of epidemiology are typically referred to as *physical activity epidemiology.*

This introduction provides a broad overview of the diverse realms of physical activity research. The first section reviews the principles and concepts in physical activity epidemiology and introduces the general terminology used to study physical activity. The next section describes the various domains in physical activity research and highlights the current state of knowledge and research priorities in each area. The final section describes the importance of improving physical activity assessments to address many of these important questions. This introduction is intended to set the stage for the further exploration of these topics in the remaining chapters of the book.

Physical Activity Epidemiology

Physical activity epidemiology examines the relationships between physical activity and various health-related states from a variety of perspectives. The description provided by Caspersen (1989, p. 425) concisely summarizes the general domains and goals of physical activity epidemiology research.

> physical activity epidemiology can be defined as a two-part process. First, it studies (a) the association of physical activity, as a health-related behavior, with disease and other health outcomes; (b) the distribution and the determinants of physical activity behavior(s); and (c) the interrelationship of physical activity with other behaviors. Second, it applies that knowledge to the prevention and control of disease and the promotion of health.

Definitions in Physical Activity Epidemiology

Physical activity has been defined as "any bodily movement produced by skeletal muscles that results in caloric expenditure" (Caspersen et al., 1985). This definition has become the accepted scientific definition of *physical activity* and has been widely adopted within the field. It broadly incorporates all forms of movement as physical activity and operationalizes these movements as contributors to overall energy expenditure. The larger the muscle mass involved in the movement, the greater the resulting energy expenditure. The related term *exercise* is viewed more narrowly and is defined as "physical activity that is planned, structured, repetitive, and results in the improvement or maintenance of one or more facets of physical fitness." *Physical fitness,* in turn, is defined as "a set of outcomes or traits that relate to the ability to perform physical activity" (Caspersen et al., 1985). Thus, physical activity is viewed as a health-related behavior that can influence the development of physical fitness.

Indicators of Physical Activity Level

As a behavior, physical activity can be quantified or characterized in a variety of ways. Physical activity is often categorized by type, intensity, frequency, and duration, but it can also be dichotomized into occupational or leisure, continuous or intermittent, weight-bearing or non-weight-bearing, as well as other classifications. Different types of activity are commonly equated through metabolic equivalents (METS). One MET is considered to represent resting energy expenditure, or approximately 3.5 ml/kg/min in terms of oxygen consumption. Because progressively more vigorous forms of activity require proportional increases in oxygen consumption, activities can be quantified in terms of multiples of this resting oxygen consumption. An activity that requires four times the oxygen consumption of rest would be defined as 4 METS; another that requires 10 times the resting level would be listed as 10 METS. This system allows all activities to be compared on a standard scale. For example, walking and bicycling are very different activities in terms of the type and form of movement, but the overall aerobic challenge to the body can be compared in terms of oxygen cost or METS.

A limitation of METS in standardizing activity levels is that it doesn't take into account the adaptability of the body to physical activity. Individuals may have a similar generalized response to physical activity, but the magnitude or effect of the response depends on a person's current level of physical fitness. From a teleological perspective, the body adapts to make exercise easier on itself. Thus, a trained person can do more work or physical activity at the same level of response or exertion as an untrained or unfit person. Alternately, the same level of activity could be perceived very differently depending on a person's maximal aerobic capacity. For a highly fit individual, a moderate-intensity walk may barely qualify as "physical activity," but the same walk could be very challenging for a sedentary individual. The criterion of 6 METS is often used as a cutoff for vigorous physical activity because it corresponds to about 60% of maximal capacity for most people. However, this value is generally too low for younger (or more-fit) adults and too high for older (or less-fit) adults.

The inability to match absolute and relative indicators of physical activity has confounded comparisons of various research studies. For example, most epidemiological studies on health outcomes have focused on absolute measures of physical activity (e.g., total energy expenditure or total amount of time spent in physical activity). In contrast, most research examining the acute or chronic responses to exercise or mechanisms associated with benefits have tended to use relative measures of exercise (e.g., % $\dot{V}O_2$max). Better standardization of the categories and levels used to define activity would make it easier to reconcile results from these different bodies of literature.

An international symposium was recently held to discuss dose-response issues related to physical activity and health (Bouchard, 2001) and considerable attention was given to the standardization of terminology (Kesaniemi et al., 2001). A table from the Surgeon General's Report (SGR) *Physical Activity and Health* (U.S. Department of Health and Human Services, 1996, table 2.4, p. 33) had provided some general consensus for physical activity classifications, but additional clarification was needed to permit consistent applications. The original table showed how absolute intensity values would vary by age across the different, relative intensity classifications. The table was based on the expected declines in maximal aerobic capacity that would occur for most "healthy adults" but didn't account for the inter-individual variability in fitness across the age continuum. A revised chart presented by Howley (2001) expresses the absolute intensity levels in terms of maximal aerobic capacity rather than age. This allows the different exercise intensity classifications (e.g., very light, light, moderate, hard, very hard, and maximal) to be compared for individuals with varying levels of fitness (see table I.1). The values from the revised chart are also plotted in figure I.1 to more clearly illustrate these distinctions. For moderate physical activity, the intensity ranges from 2.6 to 3.3 METS for a person with a $\dot{V}O_2$max of 5 METS and from 5.4 to 7.5 METS for a person with a $\dot{V}O_2$max of 12 METS. The classification of exercise intensity depends on a person's individual level of fitness. Consult the article by Howley (2001) for a thorough explanation of the assumptions and algorithms used to generate these values.

Role of Physical Activity Assessments in Epidemiology Research

An underlying measurement challenge within the physical activity epidemiology field is the need for valid and reliable measures of physical activity (Caspersen, 1989). This has long been recognized as an important research priority, but it is becoming increasingly evident that the development and refinement of physical activity assessment techniques are critical for the continued advancement of the field (Lamonte & Ainsworth, 2001; Wareham & Rennie, 1998).

TABLE I.1 Classifications of Physical Activity

Intensity	% $\dot{V}O_2$R	%HRmax	RPE	METS for different levels of maximal aerobic capacity			
				12 METS	10 METS	8 METS	5 METS
Very light	<20	<50	<10	<3.2	<2.8	<2.4	<1.8
Light	20–39	50–63	10–11	3.2–5.3	2.8–4.5	2.4–3.7	1.8–2.5
Moderate	40–59	64–76	12–13	5.4–7.5	4.6–6.3	3.8–5.1	2.6–3.3
Hard	60–84	77–93	14–16	7.6–10.2	6.4–8.6	5.2–6.9	3.4–4.3
Very hard	>85	>94	17–19	>10.3	>8.7	>7	>4.4
Maximal	100	100	20	12	10	8	5

Adapted from Howley (2001).

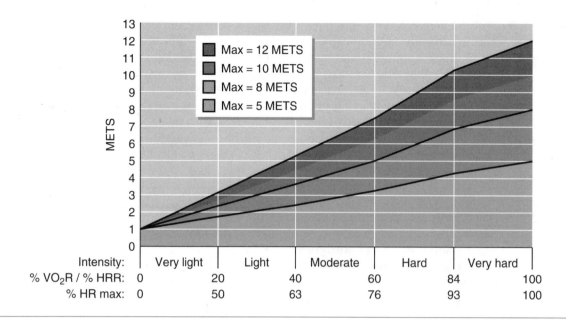

Figure I.1 Comparison of absolute exercise intensities for four different levels of cardiorespiratory fitness.

With respect to health, the effect of physical activity is strong enough that, even with fairly crude measurement techniques, researchers have been able to consistently document the strong health benefits available through physical activity. Links between fitness and health have been found to be stronger in many cases, but this is likely due to the greater accuracy with which fitness can be measured (Kesaniemi et al., 2001). As originally stated in the SGR,

> efforts to stratify studies of physical activity and coronary heart disease [CHD] by the quality of measurement have found that the methodologically better studies showed larger associations than those with lower quality scores. In addition, cardiorespiratory fitness, which is more objectively and precisely measured than the reported level of physical activity often is also more strongly related to cardiovascular disease [CVD] and mortality. Measures of association between physical activity and health outcomes thus might be stronger if physical activity measurements were more accurate (U.S. Department of Health and Human Services, 1996, pp. 144-145).

While the importance of physical activity is well established, more accurate measures are needed to better understand the specific amounts of physical activity that are needed for health benefits as well as to better understand the mechanisms through which these benefits are conferred. With respect to behavioral research, accurate assessments are needed to better test theories of physical activity or to determine if a particular behavioral intervention

was successful in changing activity behavior. If there is considerable error in the assessment, the power to detect change is reduced and large sample sizes are required to test these relationships. Clearly, advances in measurement techniques would improve our ability to more effectively study, predict, and promote exercise behavior. The subsequent section provides a more systematic review of the different areas of research in physical activity epidemiology.

Research on Physical Activity Epidemiology

Health-related research on physical activity encompasses a variety of related research applications. Although the areas are often pursued independently by different research groups, the body of knowledge in each domain has direct relevance and/or influence on the other research areas. Figure I.2 reveals some general links between these different research areas.

This section provides a brief overview of the state of research in each of the components of the model. The intent is not to provide a detailed review of each body of literature but rather to highlight the current issues and challenges in each area and to describe interdependent relationships among the areas. Recommendations from the SGR (U.S. Department of Health and Human Services, 1996) are provided to illustrate some specific research priorities

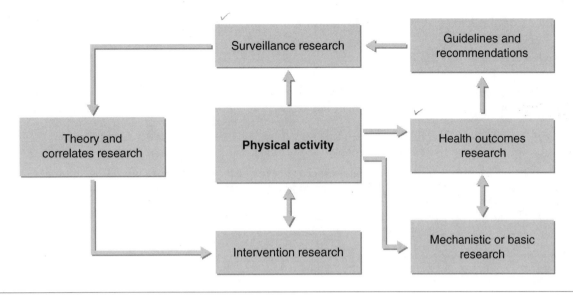

Figure I.2 Conceptual model exhibiting links between domains of physical activity research.

for the field. Emphasis is placed on the role that improved physical activity assessment techniques would play in helping to address these priorities. Readers interested in more specific details about the methods and principles of physical activity epidemiology are encouraged to consult a variety of excellent reviews available on the subject (Caspersen, 1989; Caspersen et al., 1998; Walter & Hart, 1991).

Health Outcomes Research (Physical Activity Epidemiology)

Health outcomes research focuses on understanding the relationships between physical activity and health. Research in this domain has increased dramatically in the past 30 years, and the findings from this area of research have greatly advanced our understanding of the relationships between physical activity and various health outcomes. As depicted in the model in figure I.2, these results have had a major influence on the establishment of guidelines and recommendations for physical activity in the population.

A major development in the past 10 years was the recognition that moderate levels of physical activity provide important health benefits. Previous recommendations or guidelines for physical activity were based primarily on the amount of activity needed to develop and maintain cardiorespiratory fitness. The general assumption was that improvements in fitness were needed to improve health outcomes. However, consistent observations from many epidemiological studies have confirmed that

significant improvements in health occur from even modest amounts of physical activity. This information led to the release of a new set of public health guidelines for physical activity (Pate et al., 1995). These new guidelines for physical activity were not intended to replace the previous American College of Sports Medicine (ACSM) guidelines but rather to provide an alternative way to help people benefit from an active lifestyle. The prescription of moderate physical activity is thought to be a more appealing behavioral target than the more structured, vigorous-intensity prescription in previous guidelines. This general public health guideline has been endorsed and supported in a number of subsequent positions and recommendations, including the National Institutes of Health (NIH) Consensus Development Panel on Physical Activity and Cardiovascular Health (National Institutes of Health, 1996), the SGR (U.S. Department of Health and Human Services, 1996), the revised ACSM guidelines for exercise in healthy adults (American College of Sports Medicine, 1998), the statement by the American Heart Association on the benefits of exercise (Grundy et al., 1996), and other similar international guidelines. See table I.2 for a comparison of the specific recommendations and statements.

There are several important distinctions in interpreting and applying these guidelines. One point is that the recommendations specify "moderate-intensity" activity. There is currently little or no evidence to support the premise that the accumulation of large amounts of "light" activity may provide similar benefits. The Canadian guidelines

TABLE I.2   Comparison of Various Public Health Recommendations for Physical Activity

Agency (reference)	Major statement
American Heart Association (Fletcher et al., 1996)	"Persons of all ages should include physical activity in a comprehensive program of health promotion and disease prevention and should increase their habitual physical activity to a level appropriate to their capacities, needs, and interest . . . even low- and moderate-intensity activities performed daily can have some long-term health benefits and lower the risk of cardiovascular disease." (pp. 858–859)
American College of Sports Medicine: Position Stand on The Recommended Quantity and Quality of Exercise (ACSM, 1998)	". . . there is a dose response to exercise by which the benefits derived through varying quantities of physical activity ranging from approximately 700–2000 plus kilocalories of effort per week. Many significant health benefits are achieved by going from a sedentary state to a minimal level of physical activity; programs involving higher intensities and/or greater frequency/durations provide additional benefits." (p. 976)
Centers for Disease Control and Prevention (CDC) and the American College of Sports Medicine (Pate et al., 1995)	"Every U.S. adult should accumulate 30 minutes or more of moderate-intensity activity on most, preferably all, days of the week." (p. 404)
National Institutes of Health (NIH) Consensus Development Panel on Physical Activity and Cardiovascular Health (NIH, 1996)	"All Americans should engage in regular physical activity at a level appropriate to their capacity, needs, and interests. Children and adults alike should set a goal of accumulating 30 minutes of moderate intensity physical activity on most, preferably all days of the week." (p. 241)
Surgeon General's Report on Physical Activity and Health (U.S. Department of Health and Human Services, 1996)	"People of all ages, both male and female, benefit from regular physical activity. Significant health benefits can be obtained by including a moderate amount of physical activity (e.g., 30 minutes of brisk walking or raking leaves) on most, if not all days of the week. Through a modest increase in daily activity, most Americans can improve their health and quality of life." (p. 4)

specifically recommend 60 minutes of activity because it is assumed that most people misinterpret "light" activity as "moderate" (Bouchard, 2001). Another point is that the health benefits from physical activity should be viewed on a continuum. Participation in more vigorous activity or a program with greater frequency or duration is likely to provide a number of additional benefits. A final point is that the specific benefits gained from increased physical activity depend on a person's initial fitness level. Individuals with low current levels of fitness will benefit the most from physical activity. Some evidence suggests that benefits may be obtained from both the acute aspects of a bout of activity as well as from the chronic adaptations. The modest amount of activity endorsed by the general public health guidelines may provide both an acute benefit and a training stimulus to an untrained individual but mainly an acute benefit to a more physically fit individual. This requires further study, however.

In summary, epidemiological research in this area has clearly documented the important role that physical activity plays in health. Because of its direct relationship to the cardiovascular system, the strongest documented effect of physical activity is on reducing risks for various cardiovascular diseases, including heart attacks and strokes. Links to a number of other chronic conditions are becoming increasingly evident. The SGR (U.S. Department of Health and Human Services, 1996) provides detailed reviews of the available literature on physical activity and a variety of health outcomes. See appendix A for a listing of the major conclusions from the report.

Measurement Issues in Health Outcomes Research. Emphasis in most epidemiological research is placed on the classification of individuals into distinct groups or categories. For physical activity research, this typically involves using criteria to divide a sample into different levels of activity (e.g., active and inactive populations). The types of assessments for this type of research vary depending on the specific research question but are typically from self-report instruments because large sample sizes are usually required. A number of major epidemiological studies have also successfully documented important health benefits using relatively simple classifications of physical activity based on occupational status or general ratings of physical activity. A rough estimation of the exposure variable is generally considered acceptable for epidemiological applications, since any documented link between the exposure and outcome can be considered an underestimate of the true effect that would be expected in the population. This caveat provides for conservative and cautious interpretations and allows the research in epidemiology to advance in a systematic but progressive manner.

While rough estimates have been sufficient to document the general health benefits associated with physical activity, more precise assessments are needed to address more complex questions. As previously described, better standardization in the terminology used to describe activity will allow for closer matching between epidemiology studies and basic or mechanistic research. This will allow for a better understanding of the dose-response characteristics between physical activity and different health outcomes.

Additional work is also needed to sort out the possible independent effects of physical activity and physical fitness on health. Williams (2001) recently questioned the current view of a curvilinear, inverse relationship between physical activity and health on the grounds that the relationship may be different for physical activity than for fitness. Blair (2001), however, refuted these considerations and emphasized that differences in the nature and accuracy of the measurements can account for the results. Additional comparisons and standardization will clearly help to resolve this issue. Research is also needed to address the benefits associated with the fractionalization of physical activity into shorter bouts (Hardman, 2001), as this issue has implications for physical activity guidelines and the promotion of physical activity in intervention research.

Priorities in Health Outcomes Research. The specific research priorities outlined in the SGR (U.S. Department of Health and Human Services, 1996, p. 150) for this line of research are the following:

- Delineate the most important features or combinations of features of physical activity (total amount, intensity, duration, frequency, pattern, or type) that confer specific health benefits.

- Determine specific health benefits of physical activity for women, racial and ethnic minority groups, and people with disabilities.
- Examine the protective effects of physical activity in conjunction with other lifestyle characteristics and disease-prevention behaviors.
- Examine the types of physical activity that preserve muscle strength and functional capacity in the elderly.
- Further study the relationship between physical activity in adolescence, early adulthood, and the later development of breast cancer.
- Clarify the role of physical activity in preventing or reducing bone loss after menopause.

Surveillance Research

Research on the surveillance of physical activity is aimed at identifying patterns and trends in physical activity in the population. This research provides information about important targets for behavioral interventions designed to promote activity. It also informs determinants (i.e., correlates) of research regarding possible demographic predictors of physical activity. Information about activity patterns in the population are also used to guide promotional efforts, to allocate resources for public health funding, and to develop policies that may influence health in the population (Caspersen et al., 1998). Surveillance efforts that are directly linked to public health activities are often referred to specifically as *public health surveillance* (Thacker & Berkelman, 1988; Caspersen et al., 1994).

Surveillance research is a major objective within the Centers for Disease Control and Prevention (CDC; **www.cdc.gov**). The National Center for Chronic Disease Prevention and Health Promotion (NCCDPHP) within the CDC coordinates several ongoing surveillance studies to track participation in physical activity and other health behaviors across the country. See table I.3 for a comparison of the different surveillance tools used to track activity patterns and trends in the population. Additional detail on the sampling issues and methodologies used with these instruments is available in other published sources or on the Web sites devoted to each of the instruments.

The SGR (U.S. Department of Health and Human Services, 1996) provides summaries of the results from a number of surveillance instruments. Some differences exist among the instruments, but the general patterns tend to be fairly consistent. For example, the prevalence of inactivity ranged from 21.7 to 28.7 for the National Health Interview

TABLE I.3   Public Health Surveillance Instruments Used to Examine Population Trends in Physical Activity

Name of survey instrument	Description and data collection methods
Behavioral Risk Factor Surveillance System (BRFSS): **www.cdc.gov/nccdphp/brfss**	An annual state-by-state survey of health-risk behaviors of adults aged 18 years and older.
National Health and Nutrition Examination Survey (NHANES): **www.cdc.gov/nchs/nhanes.htm**	A periodic survey designed to collect information about the health and nutrition habits of Americans. The current assessment, which combines a home interview with lab-based measures, is the eighth version since 1960.
Youth Risk Behavioral Survey (YRBS): **www.cdc.gov/nccdphp/dash/yrbs/index.htm**	A national, school-based survey of high school students in grades 9 through 12 that covers a variety of health behaviors.

Survey (NHIS), the Behavioral Risk Factor Surveillance System (BRFSS), and the National Health and Nutrition Examination Survey (NHANES) III sample. The report concluded that approximately 25% of the population is completely inactive. For regular, vigorous leisure activity, the prevalence ranged from 14.2% in the BRFSS to 16.4% for the NHIS.

Another goal of surveillance instruments is to document long-term trends in behavior over time. Because this can only be examined for questions that have been assessed similarly over time, considerable care is given to ensure that the wording can allow for some tracking over time. Macera and Pratt (2000) provide an illustration of how the BRFSS was modified to still allow for long-term tracking of

certain physical activity questions. Figure I.3 presents the trends in levels of inactivity over the past 10 years as assessed through the BRFSS.

This graph reveals that levels of inactivity across the United States have remained fairly stable on a population basis. Thus, despite considerable public health efforts, a large percentage of the population remains physically inactive. Differences in activity levels have been identified between genders and among different racial and socioeconomic groups, and these differences have important public health significance. Because the BRFSS was designed to provide state-by-state comparisons, trend and prevalence data can be processed separately for each state (see the interactive BRFSS Web site for examples: **http://apps.nccd.cdc.gov/brfss/index.asp**).

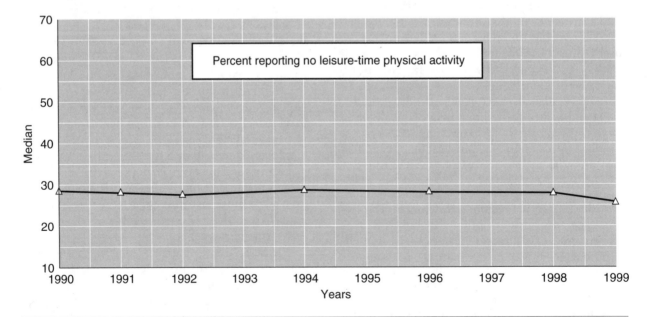

Figure I.3   Trends in levels of physical inactivity in the past 10 years. Data collected from the Behavioral Risk Factor Surveillance System (BRFSS).

These readily accessible data provide useful information to state and national health officials about patterns of activity (and other health behaviors) in the population.

The public health service recently released Healthy People 2010 (HP2010) to establish national public health goals for the year 2010 (**www.health. gov/healthypeople**). The focus area devoted to physical activity and fitness contains 15 goals and specifies variables from different surveillance instruments to track progress toward these goals. The two goals listed as "leading indicators" in the HP2010 framework are to (1) increase the proportion of adults who engage regularly, preferably daily, in moderate physical activity for at least 30 minutes per day (22.2) and (2) increase the proportion of adolescents who engage in vigorous physical activity that promotes cardiorespiratory fitness three or more days per week for 20 or more minutes per occasion (22.7). A complete listing of the HP2010 goals and current baseline values is included in appendix B.

Measurement Issues for Surveillance Research. Surveillance research requires the use of physical activity measures that can be administered to a large population in an efficient manner. Therefore, surveillance research relies almost exclusively on self-report or interviewer-administered data on physical activity. An important issue with surveillance research is the way that physical activity is operationalized and assessed. Small differences in the wording of questions can lead to large differences in prevalence estimates of physical activity. Therefore, improving the comparability and accuracy of physical activity instruments is an important priority for surveillance research. A better understanding of the equivalence (or lack thereof) of different instruments would make it possible to reconcile differences in results across studies and to establish more accurate profiles of activity levels in the population.

Sarkin et al. (2000) recently evaluated the agreement between three different self-report measures in 575 college students. They compared the seven-day physical activity recall, YRBS, and NHIS on the prevalence of meeting the traditional ACSM fitness guideline and the CDC/ACSM health-related guideline (Pate et al., 1995). The proportion meeting the fitness guideline ranged from 32 to 59%, while the proportion meeting the health-related guideline ranged from 4 to 70%. This study highlights the challenges associated with comparing prevalence rates across studies using different measures. Additional work is needed to develop procedures to allow for the triangulation of results from multiple studies. As pointed out in a recent review of self-report instruments (Sallis & Saelens, 2000), studies are also needed to begin evaluating the ability of instruments to assess absolute amounts of physical activity rather than just continuing to compare the relative validity of different instruments.

Efforts are needed to expand surveillance efforts to reach other understudied populations. Considerable work has been done to develop and evaluate the utility of a new survey tool, the International Physical Activity Questionnaire (IPAQ), that would be used to assess physical activity patterns in different countries. Extensive pilot testing was done with various survey formats and lengths to examine the reliability and validity of the instrument. For this instrument, it was important to take into account distinct sociocultural differences in the way that physical activity is perceived as well as inherent differences in language. Data has been obtained from a number of countries, but the results have not been compiled or released at this time. Continued research with this instrument will allow for broader study of international patterns of physical activity. Specific efforts are also needed to track physical activity and fitness trends among youth. While a number of data sets are available to examine patterns of activity in adults, less is known about physical activity in youth. The YRBS tracks youth behaviors but is aimed at an adolescent population. Little is known about the levels of physical activity among young (i.e., elementary school) children.

Priorities for Surveillance Research. The following are research priorities cited in the SGR (U.S. Department of Health and Human Services, 1996, p. 201) in physical activity surveillance:

- Develop methods to monitor patterns of regular, moderate physical activity.
- Improve the validity and comparability of self-reported physical activity in national surveys.
- Improve methods for identifying and tracking physical activity patterns among people with disabilities.
- Routinely monitor the prevalence of physical activity among children under the age of 12.
- Routinely monitor school policy requirements and students' participation in physical education classes in elementary, middle, and high schools.

Correlates/Determinants of Physical Activity

A wide array of research has examined the many factors that correlate with and/or influence physical activity. In relationship to physical activity epidemiology, the primary goal of correlates/determinants research is to describe and understand the factors that influence physical activity behavior across the life span. General information about patterns of activity in the population help to clarify the demographic correlates of physical activity, but there is considerable inter-individual variability with respect to physical activity. A variety of variables has been proposed to account for these patterns. To simplify or summarize these influences, the correlates are often grouped under common categories, or domains, such as personal, interpersonal, social, or environmental influences.

Studies on the correlates/determinants of physical activity are typically studied within existing theories and models of behavior. The predictive utility of the different models is evaluated by determining how much variance in physical activity can be explained with a particular set of correlates. Variables accounting for a significant percentage of the variance are deemed to represent important correlates of physical activity. In some cases, variables are listed as correlates even if they account for a small percentage of actual variance. Also, because of considerable covariation among the measures, important predictors can be easily overshadowed by other variables that entered the model first. These issues have contributed to the veritable quagmire of constructs that currently exists in the determinants literature. A number of excellent reviews have attempted to sort through the different factors influencing physical activity behavior (Dishman & Sallis, 1994; Dishman et al., 1985; King et al., 1992; Sallis & Hovell, 1990; Sallis et al., 2000). Consult these and other references for information on physical activity determinants.

To provide value for intervention research, one must delineate which measures are considered mediators of physical activity. By definition, a *mediator* is a variable (or group of variables) that explains the mechanism through which the intervention influences the outcome (Baron & Kenny, 1986). In the case of physical activity interventions, a mediator is a variable in the causal path that is specifically targeted to help promote changes in physical activity. Statistically, a mediating variable would account for the effect of an intervention if a positive relationship between the intervention and

outcome becomes nonsignificant after controlling for the effect of the mediator(s). Baranowski clearly described the importance of studying mediating variables in a commentary titled "Theory as mediating variables: Why aren't community interventions working as desired?" (Baranowski et al., 1997). He argues that efforts are needed to (1) better describe the theoretical relationship between mediating variables and behavior, and (2) document the ability of interventions to change the mediating variables. A more systematic approach to the study of mediating variables is likely to improve the effectiveness of behavioral interventions because the primary influences can be more specifically targeted. For more detail, consult a comprehensive review of mediating variables used in physical activity interventions (Baranowski et al., 1998).

Measurement Issues in Correlates/Determinants Research. Research on the determinants and correlates of physical activity has increased tremendously in recent years. While considerable progress has been made in this type of behavioral research, it still has proven difficult to accurately predict physical activity behaviors. The variance in physical activity behavior accounted for by models of physical activity have typically been less than 35% (Dishman, 1994). The limitations in the prediction of physical activity stem, in part, from a lack of specificity in the current behavioral models and/or limitations in the operationalization and assessment of the relevant constructs. Psychometric work is clearly needed to improve the specificity with which various psychological, social, or environmental constructs are assessed. Additional work is also needed to standardize the way physical activity is assessed. If the assessment of physical activity is weak, then it becomes more difficult to predict physical activity to any appreciable extent.

A particular challenge in this regard is the incongruity of different physical activity assessment techniques. Several studies have reported differences in the reported correlates of physical activity depending on the type of measure that was used. For example, Dishman et al. (1992) reported different results when correlates were evaluated with two different measures of physical activity. A composite measure of correlates predicted physical activity when it was assessed with a 7-day diary method but not when it was estimated by an objective activity monitor. The lack of convergence between the two measures is problematic because it limits the ability to fully evaluate the utility of the correlate measures. The authors recommended that further studies of determinants employ concurrent estimates of activity from different methods. They also suggested that measures of determinants must be viewed as "premature until verified by a convergence of data from self-report and concurrent objective monitoring" (p. 909). Epstein et al. (1996) reported similar challenges in evaluating the determinants for obese youth when assessed by accelerometer and self-report measures. In this study, variables were added in a specific order to determine the additional amount of variance accounted for by the other measures. At the first step, child and parent psychological variables explained 8.3 and 3.4% of the variance in the accelerometer and self-reported activity, respectively. Socioeconomic status and parent activity level added to the prediction of physical activity (i.e., 14.8% increment) for the accelerometer prediction, but fitness was the only additional predictor (i.e., 23.5% increment) for the self-report data. These discrepancies exist because there is a substantial amount of variance not shared among the different measurement approaches. Some of the incongruity is due to the way the determinants measures are assessed, but much is likely due to the different physical activity measures used. Because no definitive gold standard is available, the differences in these results cannot be fully reconciled.

Advances in physical activity assessment techniques will make it easier to study the various factors that influence physical activity behavior. Specific studies are also needed to directly compare the predictive utility of different constructs and theoretical approaches used in behavioral research.

Priorities in Correlates/Determinants Research.
The following are specific priorities highlighted in the SGR (U.S. Department of Health and Human Services, 1996, p. 249) for determinants research:

- Assess the determinants of various patterns of physical activity among those who are sedentary, intermittently active, routinely active at work, and regularly active.
- Assess determinants of physical activity for various population subgroups (e.g., by age, sex, race/ethnicity, socioeconomic status, health/disability status, geographic location).
- Examine patterns and determinants of physical activity at various developmental and life transitions, such as from school to work, from one job or city to another, from work to retirement, and from health to chronic disease.

Physical Activity Intervention Research

Physical activity intervention research involves the development and evaluation of programs that can promote physical activity within different segments of the population. Primary targets for interventions are often those populations that have been identified in surveillance research to be at higher risk for inactivity. For example, the especially low levels of physical activity observed for minority and lower socioeconomic populations have led to greater emphasis on these populations in grant funding and behavioral research. Similarly, there has been considerable interest on work with children and adolescents since a number of studies documenting sharp declines in activity levels during the adolescent years (Pate et al., 1994).

As illustrated in the model in figure I.2 (see page 7), the design and development of interventions are informed by the determinants literature and theories from behavioral research. The inherent challenge in intervention research is to conceptualize and prioritize the various factors that have been shown to influence physical activity behavior and apply them in an integrated and organized intervention. According to a quantitative review on physical activity interventions by Dishman and Buckworth (1996), studies that have employed established principles of behavior modification have had larger effect sizes. The evolving links between the exercise science and behavioral science communities have led to a greater acceptance of physical activity as a complex behavior worthy of its own study. The effectiveness of behavioral interventions will continue to improve as information accumulates on the mediators and moderators of behavior change.

The nature and scope of physical activity interventions have changed dramatically in recent years as public health guidelines and behavioral theories have evolved. Because more vigorous forms of activity were viewed as essential for good health, early intervention studies emphasized structured exercise and were conducted more as training studies. The current consensus supporting a curvilinear relationship between physical activity and health outcomes implies that physical activity will have a greater impact on health at the lower end of the physical activity continuum. Therefore, because sedentary individuals have the greatest potential for improvement in health, public health efforts now tend to focus on promoting moderate-intensity activity and on reducing inactivity. The recognition that sedentary individuals may be more responsive to moderate forms of activity has led to the development of "lifestyle physical activity interventions" (Dunn et al., 1998). The distinguishing characteristics of lifestyle physical activity interventions is that they are aimed at helping people to develop an individualized program that allows them to accumulate the recommended amount of moderate-intensity activity throughout the day (Dunn et al., 1998).

Another characteristic common in many contemporary interventions is the use of social-ecological models of health promotion (Sallis & Owen, 1997). These models acknowledge that people's behavior is strongly influenced by the physical, social, and cultural environment in which they live. The domains of influence are categorized by levels into individual, intrapersonal, institutional/organizational, community, and the societal/policy categories (Sallis & Owen, 1997). Recommendations generally call for efforts to address multiple levels of influence in the hopes of increasing the likelihood of behavior change (Smedley & Syme, 2001; The Partnership to Promote Healthy Eating and Active Living, 2001). Physical activity interventions that adopt a social-ecological perspective offer considerable advantages over narrower interventions aimed at changing individual attitudes or behavior because they can target multiple influences on a person's life (Booth et al., 2001).

A review by Marcus and Forsyth (1999) provides an illustration of how physical activity interventions may be influenced by this impending paradigm shift. In their article, the different levels of behavioral interventions were categorized based on their predominant focus and their potential reach. The three categories were downstream, midstream, and upstream. Downstream approaches represent the established methods of intervention that include cognitive behavioral interventions, clinical exercise interventions, and exercise training studies. As a group, these approaches have demonstrated some level of effectiveness but limited potential to reach large segments of the population. Midstream approaches include training primary care physicians to counsel patients, developing innovative work-site programs, and conducting media campaigns and community events that promote activity within whole communities. These methods offer the potential to reach a larger segment of the population but are currently underutilized. Upstream approaches emphasize broader environmental or policy approaches at promoting physical activity. Possible examples include modifications in insurance policies, tax incentives,

or the development of more accessible and integrated systems of bike paths and walking trails as well as more prominent stairways in buildings. These approaches may offer the most promise for population change, but little work has been done in this area.

The recognition of the importance of environmental and policy influences on behavior has also contributed to some alternative perspectives regarding the evaluation and interpretation of intervention research. The typical progression of intervention research begins with tightly controlled efficacy trials that are conducted under optimal conditions and with strict programmatic controls. If an intervention is found to be efficacious, then trials are conducted to see if the program can be conducted in other settings or under different conditions. If the program is effective under less-controlled conditions, then it may be appropriate for dissemination to the general public on a broader scale. A limitation of this progressive approach through efficacy, effectiveness, and dissemination is that it doesn't consider the overall impact an intervention can have from a public health perspective. A study can have a major impact on public health if it can induce even a small effect on a large sample of the population. Glasgow et al. (2001) recently proposed an alternative system, the RE-AIM Framework, for evaluating the effects of interventions. In this framework, greater consideration is given to the overall impact of an intervention on the population rather than solely on the magnitude of the treatment effects. They reason that small effects may be important if they can be delivered to large segments of society. This may be a particularly important distinction for physical activity research (particularly with respect to policy and environmental research), since it is now established that only moderate amounts of physical activity are needed to improve current health status.

Measurement Issues for Intervention Research. The objective of intervention research is to test approaches that may prove useful in promoting physical activity in a specific target population. Thus, a key consideration in intervention research is to ensure that the outcome measure can be assessed accurately and is sensitive to change. Due to the inherent challenges associated with assessing physical activity, researchers have often opted to use related outcome measures that can be assessed more objectively and with less error (e.g., assessments of physical fitness or changes in body composition).

Although this approach provides a "harder end-point," it is limited in that the outcome may not correspond to the activities promoted in the intervention. Many alternative outcomes such as fitness and body composition are also not highly correlated with corresponding increases in physical activity.

Advances in assessment techniques and research design have allowed more interventions to use physical activity as an outcome measure. More consideration is now given to assess physical activity over multiple days to reduce measurement error. Greater emphasis is also placed on using larger sample sizes that provide sufficient amount of power to detect a meaningful change in activity. In TAAG (Trial for Activity in Adolescent Girls), a major multisite clinical trial funded by NIH, the decision was made to use objective assessments of physical activity provided by an accelerometry-based activity monitor as the primary outcome measure. The use of an objective measure of physical activity (as opposed to a self-report instrument) is expected to reduce measurement error and increase the power to detect change. Logistical challenges remain for using activity monitors in this capacity, but they offer strong promise for measuring free-living physical activity in intervention research.

Priorities in Intervention Research. A consensus conference was held at the Cooper Institute in 1997 to discuss the current state of research on physical activity interventions. Sessions highlighted the application of interventions in different settings and with different populations and provided recommendations for future research. For more detail about the status of physical activity interventions for different target populations and groups, review the supplement in the *American Journal of Preventive Medicine* (vol. 15, no. 4) that summarized the results of this conference. The stated priorities in the SGR (U.S. Department of Health and Human Services, 1996, p. 249) on physical activity interventions also provide useful guidelines.

- Develop and evaluate the effectiveness of interventions that include policy and environmental supports.
- Develop and evaluate interventions designed to promote adoption and maintenance of moderate physical activity that addresses the specific needs and circumstances of population subgroups, such as racial/ethnic groups, men and women, girls and boys, the elderly, the disabled, the overweight, low-income groups; and persons at life transitions, such

as in adolescence, early adulthood, family formation, and retirement.
- Develop and evaluate the effectiveness of interventions to promote physical activity in combination with healthy dietary practices that can be broadly disseminated to reach large segments of the population and can be sustained over time.

Mechanistic or Basic Research on Physical Activity

Mechanistic research on physical activity refers to any disciplinary area of research that seeks to understand and describe the way the body responds to, adapts to, or is influenced by physical activity. The breadth of this research cannot be sufficiently characterized in any systematic way. Suffice it to say that for every unanswered question, there are likely researchers interested in figuring out the answer.

From a public health perspective, this area of research plays an important role in helping to understand the mechanisms through which physical activity influences health. Findings from basic research are also used as the basis for hypotheses in epidemiology studies and clinical trials. The reciprocal relationships between basic science and health outcomes research are indicated in figure I.2 with bidirectional arrows (see page 7).

An example of these reciprocal links is the integration of research on the health risks associated with obesity. Several epidemiological studies revealed that high levels of body fatness increased a person's risk for coronary heart disease, stroke, diabetes, and other conditions. Basic research examining possible mechanisms suggested that many of the risks may be part of the same underlying metabolic condition, now known as insulin-resistance syndrome or Syndrome X. This syndrome is characterized by the clustering of insulin resistance and hyperinsulinemia, dyslipidemia, essential hypertension, glucose intolerance, and an increased risk of non-insulin-dependent diabetes mellitus and cardiovascular disease (Landsberg, 1996). Physical activity has been shown to produce beneficial metabolic changes that limit the progression of this underlying disease process (Kelley & Goodpaster, 1999; Ross & Janssen, 1999), but some of this beneficial effect was also found to be independent of changes in body composition or weight loss. Subsequent epidemiology studies have used body fatness or body composition as an additional exposure variable to examine the combined effect of activity and body composition on morbidity and mortality outcomes (Lee et al., 1999; Wei et al., 1999). Consistent with the other findings, the beneficial effects of physical activity were found to persist across the range of body composition values. In fact, relative risks of cardiovascular disease were found to be greater for thin individuals who are inactive than for overweight or obese individuals who are fit. These recent results suggest that physical activity provides protection against the general health risks of obesity (Welk & Blair 2000) and indicate that physical activity may have an even greater impact on health than was previously realized.

The progression of knowledge in health outcomes research is dependent on continued advances in basic science research. For further detail on links between physical activity and obesity, see the supplement of *Medicine and Science in Sports and Exercise* (vol. 33, no. 6) that summarizes results from an international symposium. A similar compilation of results on the dose-response characteristics of physical activity on other conditions is also provided in the supplement, which was developed from the recent Hockley Valley symposium (Kesaniemi et al., 2001).

Priorities in Basic or Mechanistic Research. Research priorities in basic research are driven primarily by lingering questions that demand progressively more advanced research designs. More detailed assessments and better standardization of terminology are likely to help advance research on a variety of basic science questions related to physical activity. The specific research recommendations in the SGR (U.S. Department of Health and Human Services, 1996, p. 77) that relate to basic research include the following:

- Explore individual variations in response to exercise.
- Better characterize mechanisms through which the musculoskeletal system responds differentially to endurance and resistance exercise.
- Better characterize the mechanisms through which physical activity reduces the risk of cardiovascular disease, hypertension, and non-insulin-dependent diabetes mellitus.
- Determine the minimal and optimal amount of exercise for disease prevention.

Summary

Health care expenditures currently account for over 20% of the gross national product, but recent estimates suggest that only 5% of the annual health care budget is allocated to prevention efforts (Smedley & Syme, 2001). Public health efforts are needed for improving lifestyle behaviors, and efforts to promote physical activity may provide the largest overall benefit for public health. Physical activity has positive effects on a variety of different conditions, each of which contributes independently to the public health burden of chronic disease. Also, because the percentage of sedentary individuals is greater than the percentage of individuals with other risk factors, the attributable risk for inactivity may be greater than for other established health risks.

Physical activity promotion may be especially critical to combat the well-chronicled obesity epidemic (Mokdad et al., 1999; Troiano & Flegal, 1998; Troiano et al., 1995). Although obesity is clearly a multifactorial problem, several ecological reviews have suggested that the recent trends in obesity are more strongly related to decreases in energy expenditure than to increases in energy intake (Jebb & Moore, 1999). Efforts to combat the increasing prevalence of obesity must therefore involve widespread efforts to promote physical activity across the whole population.

Continued research is needed to further the current knowledge base on the links between physical activity and health. Research is also needed to increase the understanding of the factors that influence physical activity in the population. (See appendix C for the complete listing of the physical activity recommendations in the Surgeon General's Report.) A major goal of these sections was to describe the collective efforts in each of the described areas of physical activity research (health outcomes, surveillance, correlates/determinants, intervention, and basic or mechanistic). Efforts were made to highlight the ways in which the different areas of research were related but interdependent. A key theme described throughout these sections was the important role that physical activity assessments have in advancing health-related research on physical activity. The remaining chapters in this book provide more detailed information on the appropriate use of various physical activity assessment techniques. Measurement issues associated with physical activity assessment and the implications for the design and interpretation of research studies are also emphasized throughout the book. Continued efforts to improve physical activity assessment techniques will contribute to systematic advances in health-related research on physical activity.

References

American College of Sports Medicine. (1998). Position stand: The recommended quantity and quality of exercise for developing and maintaining cardiorespiratory and muscular fitness, and flexibility in healthy adults. *Medicine and Science in Sports and Exercise, 30,* 975-991.

Baranowski, T., Anderson, C., & Carmack, C. (1998). Mediating variable framework in physical activity interventions. *American Journal of Preventive Medicine, 15,* 266-297.

Baranowski, T., Lin, L.S., Wetter, D.W., Resnicow, K., & Hearn, M.D. (1997). Theory as mediating variables: Why aren't community interventions working as desired? *Annals of Epidemiology, 7,* S89-S95.

Baron, R.M., & Kenny, D.A. (1986). The moderator-mediator variable distinction in social psychological research: Conceptual, strategic and statistical considerations. *Journal of Personality and Social Psychology, 51,* 1173-1182.

Blair, S.N. (2001). Guest editorial to accompany physical fitness and activity as separate heart disease risk factors: A meta analysis. *Medicine and Science in Sports and Exercise, 33,* 762-764.

Booth, S.L., Mayer, J., Sallis, J.F., Ritenbaugh, C., Hill, J.O., Birch, L.L., Frank, L.D., Glanz, K., Himmelgreen, D.A., Mudd, M., Popkin, B.M., Rickard, K.A., Jeor, S.T., & Hays, N.P. (2001). Environmental and societal factors affect food choice and physical activity: Rationale, influences and leverage points. *Nutrition Reviews, 59,* 21-39.

Bouchard, C. (2001). Physical activity and health: Introduction to the dose-response symposium. *Medicine and Science in Sports and Exercise, 33,* S347-S350.

Caspersen, C.J. (1989). Physical activity epidemiology: Concepts, methods, and applications to exercise science. In J.O. Holloszy (Ed.), *Exercise and sport sciences reviews* (17th ed., pp. 423-473). Indianapolis: American College of Sports Medicine.

Caspersen, C.J., Merritt, R.K., & Stephens, T. (1994). International physical activity patterns: A methodological perspective. In *Advances in exercise adherence* (pp. 73-110). Champaign, IL: Human Kinetics.

Caspersen, C.J., Nixon, P.A., & Durant, R.H. (1998). Physical activity epidemiology applied to children and adolescents. In J.O. Holloszy (Ed.), *Exercise and sport sciences reviews* (vol. 26, pp. 341-403). St. Louis: Williams and Wilkins.

Caspersen, C.J., Powell, K.E., & Christenson, G.M. (1985). Physical activity, exercise, and physical fitness: Definitions and distinctions for health-related research. *Public Health Reports, 100,* 126-131.

Corbin, C.B., Lindsey, R., & Welk, G.J. (2001). *Concepts of fitness and wellness: A comprehensive lifestyle approach* (3rd ed.). St. Louis: McGraw Hill Higher Education.

Dishman, R.K. (1994). Introduction: Consensus, problems, and prospects. In R.K. Dishman (Ed.), *Advances in exercise adherence* (pp. 1-27). Champaign, IL: Human Kinetics.

Dishman, R.K., & Buckworth, J. (1996). Increasing physical activity: A quantitative synthesis. *Medicine and Science in Sports and Exercise, 28,* 706-719.

Dishman, R.K., Darracott, C.R., & Lambert, L.T. (1992). Failure to generalize determinants of self-reported physical activity to a motion sensor. *Medicine and Science in Sports and Exercise, 24,* 904-910.

Dishman, R.K., & Sallis, J.F. (1994). Determinants and interventions for physical activity and exercise. In C. Bouchard,

R.J. Shephard, & T. Stephens (Eds.), *Physical activity, fitness, and health: International proceedings and consensus statement* (pp. 214-256). Champaign, IL: Human Kinetics.

Dishman, R.K., Sallis, J.F., & Orenstein, D.R. (1985). The determinants of physical activity and exercise. *Public Health Reports, 100,* 158-171.

Dunn, A.L., Andersen, R.E., & Jakicic, J.M. (1998). Lifestyle physical activity interventions: History, short- and long-term effects and recommendations. *American Journal of Preventive Medicine, 15,* 398-412.

Epstein, L.H., Paluch, R.A., Coleman, K.J., Vito, D., & Anderson, K. (1996). Determinants of physical activity in obese children assessed by accelerometer and self-report. *Medicine and Science in Sports and Exercise, 28,* 1157-1164.

Fletcher, G.F., Balady, G., Blair, S.N., Blumenthal, J., Caspersen, C., Chaitman, B., Epstein, S., Sivarajan Froelicher, E.S., Froelicher, V.F., Pina, I.L., & Pollock, M.L. (1996). Statement on exercise: Benefits and recommendations for physical activity programs for all Americans. A statement for health professionals by the Committee on Exercise and Cardiac Rehabilitation of the Council on Clinical Cardiology, American Heart Assocation. *Circulation, 94,* 857-862.

Glasgow, R.E., Vogt, T.M., & Boles, S.M. (2001). Evaluating the public health impact of health promotion interventions: The RE-AIM framework. *American Journal of Public Health, 89,* 1327.

Grundy, S.M., Balady, J.G., Criqui, M.H., Fletcher, G., Greenland, P., Hiratzka, L.F., Houston-Miller, N., Kris-Etherton, P., Krumholz, H.M., LaRosa, J., Ockene, I.S., Pearson, T.A., Reed, J., Washington, R., & Smith, S., Jr. (1996). Guide to primary prevention of cardiovascular diseases. A statement for health care professionals from the task force on risk reduction. *Circulation, 95,* 2329-2331.

Hardman, A.E. (2001). Issues of fractionalization of exercise (short vs. long bouts). *Medicine and Science in Sports and Exercise, 33,* S421-S427.

Howley, E.T. (2001). Type of activity: Resistance, aerobic and leisure versus occupational physical activity. *Medicine and Science in Sports and Exercise, 33,* S364-S369.

Jebb, S.A., & Moore, M.S. (1999). Contribution of a sedentary lifestyle and inactivity to the etiology of overweight and obesity: Current evidence and research issues. *Medicine and Science in Sports and Exercise, 31,* S534-S541.

Kelley, D.E., & Goodpaster, B.H. (1999). Effects of physical activity on insulin action and glucose tolerance in obesity. *Medicine and Science in Sports and Exercise, 31,* S619-S624.

Kesaniemi, Y.A., Danforth, E., Jr., Jensen, M.D., Kopelman, P.G., Lefebvre, P., & Reeder, B.A. (2001). Dose-response issues concerning physical activity and health: An evidence-based symposium. *Medicine and Science in Sports and Exercise, 33,* S351-S358.

King, A.C., Blair, S.N., Bild, D.E., Dishman, R.K., Dubbert, P.M., Marcus, B.H., Oldridge, N.B., Paffenbarger, R.S., Powell, K.E., & Yeager, K.K. (1992). Determinants of physical activity and interventions in adults. *Medicine and Science in Sports and Exercise, 24,* 221-236.

Lamonte, M.J., & Ainsworth, B.E. (2001). Quantifying energy expenditure and physical activity in the context of dose response. *Medicine and Science in Sports and Exercise, 33,* S370-S378.

Landsberg, L. (1996). Obesity and the insulin resistance syndrome. *Hypertension Research, 1,* S51-S55.

Last, J.M. (1983). *A dictionary of epidemiology*. New York: Oxford University Press.

Lee, C.D., Blair, S.N., & Jackson, A.S. (1999). Cardiorespiratory fitness, body composition, and all-cause and cardiovascular disease mortality in men. *American Journal of Clinical Nutrition, 69*, 373-380.

Macera, C.A., & Pratt, M. (2000). Public health surveillance of physical activity. *Medicine and Science in Sports and Exercise, 71*, S97-S103.

Marcus, B.H., & Forsyth, L.H. (1999). How are we doing with physical activity? *American Journal of Health Promotion, 14*, 118-124.

McGinnis, J.M., & Foege, W.H. (1993). Actual causes of death in the United States. *JAMA, 270*, 2207-2212.

Mokdad, A.H., Serdula, M.K., Dietz, W.H., Bowman, B.A., Marks, J.S., & Koplan, J.P. (1999). The spread of the obesity epidemic in the United States, 1991-1998. *JAMA, 282*, 1519-1522.

National Institutes of Health. (1996). Physical activity and cardiovascular health. *JAMA, 276*, 241-246.

Pate, R.R., Long, B.J., & Heath, G. (1994). Descriptive epidemiology of physical activity in adolescents. *Pediatric Exercise Science, 6*, 434-447.

Pate, R.R., Pratt, M., Blair, S.N., Haskell, W.L., Macera, C.A., Bouchard, C., Buchner, D., Ettinger, W., Heath, G.W., King, A.C., Kriska, A.L.A.M.B.H., Morris, J.P.R.S., Patrick, K., Pollock, M.L., Rippe, J.M., Sallis, J., & Wilmore, J.H. (1995). Physical activity and public health: A recommendation from the Centers for Disease Control and Prevention and the American College of Sports Medicine. *JAMA, 273*, 402-407.

Powell, K.E., & Blair, S.N. (1994). The public health burdens of sedentary living habits: Theoretical but realistic estimates. *Medicine and Science in Sports and Exercise, 26*, 851-856.

Ross, R., & Janssen, I. (1999). Is abdominal fat preferentially reduced in response to exercise-induced weight loss? *Medicine and Science in Sports and Exercise, 31*, S568-S572.

Sallis, J.F., & Hovell, M.F. (1990). Determinants of exercise behavior. *Exercise and Sport Sciences Reviews, 18*, 307-330.

Sallis, J.F., & Owen, N. (1997). Ecological models. In K. Glanz, F.M. Lewis, & B.K. Rimer (Eds.), *Health behavior and health education* (2nd ed., pp. 403-424). San Francisco: Jossey-Bass.

Sallis, J.F., Prochaska, J.J., & Taylor, W.C. (2000). A review of correlates of physical activity of children and adolescents. *Medicine and Science in Sports and Exercise, 32*, 963-975.

Sallis, J.F., & Saelens, B.E. (2000). Assessment of physical activity by self-report: Status, limitations and future directions. *Research Quarterly for Exercise and Sport, 71*, S1-S14.

Sarkin, J.A., Nichols, J.F., Sallis, J.F., & Calfas, K.J. (2000). Self-report measures and scoring protocols affect prevalence estimates of meeting physical activity guidelines. *Medicine and Science in Sports and Exercise, 32*, 149-156.

Smedley, B.D., & Syme, S.L. (2001). Promoting health: Intervention strategies from social and behavioral research. *American Journal of Health Promotion, 15*, 149-166.

Thacker, S.B., & Berkelman, R.L. (1988). Public health surveillance in the United States. *Epidemiologic Reviews, 10*, 164-190.

The Partnership to Promote Healthy Eating and Active Living. (2001). Summit on promoting healthy eating and active living: Developing a framework for progress. *Nutrition Reviews, 59*, S1-S74.

Troiano, R.P., & Flegal, K.M. (1998). Overweight children and adolescents: Description, epidemiology, and demographics. *Pediatrics, 101*, 497-504.

Troiano, R.P., Flegal, K.M., Kuczmarski, R.J., Campell, S.M., & Johnson, C.L. (1995). Overweight prevalence and trends for children and adolescents. *Archives of Pediatric and Adolescent Medicine, 149*, 1085-1091.

U.S. Department of Health and Human Services. (1996). *Physical activity and health: A report of the Surgeon General.* Atlanta, GA: U.S. Department of Health and Human Services, Centers for Disease Control and Prevention.

Walter, S.D., & Hart, L.E. (1991). Application of epidemiological methodology to sports and exercise research. *Exercise and Sport Sciences Reviews, 19*, 417-448.

Wareham, N.J., & Rennie, K.L. (1998). The assessment of physical activity in individuals and populations: Why try to be more precise about how physical activity is assessed? *International Journal of Obesity and Related Metabolic Disorders, 22*, S30-S38.

Wei, M., Kampert, J.B., Barlow, C.E., Nichaman, M.Z., Gibbons, L.W., Paffenbarger, R.S., Jr., & Blair, S.N. (1999). Relationship between low cardiorespiratory fitness and mortality in normal-weight, overweight, and obese men. *JAMA, 282*, 1547-1553.

Welk, G.J., & Blair, S.N. (2000). Physical activity protects against the health risks of obesity. *President's Council on Physical Fitness and Sports Research Digest, 3*.

Williams, P.T. (2001). Physical fitness and activity as separate heart disease risk factors: A meta analysis. *Medicine and Science in Sports and Exercise, 33*, 754-761.

1

Methods for Assessing Physical Activity and Challenges for Research

Darren Dale, PhD
Eastern Connecticut State University

Gregory J. Welk, PhD
Iowa State University

Charles E. Matthews, PhD
University of South Carolina and Iowa State University

The recognition of the importance of physical activity for optimal health has led to an increased interest in assessing physical activity behavior (U.S. Department of Health and Human Services, 1996). Physical activity assessments serve as independent variables in studies examining relationships with health outcomes and behavioral or physiological correlates of physical activity. Physical activity assessments also serve as dependent variables in studies designed to promote physical activity. To address progressively more complex questions in these studies, it is essential to obtain an accurate measure of physical activity.

A variety of methods have been used to assess physical activity behavior. In 1985, Laporte pro-vided a detailed review of seven different methods that included more than 30 different instruments (LaPorte et al., 1985). These instruments were reviewed on the basis of validity, reliability, practicality, and nonreactivity and highlighted the options available for different types of research. No effort has been made to tally the number of additional techniques that have been developed since then, but it is clear that the number and sophistication of approaches have increased markedly as the need for better techniques has become more apparent. Unfortunately, despite considerable research efforts and major advances in technology, many of the lingering challenges associated with assessing free-living physical activity still remain.

The purpose of this chapter is to provide a conceptual basis for the study of physical activity assessment techniques. The first section will review the general advantages and disadvantages of the most common techniques used for assessing physical activity. The subsequent section builds from this base and describes how to compute and interpret various outcome measures of physical activity. The third section presents a measurement error model to explain sources of error associated with physical activity assessments. The challenges posed by intra-individual variation in physical activity behaviors are highlighted in this section because this type of error is especially problematic for physical activity research. The last section of the chapter provides guidelines for selecting the appropriate method of measurement to maximize statistical power. Collectively, the material presented in this chapter is intended to serve as a broad introduction to the various measurement issues and techniques presented in the rest of the book.

Comparisons of Physical Activity Assessment Techniques

Many different methods have been used in research to assess physical activity. Some methods are adaptable for use with a variety of populations (self-report, activity monitors, pedometers, heart rate monitors, doubly labeled water, and indirect calorimetry) and others have been developed specifically for children (direct observation). Several excellent reviews of physical activity assessments have described the relative merits of the various techniques (Ainsworth et al., 1994; Melanson & Freedson, 1996; Wareham & Rennie, 1998). Because of unique measurement challenges associated with activity assessment in children, several reviews have also specifically focused on assessments in children (Baranowski & Simons-Morton, 1991; Baranowski et al., 1992; Pate, 1993; Riddoch & Boreham, 1995; Rowlands et al., 1997; Welk, Corbin, & Dale, 2000).

Each of the measures has specific advantages and disadvantages that one must consider when selecting an instrument. The basic advantages and disadvantages of the different techniques have been fairly well described and are summarized in table 1.1. Because later chapters are devoted to each of the different techniques, the purpose here is to briefly introduce the general concept for each technique.

Self-Report Techniques

Self-report techniques are the most common measure of physical activity because of low financial cost and low participant burden. The term *self-report* encompasses a variety of assessment methods: physical activity diaries, interviewer-administered questionnaires, self-administered questionnaires, and reports by proxy (e.g., information provided by parents on activity patterns of children). The limitations to using self-report techniques to assess physical activity are well documented. Participants may misinterpret the questions being asked, may have difficulty recalling the time or intensity of physical activity, or may deliberately misrepresent information. Additionally, the self-report instrument may not detect one or more dimensions of physical activity (frequency, type, intensity, and duration). Despite these limitations, self-report techniques are an acceptable method of assessing physical activity behavior. In a recent review article, Sallis and Saelens (2000) concluded that there are several self-report measures with adequate reliability, content validity, and relative criterion validity for adolescent populations, adults, and older adults. They indicated that interview measures had stronger psychometric characteristics than self-administered measures and noted that few empirically based principles exist for designing self-report measures or selecting recall formats. The cognitive aspects associated with the self-report of physical activity are described in chapter 7, along with general guidelines for using these instruments in research.

Activity Monitors

Activity monitors have become one of the most popular techniques to use for assessing physical activity in free-living conditions. These devices typically use an electronic component within the device to assess the acceleration of the body in a specific dimension (uniaxial) or in multiple dimensions. The monitors provide information on the intensity of acceleration or movement at user-specified intervals and can be downloaded to a computer for data processing. Therefore, researchers are provided with considerable detail on the extent to which participants are moving around in their environment. With accelerometers, it is important to note that the acceleration of the hip (or other body part if attached elsewhere) is assessed by the accelerometer, not the absolute acceleration of the person. Thus, the accelerometer does not assess the acceleration of a person as he or she moves in a car

TABLE 1.1   Advantages and Disadvantages of Various Assessment Methods

Measure	Advantages	Disadvantages
Self-report	• Captures quantitative and qualitative information • Inexpensive, allowing large sample size • Usually low participant burden • Can be administered quickly • Information available to estimate energy expenditure from daily living (i.e., Compendium of Physical Activities)	• Reliability and validity problems associated with recall of activity • Potential content validity problems associated with misinterpretation of physical activity in different populations
Activity monitors	• Objective indicator of body movement (acceleration) • Useful in laboratory and field settings • Provides indicator of intensity, frequency, and duration • Noninvasive • Ease of data collection and analyses • Provides minute-by-minute information • Allows for extended periods of recording (weeks)	• Financial cost may prohibit assessment of large numbers of participants • Inaccurate assessment of a large range of activities (e.g., upper-body movement, incline walking, water-based activities) • Lack of field-based equations to accurately estimate energy expenditure in specific populations • Cannot guarantee accurate monitor placement on participants during long, unobserved periods of data collection
Heart rate monitor	• Physiological parameter • Good association with energy expenditure • Valid in laboratory and field settings • Low participant burden for limited recording periods (30 min to 6 hr) • Describes intensity, frequency, and duration well (adults) • Easy and quick for data collection and analyses • Potential to provide participants with educational information	• Financial cost may prohibit assessment of large numbers of participants • Some discomfort for participants, especially over extended recording periods • Useful only for aerobic activities • Heart rate (HR) characteristics and the training state affect HR-$\dot{V}O_2$max relation • Some uncertainty as to the best way of using HR data to predict energy expenditure
Pedometers	• Inexpensive, noninvasive • Potential for use in a variety of settings including workplace and schools • Easy to administer to large groups • Potential to promote behavior change • Objective measure of common activity behavior (i.e., walking)	• Loss of accuracy when jogging or running is being assessed • Possibility of participant tampering • Are specifically designed to assess walking only
Direct observation	• Provides excellent quantitative and qualitative information • Physical activity categories established a priori, allowing specific targeting of physical activity behaviors • Software programs now available to enhance data collection and recording	• Time-intensive training needed to establish between-observer and within-observer agreement • Labor-intensive and time-intensive data collection, which limits the number of study participants • Observer presence may artificially alter normal physical activity patterns • Limited research reporting on validation of direct observation coding systems against physiological criteria
Indirect calorimetry and doubly labeled water	• Precision of measure • Ability to assess energy expenditure	• Invasive • Challenges associated with assessing patterns of physical activity • High relative cost

but rather the changing acceleration of the hip as it moves during physical activity.

Many studies have validated accelerometers such as the Caltrac, the Tritrac, the Biotrainer, and the Computer Science and Applications (CSA) monitor in both laboratory and field settings (Freedson & Miller, 2000). The limitations to activity monitors include their inability to detect with accuracy a wide array of physical activities, including upper-body movement (if the unit is worn at the hip), walking on an incline, carrying a load, or other nonlocomotor movements. Obviously, water-based activities cannot be assessed. Nevertheless, activity monitors are attractive to researchers for several reasons. They are very easy to operate (i.e., initializing to begin recording, downloading data), can collect data over extended periods (i.e., several weeks), and they are small and thus noninvasive for participants. Unfortunately, the cost of commercially available monitors ($200 to $500) still prohibits their use for most studies with large sample sizes. Detailed information on the application of activity monitors for physical activity research is included in chapter 8.

Heart Rate Monitoring

Heart rate monitoring has been widely used in physical activity research. A person's heart rate provides a direct indicator of the physiological response associated with physical activity. Most heart rate watches can be programmed to record and collect data at specified intervals (e.g., 60 s, 5 s) allowing a good description of the intensity, frequency, and duration of physical activity. Data from the watch can then be downloaded to a computer for data processing and analysis.

The use of heart rate monitoring does, however, present some unique challenges to investigators. First, conditions unrelated to a bout of physical activity can cause heart rate to increase without a corresponding increase in $\dot{V}O_2$. These conditions include a high ambient temperature, high humidity, hydration, emotional state, age, gender, and training status (Leger and Thivierge, 1988; Melanson & Freedson, 1996). Second, the heart rate-$\dot{V}O_2$ (HR-$\dot{V}O_2$) relationship is affected by the kinds of physical activity. More specifically, the type of muscular contraction alters the linear relation, as does the relative size of the muscle mass used during a given activity bout (Rowlands et al., 1997). A third major challenge is the temporal lag of the heart rate in response to the initiation or cessation of physical activity. It has been reported that it takes participants engaging in lifestyle activities 2 to 3 minutes for

heart rate and $\dot{V}O_2$ to increase to a level representative of the activity being performed and a similar amount of time to decrease to resting levels (Strath et al., 2000). A final challenge posed by heart rate monitoring is deciding how to most appropriately analyze the data. A variety of outcome measures have been proposed, but each can lead to different results and interpretation (Welk, Corbin, and Dale, 2000). Despite these limitations, heart rate monitors provide objective information about the intensity of physical activity under free-living or controlled conditions. Detailed information on the various outcome measures and analysis techniques for heart rate monitor data are provided in chapter 9.

Pedometers

Pedometers offer a promising alternative to activity monitors for objective assessments of movement caused by physical activity. These devices record steps taken and can estimate distance walked if stride length is known. They can also be used to estimate physical activity energy expenditure by estimating the energy cost associated with walking. The main advantages of pedometers are the low cost and ease of use. The primary disadvantages are the inability to record nonlocomotor movements and the inability to examine the rate or intensity of movement. Some accelerometry-based monitors (e.g., CSA) include the option of recording steps instead of acceleration, but this negates the primary cost advantage provided by most electronic pedometers. The potential benefits of pedometers as an objective monitoring tool are further described in chapter 10.

Direct Observation Techniques

Direct observation systems measure the behavioral aspects of physical activity. Although it is a time-intensive technique, both to train observers and collect data, it has the tremendous advantage of being able to accurately describe what took place in the physical activity setting. Therefore, both quantitative and qualitative information can be provided. Providing information on the environmental setting will become increasingly more important in physical activity research (see chapter 15). Computer software programs have been developed that allow real-time recording of physical activity and analysis of data, including researcher-specified time intervals in which the data will be stored. This will facilitate much-needed calibration studies between direct observation practices and physiological (heart rate) or movement (accelerometry) data. The primary disadvantages of direct observation techniques are the time and expense required for data collection. Accurate coding requires discipline and considerable practice. Observers must be trained and evaluated both prior to and during data collection to ensure reliable data. Because of these challenges, the use of direct observation is typically limited to smaller studies conducted in discrete settings and over short periods of time. Chapter 11 provides guidelines for conducting direct observation research and a detailed review of available direct observation instruments.

Indirect Calorimetry

Indirect calorimetry uses respiratory gas analysis to measure energy expenditure. The technique involves measurement of oxygen consumption and carbon dioxide production over periods of short or long duration. For short time intervals, participants wear a mouthpiece, facemask, or canopy during rest or exercise (including free-living). Over longer periods (i.e., approximately 24 hours or more), participants are confined to a metabolic chamber. The advantage of this approach is the ability to precisely measure energy expenditure. The disadvantages include the invasiveness of the measure, the inability to simulate true free-living situations very well, and the financial cost of the equipment. Portable backpack systems solve some of these limitations and offer considerable promise as a gold standard for field-based validation studies.

Doubly Labeled Water

The doubly labeled water technique is a biochemical procedure that can estimate energy expenditure through biological markers that reflect the rate of metabolism in the body. The procedure involves the ingestion of two stable isotopes of water and, following a 1- to 2-week period, an analysis of the difference in the rate of loss between the two isotopes from the body (i.e., urine, sweat, evaporation). The technique provides a direct measure of carbon dioxide production and an accurate estimate of energy expenditure in physical activity. The precision and noninvasive nature of this approach is a major advantage for many research applications. Disadvantages include the cost and lack of availability of the $H_2^{18}O$ isotope as well as the inability of the technique to describe qualitative patterns of physical activity. The doubly labeled water technique is increasingly being used in combination with indirect calorimetry to provide estimates of energy expenditure associated with physical activity. Indirect calorimetry can be used for the calculation of resting metabolic rate (RMR) and the thermic effect of food (TEF). The difference between total energy expenditure and this basal level of energy expenditure yields a measure of physical activity energy expenditure. The principles involved in the doubly labeled water and indirect calorimetry techniques are described in chapter 12.

Outcome Measures of Physical Activity

Because the various techniques are based on different principles (e.g., physiological, biomechanical, biochemical), they provide raw data that are not directly comparable. For example, heart rate telemetry captures heart beats, accelerometers provide movement counts, pedometers provide number of steps, direct observation provides time spent in distinct categories, and self-report commonly provides time spent in different levels of activity. The conversion of raw data from these techniques into different outcome measures is an important part of processing physical activity data.

Computing Outcome Measures of Physical Activity

Ideally, investigators want to know two things about physical activity: the type of activity engaged in (e.g., running, walking, lifestyle activities) and the amount of activity performed. Detailed information on the type, context, and setting for physical activity is typically only available through self-report or direct observation techniques, so emphasis has traditionally been placed on trying to quantify the amount of physical activity. The amount of physical activity is often assessed by time (e.g., duration) or some value reflecting the total volume of movement. For self-report or observation tools, the categories of activity are usually defined or operationalized as part of the assessment. Therefore, the amount of activity can be expressed as the total time spent in activity. With electronic monitoring devices, the overall activity is usually represented as a summary or composite measure for the amount of activity. For example, accelerometers can provide a measure of total counts and pedometers can provide an index of total number of steps. The amount of physical activity can also be expressed with other indicators, including frequency, intensity, duration, and energy expenditure. The general procedures for these conversions are summarized in table 1.2. Specific issues associated with assessing intensity and energy expenditure are described in the following sections.

Assessing the Intensity of Physical Activity

Quantifying the intensity of activity is a particularly important goal for many physical activity studies. The intensity of activity influences the physiological response and most likely the effect of physical activity on various health risks. The criterion used to designate the different intensity ranges is specific for each measurement technique. Heart rate is the most commonly used indicator of intensity because it reflects the direct physiological response to the activity. Thus, heart rate monitors can provide detailed information about the amount of time spent in different heart rate ranges. With accelerometers, the gradations of intensity are not well established. The reason for this is that each device uses different hardware and has different levels of sensitivity. Threshold values, or *cutpoints*, have been developed for some monitors to denote specific intensity categories, but the application of these cutpoints to assess free-living activities is problematic. The conversion of accelerometry data into a common unit (e.g., g forces) could help to standardize data collected from different monitors and conditions. Pedometers don't typically parse activity by time, but some new units now track total time spent moving.

TABLE 1.2   Methods of Measurement and the Characteristics of Physical Activity That Can Be Assessed

Method of measurement	Units of measurement	Dimension of physical activity	Output measure
Self-report	Bouts of physical activity	• Frequency • Intensity • Duration • Energy expenditure	• # of bouts > criterion level • # or % of bouts • # of min > criterion level • Estimates based on METS
Activity monitors or motion sensors	Movement counts	• Frequency • Intensity • Duration • Energy expenditure	• # of bouts > criterion level • Average counts per day or interval • # of min > criterion level • Estimates from calibration equation
Heart rate (HR)	Beats (per min)	• Frequency • Intensity • Duration • Energy expenditure	• # of bouts > criterion level • Average HR per day or interval • # of min > criterion level • Estimates from calibration equation
Pedometers	Step counts	• Frequency • Intensity • Duration • Energy expenditure	• NA • NA • # of steps taken • Estimates from calibration equation
Direct observation	Activity rating	• Frequency • Intensity • Duration • Energy expenditure	• # of bouts > criterion level • # or % of bouts • # min > criterion level • Estimates based on METS
Indirect calorimetry	O_2 consumption	• Frequency • Intensity • Duration • Energy expenditure	• # of bouts > criterion level • Average $\dot{V}O_2$ level • Monitored time > threshold • Total energy expenditure
Doubly labeled water	CO_2 production	• Frequency • Intensity • Duration • Energy expenditure	• NA • NA • NA • Total energy expenditure

NA = not applicable.

By dividing steps by time, a general index of intensity over the recording period can be obtained, but data have not yet been examined in this way.

The intensity of activity is often used in conjunction with frequency and duration to provide a more detailed characterization of activity. For example, once a criterion level for intensity is established, the amount of time above that criterion can be determined. Similarly, the frequency of participation in physical activity can be determined by computing the number of days that participants meet certain criteria for both intensity and duration.

Assessing Energy Expenditure of Physical Activity

Energy expenditure is also used as a composite measure of physical activity. An advantage of the use of this outcome measure is that it takes into account intensity and individual demographic variables such as body size. For example, vigorous activity requires a greater effort than moderate activity and larger people have to expend a greater amount of energy for a given amount of movement. Therefore, estimation of energy expenditure provides an indicator that can account for these different influences and reflect total movement. It can also be expressed as an absolute measure (e.g., calories expended per day or calories expended in activity) or as a relative measure (e.g., kcal/kg/day) depending on how the data are to be used.

A common unit used in calculations of energy expenditure is multiples of resting energy expenditure (METS). The accepted figure for resting energy expenditure is 3.5 ml O_2/kg/min, although there is considerable debate about the origin and accuracy of this figure for different populations. Because oxygen consumption is directly related to energy expenditure, METS can be directly converted into energy expenditure units. The standard physiological conversion for energy expenditure is approximately 5 kcal/L of oxygen consumed (assuming a mixed diet), so resting energy expenditure is approximately 0.0175 kcal/kg/min or 1.05 kcal/kg/hr. To simplify calculations, the value of 1 kcal/kg/hr is often used to represent 1 MET. Thus, an activity that is 3 METS is approximately 3 kcal/kg/hr. This conversion allows for a quick estimation of energy for various intensities of activity. For example, if the MET level for a given type and intensity of activity is known, the energy cost of activity can be estimated for individuals of any body size. A comprehensive listing of the MET levels for various forms of physical activity known as the Compendium of Physical Activities was published to provide some consistency with the way that physical activities are quantified (Ainsworth, Haskell, et al., 2000). The compendium was originally released in 1992 but was recently revised and updated with additional categories and activities (Ainsworth, Haskell et al., 2000).

Most of the methods allow for an estimation of energy expenditure, but the approach used to obtain this estimate is different for each technique. Doubly labeled water and indirect calorimetry are based directly on physiological processes and therefore provide the most accurate determination of energy expenditure. As mentioned, heart rate is highly correlated with oxygen consumption and can therefore be used as an indicator of the physiological demand associated with activity. Calibration equations developed with multiple regression techniques have been used to estimate energy expenditure (assessed with indirect calorimetry) associated with different heart rate values. Similar calibration equations have also been developed for accelerometers and pedometers to estimate the energy cost associated with a certain amount of movement. For self-report instruments, the determination of energy expenditure is usually determined by calculating the time spent in different intensities of activity. In this case, the MET levels for the different activities are estimated and used to determine the total energy cost of the activities performed. Similar approaches are used for direct observation instruments because the energy cost for each of the activity categories is typically calibrated during the initial validation of the instrument.

Although the calculation of energy expenditure provides a way to standardize the amount of exercise in a population, it does add considerable measurement error to the estimate. There are individual error differences with the way activities are performed and individual variability in the metabolic response to activity. Additional details on the calculations and interpretation of outcome measures of physical activity are provided in the individual chapters devoted to each technique.

Comparing Outcome Measures of Physical Activity

A major challenge in physical activity assessment research is the lack of a true gold standard of measurement. Without a definitive gold standard, it is impossible to determine which technique provides the most accurate assessment of physical activity. Doubly labeled water, direct observation, and indirect calorimetry have each been proposed as logical gold-standard measures, but they also have limitations that preclude their use as a definitive standard for all, or even most, applications.

The most common methodological approach for research on assessments of physical activity has been to compare the concurrent validity of the various techniques in assessing the same activity patterns. The sophistication of this line of research has increased dramatically in recent years. Advances in technology have made it easier to collect and analyze physical activity data. Improvements in research design and standardization have also helped to better understand the sources of error associated with these techniques. (See chapters 2 and 3 of this volume.)

The International Life Sciences Institute recently commissioned (and supported) a series of studies to further advance the measurement of moderate physical activity (Montoye, 2000). The results of the studies were compiled as a supplement in *Medicine and Science in Sports and Exercise.* These studies included current, state-of-the-art approaches for the assessment of physical activity but still indicate significant measurement challenges that must be overcome (see table 1.3). A consistent theme among the studies listed in the table is the difficulty in stating

TABLE 1.3   Cautionary Comments Regarding Specific Measurement Methods

Authors	Purpose	Results	Conclusion
Hendelman et al. (2000)	Evaluated the ability of pedometers and activity monitors to accurately assess free-living energy expenditure (EE) (measured by a portable metabolic unit)	Relationships with measured EE were higher for walking but underestimated lifestyle activities by 30 to 60%.	"The counts versus METS relationship for accelerometry was found to be dependent on the type of activity performed" (p. S442).
Swartz et al. (2000)	Evaluated the utility of wearing multiple accelerometers to more precisely assess free-living activity	The combination of wrist and hip accelerometer data improved the variance accounted for in measured EE from 31.7 to 33.3%.	". . . show the difficulty of developing consistent cut points and regression equations to predict the metabolic cost of all activities" (p. S450).
Ainsworth, Bassett et al. (2000)	Examined agreement between CSA monitor and two self-report instruments over a 21-day period in 83 adults	Correlations between physical activity log and various output measures of the CSA ranged from 0.24 to 0.36.	". . . motion sensors, physical activity logs and surveys reflect physical activity; however, these methods do not always provide similar estimates of the time spent in resting/light, moderate, or hard/very hard physical activity" (p. S457).
Strath et al. (2000)	Examined the relationship between HR and $\dot{V}O_2$ under field and lab conditions to improve prediction equations for energy expenditure	The correlation between HR and $\dot{V}O_2$ was r = 0.68. Adjustments for age and fitness allowed HR to provide an accurate estimate of EE (r = 0.87, SEE = 0.76 MET).	"This method of analyzing heart rate data could allow researchers to more accurately quantify activity in free-living individuals" (p. S465).
Bassett et al. (2000)	Compared the validity of four accelerometers for the assessment of free-living physical activity	Mean error scores compared with a portable metabolic unit ranged from 0.05 to 1.12 METS. Correlations ranged from r = 0.33 to r = 0.62.	"Motion sensors tended to overpredict EE during walking. However, they underpredicted the energy cost of many other activities" (p. S471).

Authors	Purpose	Results	Conclusion
Welk, Differding et al. (2000)	Evaluated the utility of the Digi-Walker pedometer for the assessment of lifestyle physical activity	Modest correlations ($r = 0.34$ to $r = 0.39$) were found between average daily step counts and estimated energy expenditure and minutes of activity from self-report.	"Pedometers provide a useful indicator of daily step counts but variability in activity patterns make it difficult to establish step count guidelines" (p. S481).
Welk, Blair et al. (2000)	Compared the accuracy of three comparable activity monitors against measured EE under both lab and field conditions	Correlations among the monitors were high for both lab ($r = 0.86$) and field conditions ($r = 0.70$). Prediction of EE was reasonable for lab conditions, but under-estimates of 42 to 67% were found for field activities.	"The observed differences among the monitors were attributed primarily to differences in the accuracy of the calibration equations rather than to the monitors themselves" (p. S489).
Ainsworth, Haskell et al. (2000)	Provided an updated series of MET codes to improve the comparability of results of studies using self-report measures of activity	The updated version extends the number of coded activities to 605 and provides updated MET levels for specific activities.	"Thus, individual differences in energy expenditure for the same activity can be large and the true energy cost for a person may or may not be close to the stated mean MET level as presented in the Compendium" (p. S502).

with certainty that a particular measurement method could provide a true indicator of free-living physical activity. Particular challenges were evident in attempts to estimate energy expenditure under free-living conditions.

Because each technique utilizes different measurement approaches to assess physical activity, the agreement among techniques is not consistently high. Data is not currently available to allow investigators to directly equate heart rate telemetry, accelerometer movement counts, direct observation, self-report, or pedometer step counts. Such data would enable the consistent categorization of moderate and vigorous intensity physical activity, independent of the method of measurement used. Until the calibration of the raw data output from the major methods of activity assessment is undertaken and reported in the research literature, it will remain problematic to directly compare the results of assessments. The variation in physical activity behaviors, within and between participants, also makes the accurate assessment of activity a challenging task for investigators. The next section provides a measurement model to describe the sources

of error that confound efforts to accurately assess physical activity under free-living conditions.

Understanding Sources of Error in Physical Activity Assessments

In contrast to many stable health indicators that are routinely measured in research settings, such as height, weight, or body composition, physical activity is a behavior. There are naturally occurring changes in physical activity from day to day (Levin et al., 1999; Matthews, Hebert et al., 2001) from one season to the next (Levin et al., 1999; Matthews, Freedson et al., 2001), from one year to the next, and over the course of one's lifetime (Paffenbarger et al., 1984). In an effort to describe the relationship between the true values of disease risk factors that have considerable natural variation and are therefore difficult to measure without error (e.g., blood pressure), epidemiologists and biostatisticians have developed measurement error models that characterize the simplified relationships between the true value of a measurable quantity and the value obtained by assessment methods typically employed in research settings (Armstrong et al., 1992; Willet, 1990).

Measurement Error Models

In measurement error models, the measured value of any parameter (Xi) is thought to vary from the true value (Ti) by both systematic (b) and random (Ei) factors. Systematic variations (b) reflect the difference between the true value T and the measured value X and can be either positive or negative in value depending on whether the measured value is an over- or underestimate of the true value. Random variation (Ei), on the other hand, does not result in a predictable or systematic increase or decrease in the measured value from the true value, so it does not contribute to systematic variation (b). In other words, random variation simply adds "noise" to the signal to be detected in measurements as is shown in equation 1.1.

$$X_i = T_i + b + E_i \qquad (1.1)$$

Random variation may be further divided into terms describing biological and analytic variation. Biological variation represents naturally occurring changes in physical activity behavior (i.e., day-to-day variation or intra-individual variation), whereas analytic variation represents differences in the way data are recorded or collected from one administration of an instrument to the next, but that are not due to a true change in physical activity.

An understanding of the nature and magnitude of the random variability in physical activity levels is necessary for the proper design, analysis, and interpretation of studies of physical activity and health. Perhaps most importantly, estimates of biological variation in physical activity behaviors are required to identify the number of days of assessment required to accurately characterize activity patterns of individuals in a given population (Beaton et al., 1979; Gretebeck & Montoye, 1992; Levin et al., 1999). In turn, the number of days of assessment required influences directly the study design (e.g., assessment methods and sample sizes) and statistical power available to detect differences between groups, or to quantify the relationships between measured values for physical activity and an outcome of interest (Beaton, 1994). This issue of intra-individual variation represents one of the biggest challenges with physical activity assessments, so specific detail is provided in the following sec-

tions to help researchers assess and interpret this type of variability.

Intra-Individual Variation

The number of days of assessment required to reliably characterize physical activity behaviors has typically been estimated using the following reliability formula for the intra-class correlation coefficient, where reliability (R) is the proportion of total variance accounted for in the measures by inter-individual sources (Snedecor & Cochran, 1989).

$$R = \sigma_B^2 / [\sigma_B^2 + (\sigma_W^2 / n)] \qquad (1.2)$$

In equation 1.2, σ_B^2 is the inter-individual (between-subject) variance, σ_W^2 is the intra-individual (within-subject) variance, and n is the number of administrations of a physical activity instrument. A rearrangement of this formula, solving for the number of administrations to achieve a desired level of reliability, is sometimes called the Spearman-Brown Prophecy Formula (Gretebeck & Montoye, 1992; Levin et al., 1999). In general, increasing the number of administrations of an instrument (n) reduces intra-individual variation, which in turn reduces total variation. As the overall denominator decreases, the proportion of the inter-individual variation increases relative to the total variance, resulting in the elevation of the reliability value. A reliability (R) of 0.80 has been considered adequate in previous physical activity applications (Baranowski et al., 1999; Gretebeck & Montoye, 1992).

Test-retest reliability estimates are commonly reported for many assessment techniques as an indicator of random variation. Among instruments seeking to capture long-term physical activity behaviors (e.g., last 12 months), short-term test-retest coefficients (i.e., 2 weeks to a month) reflect analytic variation in the instrument rather than biological variation in activity. Separation of the source of random variation (i.e., behavioral vs. analytic) is more difficult when examining activity assessments conducted over shorter timeframes (e.g., 24-hour or 7-day recalls) because variability will be due to both sources of random variation.

When one is conducting test-retest studies, it is recommended that the intra-class correlation (R) be employed rather than the standard Pearson Product Moment correlation coefficient (or Pearson r) because the former takes into account all sources of variance in the measures, whereas the latter only evaluates the variance of the two measures in their linear relationship (Patterson, 2000).

Effect of Random Variation in Physical Activity Research

The most significant effect of random variation derived from behavioral or analytic sources is the attenuation of statistical measures of effect (e.g., correlation and regression coefficients, odds ratios) (Beaton et al., 1979; MacMahon et al., 1990). That is, because of the noise in the system, we may have a reduced ability to identify the true relationship between our measures of activity and our outcomes of interest (e.g., other activity measures, health outcomes, or measures of determinants). Table 1.4 demonstrates the level of attenuation of a hypothetical perfect relationship described by the Pearson correlation coefficient of 1.0 across a range of values for random variation (Rosner & Willett, 1988) and for an increasing number of instrument administrations. The attenuated correlation coefficients in table 1.4 were obtained using the following formulae:

$$\rho_{obs} = \rho_{true} / k$$
$$k = \sqrt{1 + VR/n} \qquad (1.3)$$

where ρ_{obs} is the observed correlation, ρ_{true} is the true correlation (1.0), k is the correction factor, VR is the variance ratio (σ_W^2 / σ_B^2), and n is the number of assessments employed. Rearranging equation 1.3 allows for the calculation of a theoretical estimate of deattenuated correlation coefficients ($\rho_{true} = \rho_{obs} \times k$) using measured values of σ_W^2 and σ_B^2 (Matthews et al., 2000; Wolf et al., 1994).

For a given level of random variation, increasing the number of administrations of an instrument lessens the degree of attenuation (see table 1.4). For example, if an instrument had an intra-class correlation (ICC) of 0.50, administering the instrument three times instead of once would be predicted to increase the correlation observed from 0.71 (one administration) to 0.87 (three administrations), much closer to the hypothetical true value of 1.0. It should be noted from table 1.4 that even a modest and generally acceptable level of random variation (i.e., ICC = 0.80) results in a modest reduction in the observed values from the true (i.e., attenuation from 1.0 to 0.89). The implications of this type of variability will be described in the subsequent section.

Understanding Factors Influencing Statistical Power

A limitation in many research studies in the physical activity field is a deficient amount of statistical power. A study with adequate power allows an investigator to detect significant differences if they do indeed exist. A study with 80% power, recommended for behavioral science research by Cohen (1988), means the investigator has an 80% chance of detecting a true difference between a treatment group and a control group. A Type 2 error (β) in this case is 20%, meaning the researcher is prepared to accept a 20% chance of being wrong, in this case meaning the investigator would report nonsignificant

TABLE 1.4   Attenuation of Pearson Product Moment Correlation Coefficients by Increasing Level of Random Variation

VR	ICC	K	Number of administrations				
			1	2	3	7	14
0.05	0.95	1.0247	0.98	0.99	0.99	1.00	1.00
0.10	0.91	1.0488	0.95	0.98	0.98	0.99	1.00
0.25	0.80	1.1180	0.89	0.94	0.96	0.98	0.99
0.50	0.67	1.2247	0.82	0.89	0.93	0.97	0.98
0.75	0.57	1.3229	0.76	0.85	0.89	0.95	0.97
1.00	0.50	1.4142	0.71	0.82	0.87	0.94	0.97
2.00	0.33	1.7321	0.58	0.71	0.77	0.88	0.94
3.00	0.25	2.0000	0.50	0.63	0.71	0.84	0.91
4.00	0.20	2.2361	0.45	0.58	0.65	0.80	0.88
5.00	0.17	2.4495	0.41	0.53	0.61	0.76	0.86

Variation is expressed as variance ratios (VR) and intraclass correlations.

Correlations are computed across an increasing number of instrument administrations.

results when, in fact, the physical activity intervention actually worked. Statistical power provides the researcher with the probability of getting the data that will lead to a correct rejection of the null hypothesis.

The type of statistical tests used can have a direct effect on the statistical power of the comparisons. In general, parametric tests are more powerful than nonparametric tests, although new approaches for nonparametric techniques have now been developed to overcome this limitation (see chapter 4). Directional (one-sided) tests and tests that block or control for extraneous factors in the study are generally more powerful. In many cases, the specific type of statistical test can also influence the results. Multivariate data analysis techniques may provide more powerful ways to detect differences in physical activity research. For example, if multiple assessments of physical activity are available, they can be used in combination to provide a better overall index of physical activity. The use of canonical correlations accommodates multiple measures and use of the optimal linear combination of these measures in the statistical comparisons. Interpretations are often more difficult or less clear with these approaches, but the advantage of having a better statistical test for a given application usually makes it worthwhile. It is best to consider techniques that provide the most powerful test for their particular application.

Designing a study with adequate statistical power ($1-\beta > 0.80$) requires consideration of four factors: the level of alpha, the sample size, effect size, and measurement error. A detailed explanation of power calculations is provided in chapter 6. This section provides only a brief commentary on how each of these factors affects power.

Level of Alpha

Conventional statistical analyses assume the null hypothesis is true (e.g., the physical activity intervention will fail, or no difference between treatment and control groups). This model of analysis causes investigators to take great care in protecting against a Type 1 error, reporting statistically significant results in the situation of a true null. However, it may be surprising to learn that it is hardly ever the case in social science intervention research that the null hypothesis is true (Cohen, 1994; Hunter, 1997; Tukey, 1991). Therefore, what should be perhaps a greater concern to investigators is the Type 2 error rate (i.e., wrongly reporting statistical nonsignificance). Unfortunately, the error rate relevant to treatment re-

search in the social sciences is approximately 60%. That is, 6 out of every 10 studies wrongly conclude statistical nonsignificance, according to critics of the null-hypothesis test (Cohen, 1962; Hunter, 1997; Lipsey & Wilson, 1993; Sedlmeier & Gigerenzer, 1989). Read the excellent commentaries by Cohen (1990, 1994) and Schmidt (1992) clarifying what the null hypothesis is really testing, as well as what can and cannot be stated as fact, when investigators find p < 0.05. In light of these observations, investigators in many cases should consider increasing alpha to 0.10 (or even higher) in many studies. For example, it is hard to find a good argument for using a conservative alpha of 0.05 in intervention research. This alpha value is used in many other clinical trials to protect against making a Type 1 error. However, the question that must be addressed is, Just how serious is Type 1 error in physical activity intervention studies? It could easily be argued that a Type 2 error (reporting nonsignificance when, in fact, the treatment intervention actually worked!) is possibly more serious than a Type 1 error. Similar arguments may be appropriate for other research applications.

Sample Size

Sampling unit is a term that defines the sample used for statistical comparisons. In experimental research, the unit is one drawn independently and randomly assigned to an experimental condition. It may be one person, one work site, or one school. Sample size has a strong effect on statistical power. For a given effect size and alpha level, a larger sample will mean a statistically more powerful study. At first glance, it seems unfortunate that smaller samples will have to be used if researchers want to use the most accurate measurement methods available. After all, the most accurate methods (e.g., indirect calorimetry, doubly labeled water, heart rate monitoring, direct observation) are also the most expensive, labor intensive, or invasive. However, the improvements in accuracy available from some of these techniques may more than compensate for the loss of power resulting from a small sample. Researchers are encouraged to examine both the size and accuracy of their measures in making power calculations for their research.

Effect Size

An effect size index standardizes differences between means, adjusting for arbitrary differences in measurement scales used. In the behavioral sciences, the differences in mean scores of two groups are divided by the standard deviation of their distribution to produce a measure in standard deviation units. An effect size indicates how meaningful the findings are or the actual magnitude of the difference between two groups (Thomas et al., 1991).

Measurement Error

Measurement error reduces the ability to detect change in outcome variables. As noted earlier, sources of random variation from either behavioral or analytic sources will attenuate the statistical measures of effect and make it harder to detect significant differences (see equation 1.1). That is, because of the noise in the system, we may have a reduced ability to identify the true relationship between measures of activity and outcomes of interest (e.g., other activity measures, health outcomes, measures of determinants). Attenuation of statistical measures of effect may result in several analytic problems, including

1. an erroneous conclusion that there is no effect,
2. that observed effects are "weak," and
3. a significant limitation in our ability to carefully characterize the dose-response relationships between physical activity and the outcome of interest (Beaton, 1994).

Because of these limitations, researchers are encouraged to take steps to reduce random error in their research.

Summary and Guidelines for Selecting an Appropriate Measurement Method

In light of the different strengths and weaknesses of the various assessment techniques, care must be used to select an appropriate technique for specific research applications. Some guidelines are proposed here to help guide this selection process. These guidelines follow closely from each of the previous sections in this chapter.

1. *Select an approach that is appropriate to the research question.* The specific theory the research question is based on should have a major influence on the method that is selected (see table 1.5). For example, if theory suggests that intensity of physical activity is responsible for changes in aerobic fitness, measurement methods that reflect intensity should be chosen (e.g., heart rate monitors or activ-

ity monitors). If theory links reduced activity energy expenditure with increased prevalence of obesity, then methods such as indirect calorimetry or doubly labeled water should be selected. If investigators aim to determine a dose-response relation between physical activity and risk of depression, then a measurement method that can accurately document the relative intensity of activity would be necessary. Because different methods are able to measure the same characteristic, the decision often rests on the relative importance of accuracy and practicality (Rowland, 1996). This is a challenging tradeoff because the most accurate measures (i.e., indirect calorimetry, doubly labeled water, direct observation) are typically the least practical in terms of cost, time, and the burden placed on participants. As stated earlier, self-report is probably the most commonly used method of assessment, certainly in large populations, because it is highly practical, inexpensive, and noninvasive. Unfortunately, the trade-off is in the accuracy of the measure.

2. *Select an appropriate outcome measure.* Because physical activity can be operationalized or expressed in different ways, the selection of an appropriate outcome measure becomes an important research issue. In general, the outcome measure should provide an appropriate reflection of the type of activity being examined in the particular study. In some studies, it may be important to use an absolute measure of physical activity such as minutes of activity or physical activity energy expenditure. In other situations, it may be necessary to consider the relative dose of activity by expressing the results by body weight or by considering the individual response to the activity. In some situations, efforts to estimate energy expenditure from raw activity data may add unnecessary error to an assessment and limit the ability to answer the research question. Understanding the unique ways in which a particular population group engages in physical activity is also imperative to selecting an appropriate outcome measure of physical activity. In a recent review article on assessment issues with children, Welk, Corbin et al. (2000) described the importance of selecting an outcome measure that can accurately assess the intermittent activity patterns of children. The authors pointed out that the results of a study could be greatly influenced by the choice of outcome measure used to summarize the data. For example, using a strict criterion that examines the percentage of children who do continuous exercise (i.e., heart rate > 140 b/min) would lead to a low estimate for the activity levels of children because children don't perform this type of

TABLE 1.5   Research Questions and Possible Physical Activity Measures

Examples of research questions	Method(s) of measurement
What are the dose-response characteristics between physical activity and various fitness and health outcomes?	• Accelerometers • Direct observation • Heart rate monitoring
How does physical activity influence weight control and daily energy expenditure?	• Doubly labeled water • Indirect calorimetry
What psychological measures are related to participation in physical activity?	• Accelerometers • Heart rate • Self-report
Can physical activity interventions increase the amount of time individuals spend in moderate-intensity physical activity?	• Accelerometers • Heart rate monitoring • Direct observation
Can large-scale behavioral intervention programs decrease *the amount of* inactivity in different populations?	• Self-report • Pedometers
Do parental interests influence the types and amount of physical activity their children are involved in?	• Self-report • Direct observation • Accelerometers

activity. Using a criterion that is based on the total accumulation of minutes of activity would lead to a much higher estimate. Because children generally have sporadic and intermittent activity patterns, it is more appropriate to use a measure that reflects this pattern. The same types of distinctions are critical for assessing other population groups. A supplement in *Research Quarterly* includes recommendations for assessing women (Ainsworth, 2000), minorities (Kriska, 2000) and elderly adults (Washburn, 2000).

3. Take steps to reduce error in measurements. The accumulation of measurement error in physical activity assessment poses a considerable challenge for research. Investigators are encouraged to use the best methods available and to make concerted efforts to reduce measurement error in their research protocol. Following scripts in administering self-report instruments can help to reduce the degree of subjectivity in interpreting physical activity questions. Similarly, following consistent procedures for the placement and positioning of activity monitors can reduce measurement error associated with these devices. The individual chapters on measurement techniques provide considerable detail on the most appropriate ways to control measurement error and provide some information about the intra-individual variability to expect from the technique when used to characterize a population.

4. Use a design that allows for sufficient power. A primary consideration in any study is to ensure that there is sufficient statistical power to answer

the research question(s). Reducing measurement error is one way to increase the precision of the assessment and, therefore, power. However, as pointed out previously, other statistical considerations such as the statistical design, selection of alpha, and the sample size have a major influence on the power in a given study. In considering issues associated with power, Lipsey and Wilson (1993) used the term *design sensitivity* to urge investigators in the behavioral sciences to carefully think through all aspects of planning a study. In this context, a sensitive study design is one that will yield statistically significant results, assuming that the intervention is effective or the results are true. In most studies, according to Lipsey and Wilson, nonsignificance is more likely a result of low statistical power, an occurrence tied inextricably to the choice of a measurement method. Lipsey (1990) states that "if treatment effects of meaningful magnitude are below the threshold of what a particular experimental design can meaningfully detect, no amount of improvement in its internal, construct, and external validity will rescue it from error" (p. 13). Therefore, the use of more powerful designs and appropriate statistical analyses can counteract measurement limitations and improve the quality of physical activity research.

This chapter provided an overview of the various techniques available for the assessment of physical activity and introduced measurement issues that influence physical activity research. Subsequent chapters will further explore various measurement issues associated with physical activity assessment (chapters 2 through 6) and provide more specific information about how to collect, process, and interpret data from the different techniques (chapters 7 through 12). Some innovative measurement approaches or research applications are provided in chapters 13 through 15.

References

Ainsworth, B.E. (2000). Issues in the assessment of physical activity in women. *Research Quarterly for Exercise and Sport, 71,* S37-S42.

Ainsworth, B.E., Bassett, D.R., Strath, S.J., Swartz, A.M., O'Brien, W.L., Thompson, R.W., Jones, D.A., Macera, C.A., & Kimsey, C.D. (2000). Comparison of three methods for measuring the time spent in physical activity. *Medicine and Science in Sports and Exercise, 32,* S457-S464.

Ainsworth, B.E., Haskell, W.L., Whitt, M.C., Irwin, M.L., Swartz, A.M., Strath, S.J., O'Brien, W.L., Bassett, D.R., Schmitz, K.H., Emplaincourt, P.O., Jacobs, D.R., & Leon, A.S. (2000). Compendium of physical activities: An update of activity codes and MET intensities. *Medicine and Science in Sports and Exercise, 32,* S498-S516.

Ainsworth, B.E., Montoye, H.J., & Leon, A.S. (1994). Methods of assessing physical activity during leisure and work. In C. Bouchard, T. Stephens, & R.J. Shephard (Eds.), *Physical activity, fitness and health: International proceedings and consensus statement* (pp. 146-159). Champaign, IL: Human Kinetics.

Armstrong, B.K., White, E., & Saracci, R. (1992). *Principles of exposure measurement in epidemiology.* New York: Oxford University Press.

Baranowski, T., Bouchard, C., Bar-Or, O., Bricker, T., Heath, G., Kimm, S.Y.S., Malina, R., Obarzanek, E., Pate, R., Strong, W.B., Truman, B., & Washington, R. (1992). Assessment, prevalence, and cardiovascular benefits of physical activity and fitness in youth. *Medicine and Science in Sports and Exercise, 24,* S237-S247.

Baranowski, T., & Simons-Morton, B.G. (1991). Dietary and physical activity assessment in school-aged children: Measurement issues. *Journal of School Health, 61,* 195-196.

Baranowski, T., Smith, M., Thompson, W.O., Baranowski, J., Hebert, D., & de Moor, C. (1999). Intra-individual variability and reliability in a 7-day exercise record. *Medicine and Science in Sports and Exercise, 31,* 1619-1622.

Bassett, D.R., Ainsworth, B.E, Swartz, A.M., Strath, S.J., O'Brien, W.L., & King, G.A. (2000). Validity of four motion sensors in measuring moderate intensity physical activity. *Medicine and Science in Sports and Exercise, 32,* S471-S480.

Beaton, G. (1994). Approaches to analysis of dietary data: Relationship between planned analyses and choices of methodology. *American Journal of Clinical Nutrition, 59,* S253-S261.

Beaton, G., Milner, J., Corey, P., McQuire, V., Cousins, M., Stewart, E., Ramos, M.D., Hewitt, D., Grambsch, V., Kassim, N., & Little, J. (1979). Sources of variance in 24-hour dietary recall data: Implications for nutrition study design and interpretation. *American Journal of Clinical Nutrition, 32,* 2546-2559.

Cohen, J. (1962). The statistical power of abnormal-social psychological research: A review. *Journal of Abnormal and Social Psychology, 69,* 145-153.

Cohen, J. (1988). *Statistical power analysis for the behavioral sciences* (2nd ed.). Hillsdale, NJ: Erlbaum.

Cohen, J. (1990). Things I have learned (so far). *American Psychologist, 45,* 1304-1312.

Cohen, J. (1994). The earth is round (p < .05). *American Psychologist, 49,* 997-1003.

Freedson, P.S., & Miller, K. (2000). Objective monitoring of physical activity using motion sensors and heart rate. *Research Quarterly for Exercise and Sport, 71,* 21-29.

Gretebeck, R.J., & Montoye, H.J. (1992). Variability of some objective measures of physical activity. *Medicine and Science in Sports and Exercise, 24,* 1167-1172.

Hendelman, D., Miller, K., Bagget, C., Debold, E., & Freedson, P. (2000). Validity of accelerometry for the assessment of moderate intensity physical activity in the field. *Medicine and Science in Sports and Exercise, 32,* S442-S449.

Hunter, J.E. (1997). Needed: A ban on the significance test. *Psychological Science, 8,* 3-7.

Kriska, A. (2000). Ethnic and cultural issues in assessing physical activity. *Research Quarterly for Exercise and Sport, 71,* S47-S53.

LaPorte, R.E., Montoye, H.J., & Caspersen, C.J. (1985). Assessment of physical activity in epidemiologic research: Problems and prospects. *Public Health Reports, 100,* 131-146.

Leger, L., & Thivierge, M. (1988). Heart rate monitors: Validity, stability, and functionality. *Physician and Sportsmedicine, 16,* S3-S7.

Levin, S., Jacobs, D.R., Jr., Ainsworth, B.E., Richardson, M.T., & Leon, A.S. (1999). Intra-individual variation and estimates of usual physical activity. *Annals of Epidemiology, 9,* 481-488.

Lipsey, M.W. (1990). *Design sensitivity: Statistical power for experimental research.* Newbury Park, CA: Sage.

Lipsey, M.W., & Wilson, D.B. (1993). The efficacy of psychological, educational, and behavioral treatment: Confirmation from meta-analysis. *The American Psychologist, 48,* 1181-1209.

MacMahon, S., Peto, R., Cutler, J., Collins, R., Sorlie, P., Neaton, J., Abbott, R., Godwin, J., Dyer, A., & Stamler, J. (1990). Blood pressure, stroke, and coronary heart disease. Part 1, Prolonged differences in blood pressure: Prospective observational studies corrected for the regression dilution bias. *Lancet, 335,* 765-774.

Matthews, C.E., Freedson, P.S., Stanek, E.J., Hebert, J.R., Merriam, P.A., Rosal, M.C., Ebbeling, C.B., & Ockene, I.S. (2001). Seasonal variation of household, occupational, and leisure-time physical activity: Longitudinal analyses from the Seasonal Variation of Cholesterol Study. *American Journal of Epidemiology, 153,* 172-183.

Matthews, C.E., Hebert, J.R., Freedson, P.S., Stanek, E.J., Merriam, P.A., Ebbeling, C.B., & Ockene, I.S. (2001). Sources of variance in daily physical activity levels in the Seasonal Variation of Cholesterol Study. *American Journal of Epidemiology, 153,* 987-995.

Matthews, C.E., Hebert, J.R., Freedson, P.S., Stanek, E.J, Ockene, I.S., & Merriam, P.A. (2000). Comparing physical activity assessment methods in the seasonal variation of blood cholesterol levels study. *Medicine and Science in Sports and Exercise, 32,* 976-984.

Melanson, E.L., & Freedson, P.S. (1996). Physical activity assessment: A review of methods. *Critical Reviews in Food Science and Nutrition, 36,* 385-396.

Montoye, H.J. (2000). Introduction: Evaluation of some measurements of physical activity and energy expenditure. *Medicine and Science in Sports and Exercise, 32,* S439-S441.

Paffenbarger, R.S., Hyde, R.T., Wing, A.L., & Steinmetz, C.H. (1984). A natural history of athleticism and cardiovascular health. *JAMA, 252,* 491-495.

Pate, R.R. (1993). Physical activity assessment in children and adolescents. *Critical Reviews in Food Science and Nutrition, 33,* 321-326.

Patterson, P. (2000). Reliability, validity, and methodological response to the assessment of physical activity by self-report: Limitations and future directions. *Research Quarterly for Exercise and Sport, 71,* 15-20.

Riddoch, C.J., & Boreham, C.A. (1995). The health-related physical activity of children. *Sports Medicine, 19,* 86-102.

Rosner, B., & Willett, W.C. (1988). Interval estimates for correlation coefficients corrected for within-person variation: Implications for study design and hypothesis testing. *American Journal of Epidemiology, 127,* 377-386.

Rowland, T.W. (1996). *Developmental exercise physiology.* Champaign, IL: Human Kinetics.

Rowlands, A.V., Eston, R.G., & Ingledew, D.K. (1997). Measurement of physical activity in children with particular reference to the use of heart rate and pedometry. *Sports Medicine, 24,* 258-272.

Sallis, J.F., & Saelens, B.E. (2000). Assessment of physical activity by self-report: Status, limitations, and future directions. *Research Quarterly for Exercise and Sport, 71,* 1-14.

Schmidt, F.L. (1992). What do data really mean? Research findings, meta-analysis, and cumulative knowledge in psychology. *The American Psychologist, 47,* 1173-1181.

Sedlmeier, P., & Gigerenzer, G. (1989). Do studies of statistical power have an effect on the power of studies? *Psychological Bulletin, 105,* 309-316.

Snedecor, G.W., & Cochran, W.G. (1989). *Statistical methods.* Ames, IA: Iowa State University Press.

Strath, S.J., Swartz, A.M., Bassett, D.R., O'Brien, W.L., King, G.A., & Ainsworth, B.E. (2000). Evaluation of heart rate as a method for assessing moderate intensity physical activity. *Medicine and Science in Sports and Exercise, 32,* S465-S470.

Swartz, A.M., Strath, S.J., Bassett, D.R., O'Brien, W.L., King, G.A., & Ainsworth, B.E. (2000). Estimation of energy expenditure using CSA accelerometers at hip and wrist sites. *Medicine and Science in Sports and Exercise, 32,* S450-S456.

Thomas, J.R., Salazar, W., & Landers, D.M. (1991). What is missing in p < .05? Effect size. *Research Quarterly for Exercise and Sport, 62,* 344-348.

Tukey, J.W. (1991). The philosophy of multiple comparisons. *Statistical Science, 6,* 100-116.

U.S. Department of Health and Human Services. (1996). *Physical activity and health: A report of the Surgeon General.* Atlanta, GA: U.S. Department of Health and Human Services, Centers for Disease Control and Prevention.

Wareham, N.J., & Rennie, K.L. (1998). The assessment of physical activity in individuals and populations: Why try to be more precise about how physical activity is assessed? *International Journal of Obesity and Related Metabolic Disorders, 22,* S30-S38.

Washburn, R.A. (2000). Assessment of physical activity in older adults. *Research Quarterly for Exercise and Sport, 71,* 79-88.

Welk, G.J., Blair, S.N., Wood, K., Jones, S., & Thompson, R.W. (2000). A comparative evaluation of three accelerometry-based physical activity monitors. *Medicine and Science in Sports and Exercise, 32,* S489-S497.

Welk, G.J., Corbin, C.B., & Dale, D. (2000). Measurement issues for the assessment of physical activity in children. *Research Quarterly for Exercise and Sport, 71,* 59-73.

Welk, G.J., Differding, J.A., Thompson, R.W., Blair, S.N., Dziura, J., & Hart, P. (2000). The utility of the Digi-Walker step counter to assess daily physical activity patterns. *Medicine and Science in Sports and Exercise, 32,* S481-S488.

Willett, W. (1990). *Nutritional epidemiology.* New York: Oxford University Press.

Wolf, A.M., Hunter, D.J., Colditz, G.A., Manson, J.E., Stampfer, M.J., Corsano, K.A., Rosner, B., Kriska, A., & Willett, W.C. (1994). Reproducibility and validity of a self-administered physical activity questionnaire. *International Journal of Epidemiology, 23,* 991-999.

Measurement and Analysis Issues for Physical Activity Assessment

2

Measurement Issues for the Assessment of Physical Activity

James R. Morrow Jr., PhD
University of North Texas

This chapter focuses on interpreting the quality of the method used to assess physical activity. The guidelines presented in this chapter are appropriate for each of the physical activity methods presented in this book. Specific examples will be used that represent issues important in assessing physical activity. To facilitate the application of this information, a hypothetical research scenario is presented first. The issues associated with measuring physical activity in this study will be used for illustrative purposes throughout this chapter.

Research Scenario

Assume that you have an intervention that you believe will help people become more physically active. Your intervention might consist of a wide variety of actions (e.g., discussion groups, group physical activities, discussions about nutrition, cognitive information about the relationship between physical activity and health, a behaviorally based

program, a program based on the stages of change model, etc.). You get approval to conduct the study and you randomly assign 30 participants to each of two groups. One is the experimental group that gets the intervention and one is the control group that does not get the intervention. The intervention lasts 10 weeks, and you then want to measure the amount of weekly physical activity engaged in by the study participants. You might use any of the methods discussed in this book to assess physical activity. You decide to create and utilize a 10-item questionnaire that assesses the amount of physical activity that one engages in each week as your measure of physical activity. You might create an instrument like the Sample Assessment of Weekly Physical Activity (SAWPA) presented in form 2.1.

Assume that we actually know that the intervention works. That is, those in the experimental group who are exposed to the intervention will actually increase their amount of weekly physical activity. This is a theoretical example, clearly. You can never be certain that you know the intervention works in

Form 2.1 Sample 10-Item Self-Report Instrument to Assess Weekly Physical Activity[a]
(SAWPA—Sample Assessment of Weekly Physical Activity).

Answer the following items regarding the amount of physical activity in which you typically engage weekly:
The word *moderate* refers to activities like brisk walking, gardening, slow cycling, dancing, or hard work around the house. The word *vigorous* refers to activities like basketball, jogging, running, fast cycling, aerobics class, swimming laps, singles tennis, racquetball, etc.

	Days per week							
	0	1	2	3	4	5	6	7
1. How many days per week do you typically engage in *almost no* physical activity? That is, you are sedentary throughout most of the day.								
2. How many days per week do you typically engage in *moderate* physical activity that accumulates to at least 30 minutes?								
3. How many days per week do you typically engage in *vigorous* physical activity that accumulates to at least 30 minutes?								
4. How many days per week do you typically engage in conditioning training exercises to improve your *strength*?								
5. How many days per week do you typically engage in activities to improve your *flexibility*?								
6. How many days per week do you purposefully take *action to increase* your amount of physical activity (e.g., walk rather than ride the bus; park farther from your worksite; etc.)?								
7. How many days per week do you purposefully increase your physical activity through *lifestyle activities* (e.g., take the steps rather than the escalator or elevator; walk quickly rather than slowly; wash the car by hand; etc.)?								
8. How many days per week do you typically engage in *household tasks or gardening or yard work* that result in an increased breathing rate, increased heart rate, or cause you to sweat?								
9. How many days per week does your *employment* typically result in you engaging in physically taxing activities of a moderate or vigorous nature? Mark zero if you are unemployed or retired.								
10. How many days per week do you typically engage in *leisure activities* other than physical exercise that result in increasing the amount of moderate or vigorous physical activity you get?								

[a]Note this instrument has not been tested for reliability and validity. It is presented here for illustrative and didactic purposes only. A total score is obtained by summing the number of days one engages in each of the 10 listed activities. Note that item #1 must be reversed before summing the items because a lower score on item #1 indicates more physical activity. Thus, the theoretical range of scores is 0 (no physical activity at all throughout the week) to 70 (completes each of the listed activities on each day of the week).

From *Physical Activity Assessments for Health-Related Research*, edited by Gregory·J. Welk, 2002, Champaign, IL: Human Kinetics.

an experimental study. However, when you compare the groups, you learn that those in the experimental group are no more active than those in the control group. Assuming that you put a great deal of time, effort, money, and resources into the intervention and that the intervention truly is effective, how could you have received such false results? Technically, you have committed an error of the second type, or a Type II error. A Type II error occurs when a relationship exists between the variables you are investigating (in this case physical activity and the intervention) and your study fails to conclude that there is a relationship.

There are several reasons that might cause you to obtain such faulty results. Perhaps you did not have enough participants. That is, your study was not "powerful" enough to detect actual differences in physical activity that occurred as a result of your intervention. There are a number of methodological artifacts that might cause you to draw inaccurate conclusions. Perhaps you used the wrong type of statistical analysis and thus arrived at the wrong conclusion. Perhaps the participants did not correctly follow your intervention procedures. Perhaps those delivering the intervention did a poor job of administering it. Finally, it might be that your 10-item self-report method of assessing physical activity is flawed. That is, it does not accurately measure the physical activity in which your participants engage. This last issue involves the psychometric characteristics of the method used to assess physical activity. The psychometric characteristics of physical activity measurement will be presented and illustrated in this chapter. *Psychometric* refers to the ability to accurately measure and report the characteristic being described, in our case physical activity.

General Concepts and Key Terminology

The key terms in this chapter are *reliability* and *validity*. Each of these terms can be further subdivided into specific types of reliability or validity. In the simplest sense, reliability is consistency of measurement. An instrument is reliable in the sense that it results in the same score or measure when repeated a second or subsequent time. A special type of reliability is objectivity. Objectivity, or inter-rater reliability, is the agreement between two or more raters or observers when scoring or observing the same characteristic, attribute, or behavior.

Validity is the truthfulness of the measurement obtained when something is assessed. Note that reliability is important because it is a vital component of validity. It is illogical to think of a measurement as truthfully measuring the characteristic, attribute, or behavior that it purports to measure unless it can consistently measure the characteristic, attribute, or behavior. The second component of validity is relevance. A measurement will be valid if it is both reliable (i.e., consistent) and relevant (i.e., related). Validity is the most important characteristic of measurement.

It is important to realize that any method of assessing physical activity (or any other characteristic, attribute, or behavior) should not be viewed as generally reliable or valid. The act of assessment (i.e., measurement) is reliable or valid with specific populations, in given circumstances, measured in specific ways, and so on. That is to say, something that is reliable or valid in one setting is not necessarily reliable or valid in another setting.

The general types of reliability and validity are summarized in figure 2.1. Subsequent sections will provide further clarification on reliability and validity and describe how the different types should be used and interpreted in physical activity research. For a more extensive review of reliability and validity, consult *Standards for Educational and Psychological Testing* (American Psychological Association, 1985; American Educational Research Association, 1999).

Reliability

While validity is the most important psychometric characteristic of physical activity measurement, reliability is described first because reliability must be established before a measurement can be valid. Reliability is denoted by $r_{xx'}$ or $R_{xx'}$. The use of reliability here is more theoretical. If a value was reported it would be appropriate to use r or R. Reliability values range from 0.00 (zero), indicating no consistency of measurement, to 1.00 (unity), indicating perfect reliability. Thus, the limits of $r_{xx'}$ are $0.00 \leq r_{xx'} \leq 1.00$. To better appreciate the nature of reliability, it is useful to review the basics of reliability theory. The types of reliability will then be described along with some specific applications of reliability.

Reliability Theory

An obtained measurement consists of two parts. The obtained (observed, recorded, or total) score actually consists of the true portion and an error portion of what is being assessed. This is illustrated in

Reliability	Validity

• Interclass: based on the Pearson Product Moment Correlation
 	
  • Two trials only
 	
    • Test–retest
    • Equivalence
 	
• Intraclass: based on the Analysis of Variance (ANOVA)
 	
  • Multiple trials
 	
    • Internal consistency
	• Evidence of content-related validity
	
	  • Based on logic, theory, observation, and expert review
	
	• Evidence of criterion-related validity
	
	  • Based on the Pearson Product Moment Correlation (r) between surrogate and criterion measures
	
	• Evidence of construct-related validity
	
	  • Based on accumulated theoretical and statistical evidence

Figure 2.1   Reliability and validity.

figure 2.2 where the square represents obtained or total variance. The true portion represents the perfectly accurate value of what is being measured and the error portion can be made up of technical, personal, consistent, bias, and random errors as well as unreliable components. Technical errors could be due to instrument errors; personal errors could be due to lack of training; consistent errors could be due to a tendency for one to be lenient or restrictive in reporting or obtaining results; bias could positively or negatively affect one's score; and random and unreliable component errors are idiosyncratic characteristics that are inconsistent and add to the error component of the measurement.

The greater the proportion of the variation in the obtained score that consists of true score variability, the greater the reliability. Consider the limits on reliability. If the obtained score consists entirely of true score, the reliability is 1.00. If the obtained score consists entirely of error score, the reliability is 0.00. In reality, neither of these generally occur, and the reliability is somewhere between 0.00 and 1.00. This is illustrated in equation 2.1. (See Morrow and Jack-

son [1993] for a review of the interpretation and presentation of the reliability coefficient.)

$$r_{xx'} = \frac{\text{True score variance}}{\text{Obtained score variance}} \quad (2.1)$$

Types of Reliability

Reliability can be subdivided into several types, described as interclass reliability, objectivity, and intra-class reliability. The purpose here is not to review how to calculate the various reliability types but to demonstrate how to interpret them. For examples of how to calculate the various types, consult a measurement text in human performance (e.g., Baumgartner & Jackson, 1999; Morrow et al., 2000; Safrit & Wood, 1995).

Interclass Reliability Coefficient ($r_{xx'}$).   Interclass reliability, typically reported with $r_{xx'}$, is based on the Pearson Product Moment correlation (r), or Pearson correlation, between two administrations of the same or equivalent test. A high (i.e., $r_{xx'} \geq 0.80$) correlation between administrations of the two trials of the test indicates a reliable test. An interclass reliability coefficient could be obtained if you administered the SAWPA on two different occasions and then intercorrelated the results from each administration. An assumption is also made that there are no mean differences between trials. Mean differences can be tested statistically with a correlated *t* test or repeated measures of analysis of variance (ANOVA). See any basic statistics book for how to conduct these tests of mean comparisons.

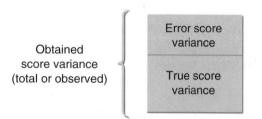

Figure 2.2   Obtained, error, and true score variance.

Objectivity Coefficient. One can use the Pearson correlation methodology to estimate the objectivity or inter-rater reliability of different raters. If the relationship is high (generally, r ≥ 0.80), one concludes that the measure reflects inter-rater reliability. A similar method could also be used to estimate intra-rater reliability where the same rater reports scores of the same event on two different occasions. Again, if the Pearson correlation is high, one can conclude that the rater or observer obtains similar results each time values are recorded. This assumes that there are no constant increases or decreases in the scores recorded from the first to the second observation. An objectivity coefficient could be obtained by having two different observers watch children engage in physical activity during recess and then correlating the amount of physical activity recorded by each observer for each child.

Intraclass Reliability Coefficient ($R_{xx'}$). The analysis of variance (ANOVA) is used to estimate the intra-class reliability, also known as the alpha coefficient (α). Actually, there are two ways that one can interpret intraclass reliability. The first is to calculate the obtained, true, and error score variances and use this to estimate the reliability of the trials from equation 2.1. An advantage of the intraclass method is that more than two trials or observers can be used. The reliability or objectivity coefficient calculated with the intraclass method results in the reliability for the average (or total) of all of the trials. This is somewhat problematic because the typical user has only one trial for which the reliability estimate is desired. In cases where more than one trial is used in intraclass reliability, the Spearman-Brown Prophecy Formula is used (see Morrow et al., 2000) to estimate the reliability for a single trial.

The second way that the intraclass reliability is estimated is called *internal consistency*. Assume that you have a questionnaire that asks people about the amount of physical activity they engage in. With internal consistency, the intent is to determine if the items on the instrument are internally consistent in the sense that they each measure the same general construct, in this case physical activity. Internal consistency values range from 0.00 to 1.00. Again, acceptable internal consistencies ($R_{xx'}$) tend to be 0.80 or higher. For example, using the SAWPA (see form 2.1), one who is very, very active would report 5, 6, or 7 days for each of the items, while someone who is quite sedentary would report 0, 1, or 2 days for nearly all of the items. Note that each of these individuals is internally consistent in the sense that their values are alike for each item.

Distinctions and Applications of Reliability Theory

While the focus of reliability is consistency of measurement, there are a variety of reliability terms, uses, and interpretations that are associated with reliability issues. It is important to understand these differences and interpretations when conducting research on physical activity. Definitions and examples are provided next to help the reader understand and interpret applications of reliability theory in the measurement of physical activity.

Stability Versus Internal Consistency Reliability. There is an important distinction between stability reliability and internal consistency reliability. Stability is the consistency of the measurement across some time period. Stability across time can be estimated with either interclass or intraclass reliability methods. Stability reliability is also often called *test-retest reliability*. However, internal consistency is determined with an intraclass model and only reports the degree to which scale respondents tend to report similar answers for the various items presented on the questionnaire. If an individual tends to be at one end of the scale for all or most items while another individual is at the other end of the scale for all or most items, the scale is said to be internally consistent (i.e., has a sufficiently high α coefficient). However, just because a scale has high internal consistency, it should not be assumed that it has high (or even adequate) stability across time.

Consider the SAWPA instrument. It is possible to reflect internal consistency across the 10 items, yet when people are asked to respond to the instrument a month later, their scores might change considerably. Thus, the SAWPA instrument would be said to reflect internal consistency reliability but not stability reliability.

Test-Retest and Equivalence Reliability. A test-retest reliability coefficient is estimated when assessment is completed at one point in time and then repeated at a subsequent time. For example, the SAWPA could be completed on Monday and then repeated the following Monday. The correlation between these measures would be considered a test-retest reliability coefficient. An equivalence reliability estimate is created when two different measurements are correlated with one another and they are both thought to measure the same construct. If the correlation is high (i.e., r ≥ 0.80), it can be said that the two instruments are essentially equivalent in the sense that they are tapping the same construct. If the SAWPA were correlated with another version of an instrument created to assess physical activity, this would be interpreted as an equivalence reliability coefficient. Either interclass or intraclass reliability methods can be used to estimate the test-retest or equivalence reliability.

Generalizability Theory. Generalizability theory (G-theory) is an extension of the intraclass reliability model that permits one to estimate a number of different reliabilities (called *G coefficients* but which are interpreted and evaluated as reliability coefficients). The benefit of generalizability theory is that many different sources of potential error can be included in the ANOVA model to determine the effect these factors (called *facets*) have on the obtained generalizability. Two types of generalizability studies are conducted. The G-study results in G coefficients for various sources of error. Thus, in a G-study, several reliability (G) coefficients can be calculated. The D-study phase of a generalizability study permits the researcher to estimate the effect of changes in the number of trials, sources of error, and so on, as well as the effects on the reliability (G) coefficient. See Brennan (1992) for a broad-spectrum presentation of G-theory and Morrow (1989) and Morrow et al. (1986) for examples in exercise science.

Factors Affecting Reliability

There are a number of factors that affect the estimated reliability. These can be generally categorized as

1. the sample and sample variability,
2. the number of items or observers,
3. the instrumentation, and
4. the time between testing.

The type of people observed affects reliability. Higher reliability is obtained when you have greater variability in the people being measured. The greater the variability in the measures, the greater the reliability will be. The number of people involved in a reliability study is not directly important in determining the size of the reliability coefficient. The type of people is the important variable affecting the obtained reliability. Certainly, having more people in a study increases confidence in the results, but generally the greater the variability in the people, the higher the expected reliability. The more familiar the tester or person being measured is with the methodology, the greater the reliability. Internal consistency and test-retest reliabilities generally

increase with an increase in the number of items on the written scale. Reliability also tends to increase with an increase in the number of observers used or in the number of occasions on which the person is assessed. Reliability generally increases when the precision of the measurement instrument or process increases. As the time between observations or measures increases, the reliability typically decreases. If the circumstances change between the measurement times, the reliability is expected to decrease.

Validity

Validity is the degree to which a measure truthfully reflects what it is intended to measure. As indicated, validity is dependent on reliability (i.e., it must be consistent) and relevance (i.e., it must be related). The types of validity are first described followed by an overview of factors that influence validity.

Types of Validity

There are three broad types of validity, defined as content related, criterion related, and construct related (see figure 2.1). Each of these types can be subdivided further depending on the nature of the comparisons being made.

Evidence of Content-Related Validity. Content-related validity is based on the logical interpretation made about the truthfulness of an instrument. For example, you might develop a series of 10 items to assess the amount of physical activity in which one typically engages. You might have experts review the questions to see if they adequately tap the information you want to obtain. If your series of 10 items is viewed as logically providing the information you want, it is said to reflect content validity. Generally, content validity is the weakest form of validation evidence. The SAWPA demonstrates content-related validity because each of the 10 items appears to assess the amount of various types of physical activity in which one might engage. If you asked experts to review the items, they would generally agree that the items tap the amount of physical activity that one engages in during a typical week. Thus, the SAWPA reflects content validity.

Evidence of Criterion-Related Validity. With criterion-related validity, the Pearson correlation coefficient is used to estimate the statistical relationship between the results of a test and a known valid criterion. The criterion is considered to be the most accurate measure of what is being assessed, but the criterion may be very difficult or expensive to measure. Thus, researchers should determine if another measurement tool, or surrogate measure, can be used to estimate the criterion because the surrogate instrument is easier or less expensive to administer. The SAWPA developed for the scenario at the beginning of this chapter is a surrogate measure of actual physical activity.

Key to criterion-related validity is having a good criterion. This is an especially challenging problem in physical activity research because there are documented limitations associated with each of the primary assessment techniques. Selecting an appropriate criterion measure for a validation study must be done carefully to provide the best standard for comparison. Assuming an appropriate criterion was identified, you might compare a 10-item self-report questionnaire like the SAWPA with results obtained from a criterion measure to see if the results obtained from the instrument correlate well with the results from the criterion. If they correlate well (e.g., r ≥ 0.80), the instrument is said to reflect criterion-related validity.

There are actually two types of criterion-related validity: concurrent validity and predictive validity. The real difference between them is the point in time at which the criterion is measured. If the criterion and the surrogate measure are obtained at the same time, this is referred to as *concurrent validity*. If the criterion is measured some time in the future, it is referred to as *predictive validity*. To differentiate between concurrent and predictive validity, an example is helpful. Assume that you assess an individual's current level of physical activity and then assess whether he or she has cardiovascular disease. Correlating these two variables yields a concurrent validity coefficient because the two measures are taken within the same time period. Now consider a second study where you measure a person's current level of physical activity but instead choose to measure whether he or she develops cardiovascular disease some time in the future (e.g., at the age of 40, 20 years from now, and so on). The correlation between physical activity level at this time and cardiovascular disease measured in the future is a predictive validity coefficient.

Criterion-related evidence, based on the Pearson correlation coefficient, ranges from −1.00 to +1.00. The closer the absolute value of the validity coefficient is to 1.00, the higher the validity and the lower the error in estimating the criterion. Thus, as the absolute value of the validity approaches 1.00, the error of estimation is reduced toward 0.00.

Evidence of Construct-Related Validity. Construct-related validity involves making a decision about the validity of a measure based on a series of studies and/or rational evidence of validity. For example, if many studies provide similar evidence, this builds support for the body of knowledge about the instrument being validated. Likewise, some validation evidence is provided when two measures that theoretically should not correlate in fact do not correlate highly. Having measures correlate that theoretically should correlate is called *convergent validity* evidence, while having measures not correlate that theoretically should not correlate is called *discriminant validity* evidence.

Essentially, construct-related evidence is provided when statistical evidence provides the support for what should logically or theoretically occur if a measure is valid. For example, if a group known to be more physically active than another group obtains a higher score on the 10-item SAWPA, this provides evidence of the construct validity of the instrument. Additional construct-related evidence would be provided for the SAWPA if people who engage in sedentary employment generally report lower scores than those who engage in physically active jobs, or if those who are overweight generally report lower scores than those who are closer to their desired weights. Additional construct-related evidence for the SAWPA can be provided by obtaining statistical evidence related to theoretically differing individuals and groups. Construct evidence would also be provided if such a comparison were made item by item for each item on the scale.

Factors Affecting Validity

A variety of factors influence the validity of an assessment. With content-related evidence, the expertise of the person making the logical judgment about the validity of an instrument is most important. Validity increases as the quality of the judge or expert increases. With criterion-related evidence, the quality of the criterion is of most importance, as it is extremely important that the criterion is as accurate as possible. Otherwise, one is correlating the surrogate measure with a criterion that could have considerable error. The degree of variability in the criterion and the surrogate measure also impacts the criterion-related validity coefficient. The greater the variability in the surrogate and criterion measures, the higher the potential validity. With construct-related validity, the amount and type of evidence provided helps to build confirmation of validity.

The number of people is important in validation studies because the confidence one can place in the obtained criterion-related validity is increased with increased sample size. It is also important to cross-validate the results obtained in a validation study (see chapter 3). That is, the results obtained with the initial study should be tested with a second sample to see if the results obtained in the initial validation work when generalized to similar people.

Criterion measures in physical activity assessment are often direct observation of the activity, electronic surveillance (e.g., motion sensors, pedometers, accelerometers), doubly labeled water, direct and indirect calorimetry, and heart rate monitoring. These are often difficult to obtain because of the time, effort, and resources needed to assess them. Thus, surrogate measures (e.g., self-reported physical activity, activity logs, activity histories) have been used to estimate physical activity.

Measurement Issues in Physical Activity Research

There are some specific distinctions in terminology and design that are important to consider in research. Several of these measurement issues are described here, including examples to help readers apply the information.

Errors of Measurement

All measures have some degree of error associated with them. The amount of error associated with reliability is referred to as the *standard error of measurement* (SEM). The amount of error associated with validity is called the *standard error of estimate* (SEE) (also called the *standard error* [SE] or the *standard error of prediction* [SEP]). Note that there are also other standard errors used in statistics. Researchers should be careful to use and interpret them correctly. The SEM and SEE are important in measurement theory. They reflect the degree to which one's obtained or observed score will fluctuate because of errors of measurement, the SEM for reliability errors and the SEE for validity errors. See Baumgartner and Jackson (1999), Morrow et al. (2000), or Safrit and Wood (1995) for examples of calculating the SEM and SEE.

Errors of measurement (SEM or SEE) can be interpreted when considering the normal distribution. Errors of measurement are theoretically normally distributed around a mean of zero. Thus, some errors are positive and add to the obtained score, and some errors are negative and result in the obtained score being lowered. But in the long run, the average of errors is assumed to be zero. All standard errors can be thought of as standard deviations (which in fact they are). Any score of ±1 SEM can be viewed as 68% likely of capturing the person's true score. A score of ±2 SEM can be viewed as approximately 95% likely of capturing the person's true score. The SEM helps to determine one's confidence in interpreting the obtained score and how closely it approximates the individual's true score. The SEE helps to determine one's confidence in interpreting the predicted score and how closely it approximates the criterion score. Think of the SEM and the SEE as the margin of error. Margin of error is often reported in national polls for presidential elections or other national questions. This margin of error is actually a type of standard error. Such national polls often report a margin of error of ±3%. This simply helps the reader place confidence in and interpret poll results.

Assume that the SAWPA has a reliability of 0.91 and an SEM of 3 units. Note from form 2.1 that the range of units for the SAWPA is 0 to 70. If the SEM is 3 units and an individual has a self-reported score of 40, you can be 68% confident that the true SAWPA score is between 37 and 43 (i.e., ±1 SEM [±3]). Additionally, you can be about 95% confident that the individual's true SAWPA score is between 34 and 46 (i.e., ±2 SEM [2 × ±3 = ±6]).

Assume the SAWPA score is related to actual energy expenditure measured with the doubly labeled water method. Additionally, assume that as a result of this relationship, the predicted weekly energy expenditure for a person is 20,000 kcal/week. However, the relationship between the SAWPA and the criterion measure (i.e., doubly labeled water) is not perfect. This results in the SEE being 2,000 kcal/week. You can be 68% confident that the person's actual energy expenditure per week is between 18,000 and 22,000 (i.e., ±1 SEM [±2,000]). Likewise, you can be approximately 95% confident that the actual energy expenditure is between 16,000 and 24,000 kcal/week (i.e., ±2 SEM [2 × ±2,000 = ±4,000]).

Notice that the SEM and the SEE are utilized in the same way, but the SEM is related to the reliability of a measure and the SEE is related to the validity of a measure.

Norm-Referenced and Criterion-Referenced Reliability and Validity

Another important distinction is between norm-referenced and criterion-referenced reliability and validity. In norm-referenced decisions, the specific score that an individual achieves on each trial is of interest for reliability and the criterion for validity. For example, you might be interested in the reliability of self-reported physical activity from a questionnaire. Assessing this information on at least two occasions leads to norm-referenced reliability information. If you were to also obtain a measure of $\dot{V}O_2$max for the individual and use it as the criterion, you could correlate the self-reported physical activity with measured $\dot{V}O_2$max and report a concurrent validity coefficient. Note that with norm-referenced decisions, the actual score is used for interpretation for both the surrogate and criterion measurements.

Criterion-referenced decisions involve categorical information. Using the example from norm-referenced reliability and validity can lead to a criterion-referenced example. Assume that you take the self-reported physical activity questionnaire and reduce it to a dichotomous variable. For example, you could reduce it to one category that includes those who are physically active enough to achieve a health benefit (i.e., vigorous physical activity for a minimum of 3 days/week, 30 min/day, or moderate physical activity for a minimum of 5 days/week, 30 min/day) and those who are not physically active enough to achieve a health benefit, thus creating two categories. If you were to repeat the self-reported questionnaire on a second occasion and again create the dichotomy, you could determine whether the individuals are placed in the same category each time. That is, are respondents reliably (consistently) reporting their physical activity so that they are placed in the same category each time the questionnaire is administered? Criterion-referenced reliability is illustrated in figure 2.3a. Bowles et al. (2001) conducted a criterion-referenced reliability study and found the criterion-referenced reliability to be quite good for two different single items assessing whether individuals are engaged in sufficient physical activity to achieve a health benefit.

There are several methods of reporting criterion-referenced reliability. These include (1) proportion of agreement between the two assessment periods, (2) the κ coefficient that adjusts the reliability estimate for chance agreements, (3) the ϕ coefficient that is simply the Pearson correlation coefficient between two dichotomously scored items, and (4) χ^2 that tests the statistical significance of the relationship between two measures. Acceptable levels of criterion-referenced reliability for these measures typically are at least 70% for proportion of agreement, >0.50 for κ; >0.80 for the ϕ coefficient; and p < 0.05 for the χ^2 test of association between trials. In the study described earlier, Bowles et al. (2001) reported that proportion of agreement was 92% and 91%, κ was 0.84 and 0.80, ϕ coefficients were 0.84 and 0.80, and χ^2 values were significant (p < 0.001), indicating that each of the two items they investigated demonstrated good criterion-referenced reliability. See Looney (1989) for an extensive discussion of criterion-referenced reliability.

Criterion-referenced validity is illustrated in figure 2.3b. Blair et al. (1989) suggested that men who have a maximum oxygen consumption ($\dot{V}O_2$max) of ≥35 ml/kg/min (i.e., 10 METS) have sufficient fitness for good health. Consider the 35 ml/kg/min criterion as the standard to be met (or not met). By creating a dichotomy based on how much physical activity is self-reported, one could determine from the SAWPA whether individuals engage in sufficient physical activity to achieve a health benefit. One could then measure participants' $\dot{V}O_2$max and create a dichotomy by identifying those who have a $\dot{V}O_2$max ≥ 35 ml/kg/min and those who do not achieve a $\dot{V}O_2$max of 35 ml/kg/min. Interest then focuses on whether there is a relationship between these two dichotomies (i.e., the self-report dichotomy and the $\dot{V}O_2$max dichotomy).

As with criterion-referenced reliability, proportion of agreement, the ϕ coefficient, the contingency coefficient (C), and the χ^2 test of association can be used to determine evidence of the validity of the self-reported physical activity questionnaire. Desired criterion-referenced validity values are ≥80% for proportion of agreement, >0.80 for the ϕ coefficient, >0.80 for the contingency coefficient, and p < 0.05 for the χ^2 test of association between the surrogate and criterion measures dichotomies. See Safrit (1989) for an extensive review of criterion-referenced validity.

Reliability and Validity Generalization

The generalizability for the obtained reliability and validity coefficients is another major issue in physical activity research. It is important to evaluate whether an instrument that is reliable and valid for one setting is reliable and valid in another setting. Setting includes participants (who vary in age, cognitive ability, gender, reading ability, and many

a

	Day 1	
Day 2	*Insufficient physical activity for a health benefit*	*Sufficient physical activity for a health benefit*
Insufficient physical activity for a health benefit	You expect people here when reliability is good	Respondents were inconsistent in reporting their physical activity levels
Sufficient physical activity for a health benefit	Respondents were inconsistent in reporting their physical activity levels	You expect people here when reliability is good

b

	Criterion	
Self-reported physical activity	$\dot{V}O_2max < 35$ ml/kg/min	$\dot{V}O_2max \geq 35$ ml/kg/min
Insufficient physical activity for a health benefit	You expect people here when validity is good	You do NOT want people here
Sufficient physical activity for a health benefit	You do NOT want people here	You expect people here when validity is good

Figure 2.3 Criterion-referenced *(a)* reliability and *(b)* validity.

other factors), time (of day, of the year, and so on), location, and any other variable that might change from one administration to another. It is best to assume that the reliability or validity of an instrument does not generalize to other people, places, or times unless there is specific evidence that it does.

Revisiting the Research Scenario

It should now be clear that having a reliable and valid instrument is paramount when attempting to assess physical activity. If the instrument (or surrogate measure) cannot be trusted as a result of poor reliability or validity, it could well give you inaccurate or misleading information and cause you to draw invalid conclusions. In the scenario presented at the beginning of this chapter, the instrument may not have been of sufficient quality to identify true differences in the groups tested. Thus, the Type II error was made and you falsely concluded that your intervention was not successful. Researchers should always describe the sample well and report the reliability and validity obtained in both the study and related studies.

Summary and Examples

Table 2.1 contains examples of reliability studies in the assessment of physical activity and table 2.2

TABLE 2.1   Examples of Reliability Studies in Physical Activity

Author(s)	Purpose	Method	Results	Comments
Albanes et al. (1990)	Provide evidence of the equivalence reliability of the Minnesota Leisure-Time Physical Activity Survey (MLTPAS) with other measures of physical activity	Correlate MLTPAS with 6 other measures of physical activity	Harvard Alumni Scale (0.54); Stanford 7-day recall (0.48); Framingham Activity Survey (0.33); Baecke Index (0.36); Lipid Research Clinics Survey (0.63); Health Insurance Plan of New York (0.52)	The actual purpose of the study was to provide evidence of the validity (see table 2.2) of the MLTPAS. However, the results could also be used as evidence of the equivalence reliability of the 6 scales.

Author(s)	Purpose	Method	Results	Comments
Crocker et al. (1997)		Self-administered 7-day recall measure with 10 items to 215 students aged 9 to 15	Internal consistency for females was 0.83 and 0.80 for males. Using the intraclass model, test-retest across 1 week, reliability was 0.82 for females and 0.75 for males.	Both internal consistency (α coefficient) and test-retest coefficients suggest that the instrument is reliable.
Jacobs et al. (1993)	Estimate test-retest of 10 physical activity questionnaires	1-month reliability estimated for each questionnaire for 78 men and women	36 test-retest reliabilities were calculated for subscales of the questionnaires. Results were $0.12 \leq r_{xx'} \leq 0.95$.	Test-retest depended on the questionnaire and specific subscale evaluated.
Rowe et al. (1997)	Determine the test-retest intraclass reliability of the System for Observing Fitness Instruction Time (SOFIT) for measuring physical activity in 173 students in grades 1 through 8	Students' heart rates obtained for 7 activities on 2 different days	Intraclass $R_{xx'}$ ranged from 0.66 to 0.84 for heart rates measured during the activities.	Heart rate can be reliably obtained with a single-day measure for lying, sitting, standing, walking, jogging, curl-ups, and push-ups.
Sallis et al. (1990)	Estimate inter-instrument reliability of the Caltrac, 1 worn on each hip	Caltracs worn by 12 children for 2 days	$r_{xx'} = 0.96$	Caltrac instruments can be used inter-changeably.
Westerterp et al. (1992)	Estimate test-retest reliability of the Zutphen Physical Activity Questionnaire	Questionnaire completed by 21 older Dutch men (age > 64) on 2 occasions, 4 months apart	$r_{xx'} = 0.93$	Reliability was very good across 4 months between measure-ments.

TABLE 2.2 Examples of Validity Studies in Physical Activity

Author(s)	Purpose	Method	Results	Comments
Albanes et al. (1990)	Provide evidence of the construct validity of the Minnesota Leisure-Time Physical Activity Survey (MLTPAS)	Correlate MLTPAS (criterion) with 6 other measures of physical activity	Harvard Alumni Scale (0.54); Stanford 7-day recall (0.48); Framingham Activity Survey (0.33); Baecke Index (0.36); Lipid Research Clinics Survey (0.63); Health Insurance Plan of New York (0.52)	The 6 measures correlate moderately with the MLTPAS. Theoretically, they should because they are tapping the same construct (i.e., physical activity). Thus, this is evidence supporting the construct validity of the MLTPAS.

(continued)

TABLE 2.2 · *(continued)*

Author(s)	Purpose	Method	Results	Comments
Freedson et al. (1998)	Estimate the validity of the CSA accelerometer	Correlate CSA activity counts with METs (criterion)	r = 0.91	The CSA counts validly estimate energy expenditure.
Klesges and Klesges (1987)	Validate the Caltrac with 9 hours of direct observation in 30 preschool children	Correlate Caltrac recordings with direct observation (criterion)	The range of r for hourly comparisons was from 0.62 to 0.95. For the all-day comparison, r was 0.54.	Concurrent validity for the Caltrac was moderate to high when using direct observation of physical activity as the criterion.
Rikli and Jones (1999)	Criterion-related validity evidence for 5 test items for older adults	Correlations calculated with specific criterion for each item	Concurrent validities ranged from 0.76 to 0.82 for men; 0.71 to 0.86 for women.	There was good evidence of criterion-related validity.
Rowe et al. (1993)	Determine validity of the System for Observing Fitness Instruction Time (SOFIT) for measuring physical activity in 173 students in grades 1 through 8	Heart rates (criterion) compared to amount of physical activity recorded with the SOFIT	Heart rates were related to the amount of physical activity in which the student was involved.	The study provided construct-related evidence of the validity of the SOFIT.
Sallis et al. (1993)	Estimate the validity of a self-administered physical activity survey with 4th-grade students	Correlate activity checklist with Caltrac (criterion)	r = 0.28	Validity with 4th-grade children was limited.
Starling et al. (1999)	Estimate validity of the Yale Physical Activity Survey (YPAS)	Compare energy expenditure estimates from the YPAS with energy expenditure from doubly labeled water (criterion)	Mean differences for men and women with the 2 measures were not significantly different.	Construct validity evidence is provided because mean values do not differ for the surrogate (YPAS) and the criterion measure (doubly labeled water)
Westerterp et al. (1992)	Estimate validity of a questionnaire with elderly Dutch men	Correlate questionnaire with activity index based on doubly labeled water (criterion) for 21 men	r = 0.61	Validity with senior adult Dutch men (mean age = 74) was moderate.

contains similar information for validity studies in physical activity. Both tables provide comments on the methods and interpretation of results. These examples should be helpful as you encounter reliability and validity information when conducting or reading research about physical activity assessment. Alternatively, the tables may provide you with examples of the method you might choose to use when conducting research on physical activity. The June 2000 *Research Quarterly for Exercise and Sport* (RQES, 2000), a special issue, contains manuscripts directly related to measuring physical activity, with specific reference to reliability and validity issues. The manuscripts are from Measurement of Physical Activity, the second of the Cooper Institute conference series. This specialty conference was held in conjunction with the Measurement and Evaluation Council of the American Association for Active Lifestyles and Fitness. The January 1993 issue and the December 2000 supplement of *Medicine and Science in Sports and Exercise* (MSSE) both contain a series of research papers on the assessment of physical activity. If you are interested in reliability and validity issues in the assessment of physical activity, review the RQES (2000) supplement and the MSSE issues.

References

Albanes, D., Conway, J.M., Taylor, P.R., Moe, P.W., & Judd, J. (1990). Validation and comparison of eight physical activity questionnaires. *Epidemiology, 1,* 65-71.

American Educational Research Association. (1999). *Standards for educational and psychological testing.* Washington, DC: American Educational Research Association.

American Psychological Association. (1985). *Standards for educational and psychological testing.* Washington, DC: American Psychological Association.

Baumgartner, T.A., & Jackson, A.S. (1999). *Measurement for evaluation in physical education and exercise science* (6th ed.). Boston: MCB McGraw-Hill.

Blair, S.N., Kohl, H.W., III, Paffenbarger, R.S., Jr., Clark, D.G., Cooper, K.H., & Gibbons, L.W. (1989). Physical fitness and all-cause mortality: A prospective study of healthy men and women. *JAMA, 262,* 2395-2401.

Bowles, H.R., Morrow, J.R., Jr., & Lochmann, A.C. (2001). Criterion-referenced reliability of single-item self-reported measures of physical activity. *Medicine and Science in Sports and Exercise, 33* (Suppl.), S119 [Abstract].

Brennan, R.L. (1992). *Elements of generalizability theory.* Iowa City: American College Testing.

Crocker, P.R.E., Bailey, D.A., Faulkner, R.A., Kowalski, K.C., & McGrath, R. (1997). Measuring general levels of physical activity: Preliminary evidence for the Physical Activity Questionnaire for Older Children. *Medicine and Science in Sports and Exercise, 29,* 1344-1349.

Freedson, P.S., Melanson, E., & Sirard, J. (1998). Calibration of the Computer Science and Applications, Inc. accelerometer. *Medicine and Science in Sports and Exercise, 30,* 777-781.

Jacobs, D.R., Ainsworth, B.E., Hartman, T.J., & Leon, A.S. (1993). A simultaneous evaluation of 10 commonly used physical activity questionnaires. *Medicine and Science in Sports and Exercise, 25,* 81-91.

Klesges, L.N., & Klesges, R.C. (1987). The assessment of children's physical activity: A comparison of methods. *Medicine and Science in Sports and Exercise, 19,* 511-517.

Looney, M.A. (1989). Criterion-referenced measurement: Reliability. In M.J. Safrit & T.M. Wood (Eds.), *Measure-ment concepts in physical education and exercise science* (pp. 137-152). Champaign, IL: Human Kinetics.

Morrow, J.R., Jr. (1989). Generalizability theory. In M.J. Safrit & T. Wood (Eds.), *Measurement concepts in physical education and exercise science* (pp. 73-96). Champaign, IL: Human Kinetics.

Morrow, J.R., Jr., Fridye, T., & Monaghen, S.D. (1986). Generalizability of the AAHPERD health related skinfold test. *Research Quarterly for Exercise and Sport, 57,* 187-195.

Morrow, J.R., Jr., & Jackson, A.W. (1993). How "significant" is your reliability? *Research Quarterly for Exercise and Sport, 64,* 352-355.

Morrow, J.R., Jr., Jackson, A.W., Disch, J.G., & Mood, D.P. (2000). *Measurement and evaluation in human performance* (2nd ed.). Champaign, IL: Human Kinetics.

Rikli, R.E., & Jones, C.J. (1999). Development and validation of a functional fitness test for community-residing older adults. *Journal of Aging and Physical Activity, 7,* 129-161.

Rowe, P.J., Schuldheisz, J.M., & van der Mars, H. (1997). Validation of SOFIT for measuring physical activity of first-to eighth-grade students. *Pediatric Exercise Science, 9,* 136-149.

RQES (2000). Special issue. *Research Quarterly for Exercise and Sport, 71.*

Safrit, M.J. (1989). Criterion-referenced measurement: Validity. In M.J. Safrit & T.M. Wood (Eds.), *Measurement concepts in physical education and exercise science* (pp. 119-135). Champaign, IL: Human Kinetics.

Safrit, M.J., & Wood, T.M. (1995). *Introduction to measurement in physical educational and exercise science* (3rd ed.). St. Louis: Mosby.

Sallis, J.F., Buono, M.J., Roby, J.J., Carlson, D., & Nelson, J.A. (1990). The Caltrac accelerometer as a physical activity monitor for school-age children. *Medicine and Science in Sports and Exercise, 22,* 698-703.

Sallis, J.F., Condon, S.A., Goggin, K.J., Roby, J.J., Kolody, B., & Alcaraz, J.E. (1993). The development of self-administered surveys for 4th grade students. *Research Quarterly for Exercise and Sport, 64,* 25-31.

Starling, R.D., Matthews, D.E., Ades, P.A., & Poehlman, E.T. (1999). Assessment of physical activity in older individuals: A doubly labeled water study. *Journal of Applied Physiology, 86,* 2090-2096.

Westerterp, K.R., Saris, W.H.M., Bloemberg, B.P.M., Kempern, K., Caspersen, C.J., & Kromhout, D. (1992). Validation of the Zutphen Physical Activity Questionnaire for the elderly with doubly labeled water. *Medicine and Science in Sports and Exercise, 24,* S68 [Abstract].

3

Construct Validity in Physical Activity Research

Matthew T. Mahar, EdD
David A. Rowe, PhD
East Carolina University

Valid measures of physical activity are essential to allow researchers to accurately answer questions about the relationship between physical activity and other health-related variables and about whether an intervention has an effect on physical activity. In addition, valid measures of physical activity must be used to accurately document the prevalence of physical activity or inactivity. Individuals who conduct research on physical activity must choose a measure with evidence of validity. Validity evidence is specific to the purpose of the measurement and to the particular population under investigation. Thus, researchers should choose a measure of physical activity that relates specifically to the particular purpose of their study and to the particular type of people with whom the measure will be used. Likewise, researchers should not develop and use new measures of physical activity until sufficient evidence of validity has been demonstrated on that measure.

Although the framework for understanding validity has changed several times (American Psychological Association et al., 1954, 1966, 1974, 1985, 1999), validity has always been regarded as the most important component of a measurement. The definition of *validity* as "the extent to which a test measures what it purports to measure" (Garrett, 1937, p. 324), although important, is no longer considered an adequate description of this central concept in psychometrics.

Validity refers to the appropriateness of inferences made from specific measures. Validation is the process of obtaining evidence to support these inferences. Thus, the specific measures of physical activity are not validated. Rather, it is the inferences from specific measures that are validated. Evidence of validity should be judged based on the intended use of the data. For example, if one wished to estimate the energy expended during bicycling from an accelerometer worn on the hip, the estimate would probably be in error. This is because little hip movement occurs during cycling, even though a great deal of energy could be expended. Thus, an accelerometer worn on the hip would indicate little motion and underestimate the energy expended during cycling. Any decisions made about people based on data that lack validity are at best questionable, and probably in error.

As we will attempt to describe in this chapter, validity is multifaceted. Thus, evidence of validity can, and should, come from various sources. Validation is the *process* of the accumulation of evidence about the meaningfulness and usefulness of a measurement. The various physical activity validation studies should be thought of as contributing evidence about the validity, or lack of validity, of a measure for a specific purpose. The accumulation of this evidence is an ongoing process that slowly evolves and eventually allows us to place confidence in the choice of instrument to assess physical activity.

Much of our understanding of validity comes from theory developed in areas other than those related to the study of human movement, particularly from psychology and educational measurement. The most commonly cited source of information about validity is a series of manuals usually referred to as the *Test Standards,* developed by the American Psychological Association, American Educational Research Association, and National Council on Measurement in Education (1954, 1966, 1974, 1985, 1999).

Validity and validation are much more complex than we perhaps realize. Because much of our understanding of validity comes from the discipline of psychological measurement (psychometrics), a brief summary of the main developments in validity gained from that field is in order. First, however, it is important to remember one important caveat. In many ways, our field (whether that be exercise science, physical education, recreation, or some other branch of what has been termed broadly *kinesiology,* or *human movement studies*) is very different from psychology. The variables we are interested in (e.g., physical activity, fitness, body composition) are quite different from those studied in psychology. There is often a lot of overlap—the behavioral characteristics of physical activity, for example, fit the psychology model of understanding quite well—but many of the variables we study are quite different from "typical" psychology variables and are measured using methods that are very different from the questionnaire method typically used in psychology. We are quite lucky in that many physical activity-related variables (e.g., heart rate, oxygen consumption, acceleration) are conceptually easier to define and understand than the abstract constructs studied by psychologists. So, although the theoretical understanding of validity developed by psychometricians is important to us, we should consider its relevance to our field carefully at all times.

In many measurement classes, validity is presented as constituting three types: construct, content, and criterion. This is often accompanied by a statement that only one type of validity is necessary, depending on the intended use of the test. In the 1985 APA standards, there was a shift away from this conceptualization toward a more unified view that *content-related, criterion-related,* and *construct-related* were terms applicable to different types of validity evidence, and that construct validity was at the heart of this model. A major shift in thinking was also that all types of validity evidence are necessary for all types of test use. This leads us to a broader, more complex set of questions than the one stated earlier: Does the test measure what it purports to measure? We should instead be asking the following types of questions: What does the instrument measure? What do the scores derived from the instrument tell us? On what populations and in what situations can the instrument be used appropriately? What are appropriate uses of the scores derived from the instrument?

These questions cannot be answered in a single study. To address all of the necessary questions effectively requires a broad range of research designs and methods. Current research in physical activity addresses some of these questions related to construct validity, especially the issue related to the nature of the construct. Although it may not be explicitly acknowledged in the literature, researchers are conducting studies that fall under the umbrella of construct validation. Unfortunately, we are faced with a Catch-22 situation in our research endeavors. One of the major problems we have in investigating, and therefore understanding, the nature of physical activity is that we do not yet know what is the best way to measure it. On the other hand, a key ingredient in designing optimal measures of physical activity is a clear understanding of the nature of the construct.

Considering the vast number of recent research publications in physical activity, it might be reasonable to expect that by now we would have at least one method of measuring physical activity that is acceptable as a stand-alone gold standard. We do not. Every major method of physical activity assessment appears to have at least one Achilles' heel, one weakness that precludes it as a global measure of physical activity. In fact, the doubly labeled water technique, which is often cited as the gold-standard method of measuring physical activity, does not provide information on any of the major dimensions routinely used to describe physical activity (i.e., frequency, intensity, duration, mode, context)!

The fact that researchers are faced with some thorny measurement dilemmas in their efforts to quantify physical activity is inescapable. We maintain that the majority of these problems are really all related to a single issue: the lack of a documented, commonly accepted, systematic program of construct validation that fits this field of study. If physical activity were a simple, concrete phenomenon, such as body weight or arm flexor strength, then this might pose little problem to the average researcher. Through a mixture of logic and good luck, a suitable measure might be found with little difficulty. However, physical activity is much more complex in nature than either of these examples.

In this chapter, we present a paradigm by which the construct validity of physical activity can be investigated. Measurement specialists in our field have benefited from the innovative ideas of leading psychometricians such as Messick (1980, 1989), Cronbach (1971, 1989), Nunnally (1978), and Loevinger (1957). Anyone wishing to fully understand the concept of validity should consult the works of these authors. However, some parts of the validity theory developed in psychology and educational measurement just do not seem to fit the different types of research conducted in the exercise and health sciences. It is with this in mind that we present a blueprint for a strong program of construct validation that will fit a wide variety of constructs and contexts, especially those relating to the study of physical activity.

In summary, validation should be thought of as a process by which evidence is accumulated to demonstrate that inferences made from test scores are appropriate and meaningful. Validation studies that provide various types of evidence about the usefulness of test score interpretation should be conducted on an ongoing basis. The remainder of this chapter is organized by the various stages and types of evidence that researchers need to collect to study the validity of the construct of physical activity. The stages of construct validation include the definitional evidence stage, the confirmatory evidence stage, and finally the theory-testing evidence stage.

Definitional Stage

As a first step in construct validation, definitional evidence of validity should be used to determine and define the domain of physical activity. Definitional evidence of validity is collected to examine how well the operational domain of the instrument represents the theoretical domain of the construct. Historically, the term *content validity* was used to represent definitional evidence of validity on written knowledge tests. In the knowledge test paradigm, content validity is evaluated based on the judgments of experts who determined whether the items on the test represent all of the important content areas and tasks that should be tested. Because of the differences between the knowledge test paradigm and the physical activity assessment paradigm, the term *definitional evidence of validity* better represents what should take place during this stage of validation.

Definitional evidence of validity differs from other aspects of validity evidence. As stated earlier, it is the inferences made from test scores that are validated and not the test itself. Definitional evidence of validity differs in that it is a property of the test rather than a property of the use of the test scores. In other words, definitional evidence of validity is focused on the measurement instrument rather than on the scores resulting from use of the measurement instrument.

Because the focus of definitional evidence of validity is on the measurement instrument, it is appropriate to begin the process of validation by considering definitional evidence. Before one can develop a measure of physical activity or before one can choose an appropriate measure of physical activity for a specific situation and purpose, the definitional validity question of what is physical activity must be examined. Physical activity researchers should use definitional evidence of validity to determine the nature of physical activity and how physical activity should be defined. *Physical activity* has been defined as "any bodily movement produced by skeletal muscle that results in caloric expenditure" (Caspersen, 1989, p. 424). Others (e.g., Freedson & Miller, 2000) have suggested that physical activity also involves a behavioral component, indicating that physical activity is voluntary. Thus, it appears that physical activity has at least three major dimensions—a behavioral dimension, a movement dimension, and an energy expenditure dimension—each of which might also be conceptualized as having subdimensions (e.g., frequency, intensity, duration, mode, context).

Researchers attempting to develop a new measure of physical activity must first determine the specific dimension of physical activity they are trying to measure. Then they must design the instrument so that it assesses the dimension of physical activity that was identified as most important. Likewise, researchers who choose from among available measures of physical activity must first define the construct of interest and then determine which measure best represents physical activity as they defined it. For example, researchers who implement a program designed to increase movement of obese children may choose to use accelerometers because accelerometers are sensitive measures of the amount of movement performed.

Similarly, a well-defined construct can help provide direction to the development or choice of a questionnaire to assess physical activity. For example, if physical activity were defined as the frequency, intensity, and duration of regular, planned exercise, the questionnaire would contain different items than it would if physical activity were defined as any body movement that results in energy expenditure.

As anyone who has tried to measure physical activity knows, valid assessment of physical activity is a daunting task. We emphasize that an important starting point is to specifically define the construct of physical activity to be examined. Researchers should first examine previous research and various theories related to physical activity to define the theoretical domain of physical activity. The resulting definition should characterize physi-

cal activity as it is understood at that point in time. Development of this definition requires a community of researchers who collaborate with each other and examine previous research in the area. Subsequent studies are used to refine this definition of physical activity.

Development of the operational domain of physical activity follows the delineation of the theoretical domain. The operational domain is where researchers specifically define how physical activity should be assessed. The operational domain should contain all potential measures of physical activity. Better operational definitions will result when the theoretical domain is well developed. The main issue here is that theory (the theoretical domain) should drive test development (the operational domain).

The theoretical domain of physical activity might include behavioral, movement, and energy expenditure dimensions. The concepts of frequency, intensity, duration, mode, and context may also be considered an important part of the theoretical domain of physical activity. This is not meant to be a complete definition of the theoretical domain of physical activity, but is used for illustrative purposes and can be used as a starting point for others to refine.

Table 3.1 provides a sample characterization of the operational domain of physical activity. Listed in the left column are the potential ways (note that this is not meant to be an exhaustive list) to measure physical activity. The other columns list the concepts that represent one characterization of the

TABLE 3.1   Sample Characterization of the Operational Domain of Physical Activity

Method	Frequency	Intensity	Duration	Mode	Context	Energy expenditure
Diary	Y	Y	Y	Y	Y	N
Questionnaire	Y	Y	Y	Y	Y	N
Accelerometer	Y*	Y*	Y*	N	N	Y*
Heart rate monitor	Y*	Y*	Y*	N	N	N
Pedometer	N	N	N	N	N	Y*
Observation	Y	Y	Y	Y	Y	N
Doubly labeled water	N	N	N	N	N	Y
Indirect calorimetry	Y*	Y*	Y*	N	N	Y
Caloric intake	N	N	N	N	N	Y

Note: Y = yes, can assess that aspect of physical activity; N = no, cannot assess that aspect of physical activity; asterisk (*) denotes that this information is available only from some versions of this type of instrument.

theoretical domain of physical activity. The notations in the body of table 3.1 indicate whether the particular measurement instrument can assess the particular aspect of physical activity listed in the column. For example, accelerometers can assess the aspects of frequency, intensity, and duration of movement but do not assess the mode or context of the movement.

Several problems can occur during specification of the theoretical and operational domains (Messick, 1989). Construct underrepresentation occurs when the instruments do not adequately assess the construct and when the operational domain is too narrowly defined. For example, instruments that assess the frequency, duration, and intensity, but not the context of physical activity, can substantially underestimate the prevalence of physical activity in certain cultural groups. Cultural groups that obtain physical activity in occupational environments, but who do not tend to perform regularly planned exercise, may have low prevalence estimates if the measure of physical activity does not assess occupational activity. In this case, the conclusion drawn from the measure of physical activity would not be valid because the researcher failed to assess the context in which physical activity took place. In other words, the instrument did not adequately assess the construct because physical activity was too narrowly defined.

Another potential problem during this stage is construct irrelevancy. Construct irrelevancy occurs when the instrument measures something other than what was intended. For example, if individuals systematically overestimate their levels of physical activity on a self-report questionnaire, then the systematic variability associated with overreporting can lead to incorrect test interpretation. The phenomenon of social desirability is a major issue to consider in all self-report measures of physical activity. People may tend to answer questionnaires in a socially desirable way so as to make themselves look better than they actually are or in an attempt to please the investigator. In the physical activity literature, Warnecke et al. (1997) noted that social desirability can result in an overreporting of physical activity. When the trait of social desirability contaminates a physical activity self-report questionnaire, the resulting scores on the questionnaire are no longer valid measures of the construct in question.

Consideration of definitional evidence of validity is of obvious importance to self-report measures (e.g., questionnaires or diaries) of physical activity. Researchers who develop these measures must con-

sider the various components of physical activity they wish to assess. A questionnaire item can assess the type, frequency, intensity, duration, and context of physical activity. Some self-report instruments attempt to assess all or a combination of these components. For example, the following question, or self-report item, from the Youth Risk Behavior Survey (YRBSS, 2001) assesses the frequency of physical activity. In addition, this item attempts to assess the frequency of physical activity only for a specific intensity (i.e., sweat and hard breathing), duration (i.e., at least 20 minutes), and type of activity (i.e., aerobic activities).

> On how many of the past 7 days did you exercise or participate in sports activities for at least 20 minutes that made you sweat and breathe hard, such as basketball, jogging, swimming laps, tennis, fast bicycling, or similar aerobic activities?

The interpretation of results from this self-report item to represent the prevalence of all types of physical activity, activity for periods longer than one week, activity at different times of the year, moderate- or low-intensity activity, or leisure physical activity would lack validity.

Consideration of definitional evidence of validity is important for all types of physical activity measurement procedures. For example, if physical activity is defined as energy expenditure, then a choice of measurement with some degree of definitional validity evidence might be doubly labeled water or indirect calorimetry (depending on the anticipated use of the measurement results). If physical activity is defined as movement, then a valid choice of an activity measure might be a pedometer or accelerometer (again depending on the anticipated use of the measurement results).

One final consideration related to definitional evidence of validity is that physical activity researchers should be cautious in the terminology they use to draw conclusions about "physical activity." Kelley (1927) recognized that it was a fallacy to assume that two instruments with the same name are necessarily measuring the same trait. He termed this the *jingle fallacy*. Kelley also cautioned that it was a fallacy to assume that two instruments with different names are necessarily measuring different traits. This he termed the *jangle fallacy*. Researchers should accurately and specifically state the nature and aspect of physical activity they assess. The extensive differences between total energy expenditure, occupational physical activity, minutes of moderate to vigorous physical activity, body movement counts, and self-reported activity over the previous week preclude the use of a global term like *physical activity* to represent each of these measures.

In summary, researchers should consider definitional evidence of validity as a starting point that will help guide them in developing new measures of physical activity or in deciding which of the available instruments is most appropriate for the particular situation and purpose. The remaining sections in this chapter provide other methods to demonstrate validity evidence after definitional evidence of validity has been considered.

Confirmatory Stage

In the second stage of construct validation, data are gathered to help us confirm our conceptualization of the construct (physical activity) arrived at in the first stage. Because the construct is inextricably linked with our operationalization of it via measurement instruments, this stage also involves evaluation of the test's ability to measure the construct. In this section, we describe four methods for obtaining and evaluating validity evidence: (1) factor analysis to evaluate internal evidence, (2) the multitrait-multimethod matrix for obtaining convergent and discriminant evidence, (3) regression techniques for obtaining criterion-related evidence, and (4) the known difference method to confirm expected relationships with the instruments.

Internal Evidence Using Factor Analysis

Internal evidence helps to confirm the theoretical dimensionality or nature of the construct, which was proposed during the construction of the test and which guided its design (see previous section on construct definition). In some of the measurement literature, this has also been described by the term *structural validity* (see Loevinger, 1957). Physical activity is often described as having dimensions of frequency, duration, and intensity, for example. It is also sometimes conceptualized as having dimensions of mode, such as walking, gardening, dancing, and so on, and context, such as occupational, leisure, household, and sport. (This is especially true with regard to the conceptualization of physical activity that underlies many questionnaire measures.) In psychology, internal dimensionality of a measurement instrument (or structure of a construct) is usually tested using analyses such as exploratory and confirmatory factor analysis. In a traditional factor analysis design in psychological research, the individual item scores from a single

questionnaire are input as the raw data. The researcher then uses the factor analysis output to determine whether items on the same subscale appear to be functioning appropriately. In other words, the analysis tells us whether the items seem to be measuring the same construct as each other (which is desirable, if they are contained in the same subscale).

It has been suggested in the psychometric literature that internal consistency estimates also provide evidence that items within the same subscale measure the same dimension of the construct. This is partly true, although internal consistency does not remove measurement error from the item variances (i.e., intercorrelations between the items may be explained partly by correlated measurement error). Factor analysis, on the other hand, reflects correlations only between that portion of the item variance that is explained by the underlying construct (for further explanation, see Gorsuch, 1983). Although the use of internal consistency estimates is therefore less appropriate from a theoretical perspective, it may provide a practical starting point for pilot work where sample size precludes the use of factor analysis. An additional weakness of internal consistency estimates are that they allow analysis of only one dimension or subscale at a time, whereas factor analysis allows analysis of all dimensions and items simultaneously.

The application of this paradigm to typical measures of physical activity has been limited. Indeed, almost no evidence of this type is available in the published physical activity measurement literature. The analogy between the item-subscale structure in the psychological questionnaire context and the structure of physical activity measures is not readily apparent. Consider first a sample questionnaire from the psychological literature, the Children's Attitude Toward Physical Activity Scale (Simon & Smoll, 1974). This questionnaire is designed to investigate the reasons why children participate in physical activity. In one subscale, Physical Activity for Health and Fitness, children are asked to use a series of semantic differential scales (bipolar adjectives representing opposite ends of a continuum: from good to bad, nice to awful, and so on) and rate what they think of the idea of doing physical activity to be fit and healthy. Each pair of adjectives represents a positive (e.g., pleasant) attitude or a negative (not pleasant) attitude. It is reasonable to assume that respondents would rate themselves similarly on all of these items. One wouldn't expect a response of "pleasant" on one subscale item and "bad" on the next, for example. Because this sample instrument is a measure of attitude toward physical activity, however, it is not a physical activity measure.

The equivalent situation in physical activity measures is not readily obvious. Even in questionnaire measures, the use of factor analysis does not seem to fit well. First, the format of items within a questionnaire is often inconsistent, ranging from dichotomous categorical items (e.g., "Do you have a garden?" from the Zutphen Physical Activity Questionnaire [Caspersen et al., 1991]) to continuous items (e.g., "How many hours, on average, do you sleep at night?" from the same questionnaire). Second, it is often not reasonable to expect that a person would respond similarly to all items contained within the same subscale. In the Yale Physical Activity Survey (DiPietro et al., 1993), for example, subscales exist for different classifications of physical activity (e.g., work, yard work, care taking, recreational). Each subscale has a list of types of activities within that classification as well as a response area where participants enter the number of hours spent in each activity in a typical week. Within a given subscale, it is unreasonable to expect that a person who indicated under the "work" section that they participate in light home repair for 1 hour each week would indicate a similar response to all of the other items in the same section. (It is, however, quite conceivable that a person might spend several hours a week on food preparation and no time on heavy home repair, both of which are in the same subsection of the questionnaire.)

Other types of instruments used to measure physical activity, such as motion sensors, heart rate monitors, and direct observation, lend themselves even less to the item-subscale configuration that seems necessary to use factor analysis appropriately. Should we, then, abandon this method for obtaining confirmatory evidence of construct dimensionality? This is not a desirable course of action. We must have some way of confirming our conceptualization of the construct domain, and the factor analysis technique has been too useful in construct validation in other areas of study (e.g., psychology and sociology) to discard it too readily. Instead, we might reassess our conceptualization of how to design studies to use this powerful technique. We suggest two possible alternatives to the item-subscale, cross-sectional design that has been used traditionally. The first is to change our idea of what might constitute an item in a factor analysis. The second is to consider using experimental designs to determine whether an instrument discriminates between different subtypes of activity.

For the first suggestion, we would broaden the design so that, rather than administer a single instrument that operationalizes several subdimensions, we would administer three or more instruments each of which is purported to measure the same subdimensions of physical activity. (In a factor analysis, there should be at least three items measuring each factor; this is why we suggest using three or more instruments here.) The subscale total scores (or dimension scores) from each instrument subsequently would be input to the factor analysis. Consider the example of three self-report questionnaires that purportedly measure time engaged in different contexts of activity (e.g., occupational, recreational, and household/family). In this example, each of the three questionnaires would yield three subscale scores (i.e., each respondent would have nine outcome scores).

A confirmatory factor analysis could be used to test the proposed three-dimensional structure (i.e., dimensions of occupational, recreational, and household/family). Each dimension is represented by three items and each of these items is a subscale score from one of the three questionnaires. The proposed model is illustrated in figure 3.1. Poor model fit, or solutions that are not interpretable, would shed doubt on either the conceptualized three-dimensional structure, which would have been defined in the first level of construct validation, or the suitability of the instruments used to measure the construct. Interpretation of specific model parameters, such as factor loadings, correlated errors, and modification indices, would help lead the researcher toward further research. (For a guide to use and an interpretation of factor analysis, see Comrey, 1978; Comrey & Lee, 1992; and Jöreskog, 1979.)

This paradigm is suitable for other types of instruments also. The conceptualization of physical activity as comprising discrete classes of physical activity intensity (e.g., inactive, low intensity, moderate intensity, vigorous intensity) could be tested using motion sensors, for example. By attaching several different models of motion sensors that record activity counts to people during regular lifetime activities, a similar set of data to that described in the previous example could be collected. After 7 days in which the study participants were instructed to follow their usual routine, data subsequently could be downloaded and separated into records of time spent in each of the four activity intensities. In this situation, the measurement model would comprise four factors (i.e., inactive, low, moderate, vigorous), each with three items (i.e., the relevant

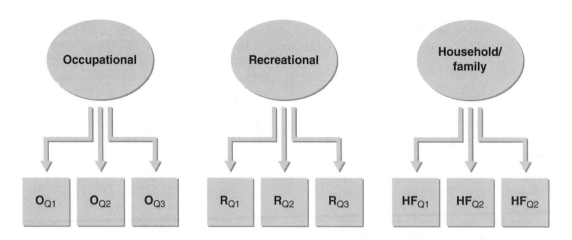

Figure 3.1   A three-factor, nine-item measurement model. Note: Q1, Q2, Q3 = questionnaire 1, 2, and 3; O = occupational subscale total score; R = recreational subscale total score; HF = household/family subscale total score.

time count resulting from analysis of each of the three motion sensor data records).

For the second suggestion, we recommend that an experimental design also could be used to obtain data for subsequent factor analysis. In a similar example, three different motion sensors could be used, or even three different types of objective instruments believed to be sensitive to intensity of activity (e.g., a heart rate monitor, a uniaxial accelerometer, and a triaxial accelerometer). A self-report physical activity questionnaire, rating of perceived exertion (RPE), and portable O_2 analyzer could even be used. In this experiment, study participants would be asked to participate in selected activities in a fairly controlled environment. Using previously established MET ratings for guidance (e.g., the compendium of Ainsworth et al., 1993), the researchers would select a small set of activities that would be described as low intensity (e.g., below 4 METS), such as making a bed (2 METS), playing darts (2.5 METS), horseshoe pitching (3 METS), and carrying a 15-pound suitcase while walking (3.5 METS). Each participant would do each activity for a brief period of time (which would be standardized for every activity). At the end of each low-intensity activity, item scores would be collected from each instrument (e.g., vector magnitude from an accelerometer, $\dot{V}O_2$ from a portable gas analyzer, average heart rate from a heart rate monitor, or self-rated intensity from an RPE scale or questionnaire). A similar arrangement would be used to collect data during a subset of moderate- and vigorous-intensity activities. Depending on how many different instruments were used, and how many sample activities were selected for each intensity class (i.e., low, moderate, and so on), there would be several items forced

to load on each factor in the confirmatory factor analysis. Factor loadings would help the researcher to determine the answer not only to general questions such as "How well does the heart rate monitor measure moderate activity?" but also to more specific questions such as "Does the accelerometer measure moderate activity of certain types (e.g., walking) better than others (e.g., cycling)?"

Designs such as this one would provide convergent evidence of validity for different physical activity measurement instruments over different modes and intensities of activity. Convergent evidence can also be obtained as part of another validity study design, the multitrait-multimethod matrix, which is described next.

The Multitrait-Multimethod Matrix

Another validation method that has escaped attention in the physical activity literature is the multitrait-multimethod (MTMM) matrix introduced by Campbell and Fiske (1959). These authors introduced a systematic procedure for evaluating validity of a measurement instrument via convergent evidence (evidence that measures of the same construct correlate highly), and discriminant evidence (evidence that measures of different constructs do not correlate as highly as measures of the same construct).

Scores on a measurement instrument can be influenced by a variety of things. One systematic influence is the person's true score on the variable we want to measure (i.e., physical activity). We hope that a person's true score will explain most of the variance in a set of scores that are supposed to measure physical activity. In other words, we hope that differences between people's physical activity scores from a measurement instrument are due mainly to differences in their true physical activity level. Another systematic influence is the method being used. For example, certain people may tend to score more highly than others on physical performance tests, regardless of what construct is being measured by the test. (Physical performance tests can be used to measure constructs such as cardiovascular fitness, manual dexterity, reaction time, decision making, or flexibility.) Scores can also be influenced by random effects, or measurement error, related to reliability of the instrument. The MTMM matrix provides a system by which we can evaluate how well a measurement instrument is measuring physical activity, and also how much it is affected by extraneous method variance and by measurement error.

The proper construction of an MTMM matrix involves the measurement of more than one trait, using more than one method of measurement. This is usually done using a fully crossed design so that every trait under consideration is measured using every type of method under consideration. Each measure of each trait is subsequently correlated with all other measures of other traits, resulting in a t (number of traits) by m (number of methods) symmetrical matrix of bivariate correlations. By measuring two different traits with two different methods, we can set up the simplest MTMM matrix, a 2 (trait) $\times$ 2 (method) design.

What constitutes a different method? Some examples from published MTMM studies in psychological and educational measurement are

- multiple-choice items versus incomplete sentences (for a test of knowledge),
- interview versus trait checklist (for a measure of attitudes toward an employer), and
- paper and pencil versus computer administered (for a test of recall ability).

In some previous studies in other areas of research, the different methods investigated were broad classes of instruments (e.g., subjective rating versus objective, or verbal versus visual), while in others the different methods were more "similar," in that they were slightly different formats of the same class of instrument (e.g., subjective self-rating versus subjective rating by teammate versus subjective rating by administrative staff, or questionnaire A versus questionnaire B). One might question whether the latter examples truly fit the original intentions of Campbell and Fiske (1959), who suggested that maximally different traits and methods be used. Perhaps the most helpful advice is that, as

with all measurement research, guidance should come from the intended use of the test. If, for example, a researcher wishes to detect differential effects of a structured after-school recreational activity program on free-living physical activity and academic success, each measured by a subjective self-rating questionnaire and some other objective measure, then the selection of trait and method for a pilot MTMM study is clear.

Table 3.2 illustrates a simple MTMM matrix. In this design two different traits (trait A and trait B) were each measured by two different methods (method 1 and method 2). This would entail administering four measurement instruments to each person in the study (i.e., trait A measured by method 1, trait B measured by method 1, trait A measured by method 2, and trait B measured by method 2) and would result in six bivariate correlations, from the six pair combinations of the different tests. Because the correlation matrix is symmetrical, only the lower half contains information relating to the set of correlations between the different tests. The cells on the upper diagonal of the matrix (denoted $R_{xx'}$) contain reliability estimates for each of the four tests. These reliability estimates should ideally be obtained from the same study sample that generated

TABLE 3.2 A Simple Multitrait-Multimethod Correlation Matrix

	Method 1 Trait A	Method 1 Trait B	Method 2 Trait A	Method 2 Trait B
Method 1				
Trait A	$R_{xx'}$			
Trait B	DV1	$R_{xx'}$		
Method 2				
Trait A	CV	DV2	$R_{xx'}$	
Trait B	DV2	CV	DV1	$R_{xx'}$

Note: $R_{xx'}$ = reliability coefficient; CV = convergent validity coefficient (monotrait-heteromethod); DV1 = divergent validity coefficient (heterotrait-monomethod); DV2 = divergent validity coefficient (heterotrait-heteromethod).

Evaluation criteria:

1. Reliability coefficients should be high.
2. Convergent validity coefficients (CV) should be significantly higher than zero, and clinically, or practically, high.
3. Convergent validity coefficients (CV) should be higher than heterotrait-monomethod discriminant coefficients (DV1).
4. Convergent validity coefficients (CV) should be higher than heterotrait-heteromethod discriminant coefficients (DV2).
5. Within each correlation matrix, the pattern of correlations should be similar (this is more evident in larger matrices from designs with several traits and several methods).

the MTMM matrix; use of reliability evidence from previous research data is less desirable.

In their original paper, Campbell and Fiske (1959) did not specify what type of reliability evidence should be obtained, although subsequent authors have claimed that internal consistency estimates should be obtained (e.g., Lowe & Ryan-Wenger, 1992). This may be appropriate for most paper-and-pencil instruments used in the behavioral sciences, but it does not suit many of the instruments used in the study of physical activity. Researchers should be guided by their intended use of the instrument. For example, if 7 days of accelerometer data are typically used in applied physical activity research, this measurement schedule should also be used in an MTMM study, and inter-day reliability should be used on the main diagonal.

The coefficients on the diagonal within the matrix (denoted CV) contain the convergent validity coefficients, which represent the monotrait-heteromethod correlations (between measures of the same trait, using different measurement methods). The cells marked DV1 contain discriminant validity coefficients of heterotrait-monomethod correlations (between different traits, measured using the same method). The cells marked DV2 contain heterotrait-heteromethod correlations, which are another type of discriminant validity coefficient, this time representing correlations between different traits, each measured using a different method.

Campbell and Fiske (1959) recommended a series of steps to follow in evaluating the correlations in an MTMM matrix. First, the convergent validity coefficients (CV) should be significantly greater than zero, and high enough to warrant further investigation. How high is high enough? This is partly determined by the area of research, the types of measures being used, and the current state of research in the area. If prior physical activity research (not necessarily MTMM studies) using a similar type of measurement instrument has produced convergent correlations of 0.5 to 0.7, then this could be used as a guideline for what is an acceptable criterion in this first stage of a subsequent MTMM study.

The second step is that the convergent validity coefficients (CV) should be higher than the discriminant validity coefficients (DV2) in the same correlation block (in the example provided, there is only one of these blocks; in more complex MTMM designs with multiple traits and methods, there would be more). This may seem like a minimal condition that should easily be met, but as Campbell and Fiske (1959) stated, there are numerous examples in the behavioral sciences literature that fail to meet this fairly elementary criterion.

The third step is that correlations between different methods used to measure the same trait (monotrait-heteromethod, or CV) should be higher than correlations between measures of different traits that have only their measurement method in common (DV1). For example, an objective mechanical measurement of physical activity (a 7-day pedometer count) should correlate more highly with a subjective measure of physical activity (a 7-day recall questionnaire) than with an objective mechanical measure of body fatness (from a bioelectrical impedance machine). If the data do not fit this pattern, this indicates that much more of the variance in scores is accounted for by the method used than is attributable to the construct being measured. This would lead us to lose faith in the validity of these physical activity measures.

The fourth step is to investigate the patterns of heterotrait correlations in each monomethod block to determine that the patterns of intercorrelations are similar (e.g., in a three-trait design, the r_{t1t3} correlation might be highest, followed by r_{t1t2}, with r_{t2t3} lowest). In the example in table 3.2, because there are only two traits, there is no way of evaluating this final criterion.

Before conducting an extensive MTMM study, the researcher would benefit from at least establishing from prior evidence that the reliability of the instruments is sufficiently high. Because low reliability has the effect of attenuating (decreasing) subsequent correlation coefficients, establishment of sufficiently high reliability should be a precursor to any MTMM study. One commonly cited criterion for evaluating reliabilities is the recommendation of Nunnally (1978), that a value of 0.70 is minimally acceptable. Within an MTMM matrix containing established measures, it would therefore be advisable to look for reliabilities above 0.80.

The MTMM procedure is a promising method for physical activity researchers who wish to obtain validity evidence for their measurement instruments. Several objective and subjective methods are available to measure physical activity. To set up an MTMM matrix, we would need to identify another, unrelated (or lowly related) trait that could be measured using instruments of a similar format. Fitness is one possibility (although it is related to physical activity, it is theoretically a separate trait), as it could be measured using objective and subjective measures (e.g., via a battery of practical fitness tests and self-rating). Also, such a study would have great practical importance for researchers who wish to investigate both physical activity and fitness, especially where research questions involve discrimination between the independent effects of physical activity and fitness on a health outcome. Another possibility is motor ability, which could be measured by a series of motor tests and a perceived motor competence questionnaire.

Table 3.3 contains an example of an MTMM matrix, with fabricated data for illustration purposes. The two traits being measured are physical activity and motor ability. The two methods are objective (e.g., a composite score from a population-appropriate motor skills test battery and total vector magnitude score from 7 days of accelerometer data) and subjective (e.g., a population-appropriate questionnaire measure of perceived motor competence and a physical activity recall questionnaire covering the same 7 days over which accelerometer data were collected). The results of this fictitious study meet all of the criteria set out by Campbell and Fiske (1959), and would support the validity of using these two measures for other physical activity research in similar settings and with similar populations to that used in the MTMM study. First, the reliability of all four measures is high (i.e., between 0.88 and 0.91). Second, the convergent validity coefficients are both statistically significant and clinically high (i.e., 0.73 and 0.77). Third and fourth, the monotrait-heteromethod correlations (i.e., 0.73 and 0.77) are higher than the DV1 heterotrait-monomethod correlations (i.e., 0.38 and 0.41) and are also higher than the DV2 heterotrait-heteromethod correlations (i.e., 0.30 and 0.33).

The advantage of the MTMM design over classical treatments of measurement error is that it provides a means by which to evaluate the influence of nonrandom error, due to systematic method variance. In situations where systematic method variance exists, a correlation between two measures of the same or different traits may be spuriously inflated by shared method variance, if the methods of measurement are identical, or similar. This would be apparent in an MTMM study where large differences exist between DV1 heterotrait-monomethod correlations and corresponding DV2 heterotrait-heteromethod correlations. In physical activity research, this type of systematic error could lead to incorrect rejection of a null hypothesis, and incorrect interpretations of theory. The last result described from the example in table 3.3 (i.e., that the DV1 correlations are only marginally higher than the DV2 correlations) indicates that, in this particular sample situation, there is little effect of method on the test scores, and that there does seem to be a low to moderate correlation between the two different traits being measured.

A brief caveat is warranted here regarding interpretation of data from the MTMM design, specifically in physical activity construct validity research. Bearing in mind the previous statements regarding the complex nature of physical activity, researchers must select the instruments for an MTMM study carefully. The global, or general, construct of physical activity incorporates a wide variety of behaviors, types of movements, contexts, and so on. All instruments selected for an MTMM study should theoretically measure the same aspects of physical activity. It is not sufficient to choose any two methods that are labeled as "physical activity" measures. Similarly, when interpreting weak monotrait correlations, researchers should be mindful of whether one or both instruments failed to measure physical activity, or whether this is evidence that the two (or more) instruments measure different aspects of physical activity. Further research is necessary to clarify such a finding.

Despite its contribution to the practice of measurement research, and the clear advantage over prior methods for investigating validity, the MTMM procedure has been criticized widely for several inherent weaknesses. Most notable of the criticisms was that the criteria were inexact and subjective, leaving little room for interpretation of "borderline" situations, also making interpretation of large matrices (incorporating several different traits and methods) unwieldy. The simple example presented here clearly met each of the criteria. However, a researcher is left in a quandary if, for example, a subset of the convergent validity coefficients in the matrix is not high, or a subset of the

TABLE 3.3　A Multitrait-Multimethod Correlation Matrix With Data

| | Objective test | | Subjective self-rating | |
	Activity	Motor ability	Activity	Motor ability
Objective test				
Activity	0.89			
Motor ability	0.38	0.91		
Subjective self-rating				
Activity	0.77	0.33	0.91	
Motor ability	0.30	0.73	0.41	0.88

discriminant validity coefficients are higher than some of the convergent validity coefficients. Interpretation of such results presents a somewhat more thorny problem than interpreting our sample matrix, where all criteria were clearly met. Apart from the prerequisite that measures should demonstrate adequate reliability, there is no method for extracting or controlling random measurement error, as all correlations involve raw scores (in contrast to the underlying factor scores that contribute to intercorrelations between factors in a factor analysis). This is particularly problematic when reliabilities are low, or vary greatly across measures, especially with regard to the final criterion of Campbell and Fiske (1959), requiring that the pattern of correlations be similar between different sections of large matrices. In such a situation, this criterion would most likely not be met, thus demonstrating the well-known statement that validity cannot be present in the absence of adequate instrument reliability.

An underlying assumption of the Campbell and Fiske (1959) MTMM procedure is that there are no trait-method correlations. This is also problematic, and it is not testable without methods for separating sources of variance. Thankfully, since the publication of the Campbell and Fiske article, researchers have developed new applications of statistical procedures such as analysis of variance and confirmatory factor analysis, for decomposing the contributions of trait variance, method variance, and measurement error from the MTMM matrix. A detailed description of these procedures is beyond the scope of this chapter, but extensive explanation and clear examples can be found in Kenny and Kashy (1992), Marsh and Hocevar (1983, 1988), Schmitt and Stults (1986), and Widaman (1985).

Criterion-Related Evidence

Criterion-related evidence of validity is demonstrated by examining the relationship between a surrogate measure of a construct and a criterion measure of that construct. In physical activity research, the surrogate measures in most studies that investigate criterion-related evidence of validity are typically self- or proxy reports of physical activity or activity monitoring devices. The criterion measure selected is very important because conclusions will be drawn based on how well the surrogate measure correlates with the criterion. Thus, the criterion should be the most accurate measure of the aspect of physical activity being assessed. Criteria that have been typically used for physical activity

research include indirect calorimetry, doubly labeled water, and direct observation. However, both self-reported physical activity and activity monitoring devices have also been used as the criterion in criterion-related validity studies (e.g., Bassett et al., 2000; Dishman et al., 1992; Garcia et al., 1997; Jacobs et al., 1993; Janz et al., 1995; Kowalski et al., 1997; Lowther et al., 1999; McMurray et al., 1998; Sallis et al., 1993, 1996; Sirard et al., 2000; Trost et al., 1999; Weston et al., 1997).

Because there is no single accepted criterion measure of physical activity, it is difficult to summarize and draw any definite conclusions from the criterion-related validity studies that have been conducted. Researchers need to use a defensible criterion measure when conducting such studies. It is essential that the criterion be a valid measure of the type of physical activity that the researcher wishes to detect with the surrogate measure being validated. If some recognized gold standard of physical activity were available, it would ideally be used as the criterion to validate other measures of physical activity. Unfortunately, no gold-standard measure of physical activity is available and the process of obtaining criterion-related evidence of validity of physical activity measures is not so straightforward. Researchers must choose from among several possible criterion measures of physical activity, and the criterion that is selected will affect the magnitude of the correlation or validity coefficient. We must think through the most important elements of physical activity that we are trying to assess with the surrogate measure and then judge which criterion is the best measure of these elements.

If the criterion measure selected is poor (e.g., unreliable, assesses different aspects of physical activity than the surrogate measure selected), then it will not correlate well with the surrogate measure of physical activity being evaluated. For example, if a self-report measure assesses activity during the previous day (e.g., Previous Day Physical Activity Recall [PDPAR]), then a measure of energy expenditure for the previous two weeks from doubly labeled water would be a poor choice of a criterion. A low correlation may result between the PDPAR and energy expenditure from doubly labeled water because the surrogate instrument and the criterion actually assess two different aspects of physical activity—not because the PDPAR is necessarily a poor measure of physical activity. In criterion-related validity studies, evidence of validity is provided if the correlation between the surrogate instrument and the criterion is substantial.

Likewise, if the correlation between the surrogate instrument and the criterion is poor, then one typically concludes that the surrogate instrument lacks criterion-related evidence of validity.

Two types of designs exist to provide criterion-related evidence of validity: concurrent evidence and predictive evidence. The designs differ based on the intended use of the data. Concurrent evidence of validity is desired when the intent is to use the surrogate measure as a substitute for the criterion measure.

One would typically want a substitute for a criterion measure when the surrogate measure is more feasible than the criterion measure. For example, if physical education teachers want to assess the movement aspect of physical activity, it is more feasible to use pedometers to measure physical activity than to use a validated observational instrument (e.g., McKenzie et al., 1991). In this example, an observational instrument would be an appropriate choice of a criterion measure because it can be used to assess the same aspect of physical activity (i.e., movement) that the physical education teacher wishes to assess with the surrogate measure. The criterion measurement instrument typically requires more training and is more time intensive than the surrogate measure. Likewise, it is not feasible for epidemiologists to use doubly labeled water on a large sample to assess energy expenditure, but it is feasible to use self-report questionnaires for this purpose. If concurrent evidence of validity were available, researchers and test users would have more confidence in the conclusions they might draw from the measurement.

Interpretation of concurrent validity depends on knowledge of correlation coefficients. A Pearson correlation (r_{xy}) is calculated to examine the relationship between the surrogate measure and the criterion. The Pearson correlation ranges between −1.00 and +1.00, and represents the degree of linear relationship between the two measures. The closer the correlation is to +1.00 or −1.00, the higher the relationship between the two measures and the greater the concurrent evidence for validity of the surrogate measure. Correlations close to zero represent a low relationship between the surrogate measure and the criterion and, thus, no concurrent evidence for the validity of the surrogate measure.

The direction of the correlation often depends on the units of measure of the instruments. If, for example, high scores on an accelerometer represent greater levels of physical activity and high scores on oxygen consumption represent greater levels of physical activity, then one would expect a positive correlation between accelerometer output and measured oxygen consumption. In the physical activity literature, few examples are available where one would expect a negative correlation between a surrogate measure of physical activity and a criterion measure of physical activity. That is because on most measures of physical activity, higher scores (e.g., accelerometer counts, caloric expenditure, number of steps per day) represent higher levels of activity. A negative correlation might result between the number of hours spent watching television and a measure of physical activity because a high score on the number of hours spent watching television may represent a low level of physical activity.

The size of the expected correlation between the surrogate measure and the criterion measure is more difficult to anticipate than the direction. The size of the correlation can potentially be affected by a small sample size, criterion unreliability, and variations in the dispersions of the samples. Thus, it is not possible to provide strict guidelines for the size of the correlation that one should expect when conducting concurrent validity studies in physical activity measurement. In areas of study where a gold standard is available, such as in body composition, validity coefficients greater than 0.80 are common (Jackson & Pollock, 1978; Jackson et al., 1980). In physical activity research, we tend to see lower correlations between the surrogate measure and the selected criterion. The coefficient of determination is the squared correlation (r^2) between the surrogate measure and the criterion measure. The coefficient of determination is a useful index to aid in interpretation of the obtained validity coefficient because it represents the amount of shared variance between the two variables or the amount of variance in the criterion variable that can be predicted by the surrogate measure. For example, if the correlation between the surrogate measure and the criterion is 0.70, then the coefficient of determination is 0.49. This indicates that 49% of the variability in the criterion measure can be predicted from the surrogate measure. It also means that 51% of the variability in the criterion measure cannot be predicted from the surrogate measure and is considered error or residual variance. Consideration of r^2 may help researchers determine whether the surrogate measure and the criterion measure are really assessing the same aspect of physical activity. Criterion-related validity coefficients lower than 0.70 indicate that the surrogate measure and the criterion measure share less than 50% of the variance in common.

The size of the sample used for concurrent validity studies is always important. It is hazardous to calculate correlations with fewer than 30 people. The suggestion that a large sample should be used in correlation studies is nearly universal in measurement and evaluation textbooks. For example, Safrit and Wood (1995) recommend a minimum of 30 people when calculating a correlation, and Baumgartner and Jackson (1999) recommend a minimum of 50 people when developing a simple regression equation (i.e., one predictor variable). When a small sample is used to calculate a correlation coefficient, the resulting value may not well represent the true value for the population because the small sample may not adequately represent the population. In addition, the effect of outliers (i.e., scores far from the mean) on correlation coefficients is greater when small samples are used.

In addition to calculation of a Pearson correlation, most concurrent validity studies will provide results from a simple or multiple regression analysis. Simple regression is conducted when one predictor variable is used to estimate the criterion. When two or more variables (e.g., activity counts and body mass) are used to estimate the criterion (e.g., activity energy expenditure), then multiple regression analysis procedures are used.

A simple linear regression equation takes on the following form.

$$Y' = bX + c$$

A multiple (more than one predictor variable) linear regression equation takes the form

$$Y' = b_1 X_1 + b_2 X_2 + c$$

where Y' is the criterion measure to be estimated, b represents the slope of the regression line, X represents a score on a predictor variable, and c represents the intercept of the regression line.

Once a prediction equation has been developed, its accuracy is estimated. If the prediction accuracy is within reasonable limits, then practitioners can use the equation for people similar to those on whom the equation was developed.

Prediction accuracy is estimated with simple and multiple correlations, the squared correlations, and the standard error of estimate (SEE). A multiple correlation is a correlation between the actual criterion score (Y) and the predicted value of the criterion (Y') that is estimated by two or more predictor variables. These accuracy statistics should be compared to other values published in similar studies.

The SEE is a very important statistic to examine prediction accuracy because it presents the accuracy of the prediction equation in the units of the criterion variable. When developing a regression equation, the following equation is used to calculate the SEE:

$$SEE = S_y \sqrt{1 - r^2} \qquad (3.1)$$

where SEE is the standard error of estimate for predicting the criterion from the predictor variables, S_y is the standard deviation of the criterion variable, and r^2 is the squared correlation between the measured criterion variable (Y) and the predicted criterion variable (Y'). Because the criterion of least squares is used to develop a regression equation, the difference between the mean measured criterion (Y) and the mean predicted criterion (Y') is necessarily zero in the sample of people on whom the regression equation is developed. Note that the standard deviation of the criterion measure is used in calculating the SEE. Thus, it is important to examine the standard deviation of the criterion scores in the sample whenever examining the accuracy of estimation. If the standard deviation of the criterion scores is lower than normal for the criterion variable in the sample used, then a spuriously low SEE can result. Researchers who do not examine the standard deviation in the sample used, along with the correlation between the measured and predicted criterion scores, may wrongly conclude that a regression equation is accurate simply because the author used a sample with an inordinately small standard deviation. Because this is most likely to occur in small samples, large samples are necessary for developing and testing regression models.

Cross-validation is the process of testing a published regression equation on a different sample of people. Cross-validation should always be done before a regression equation is used. When an equation is cross-validated, a systematic mean difference might exist between the measured and the predicted values of the criterion (i.e., a regression equation may overestimate or underestimate the criterion). Because equation 3.1 does not take into account the possibility of systematic mean differences, the standard error of estimate should be calculated differently when an already published regression equation is cross-validated. When an existing equation is cross-validated, the standard error of estimate is calculated with the formula in equation 3.2 and is often called total error (TE):

$$TE = \sqrt{\Sigma(Y - Y')^2 \div (N-1)} \qquad (3.2)$$

where TE is total error, Y is the measured value of the criterion, Y' is the predicted value of the criterion, and n is the sample size.

If TE is greater than SEE, a systematic bias in prediction exists. That is, the regression equation produces estimates of the criterion that are either systematically greater than or systematically less than the measured criterion score.

One important assumption of regression analysis is that the *observations should be independent.* Multiple observations of the same individuals have sometimes been combined and used in the same regression analysis in recent literature (e.g., Eston et al., 1998; Rowlands et al., 1999). For example, each participant may be tested at three different speeds on a treadmill with VO_2 and activity counts from an accelerometer measured at each speed. Then the observations at each speed are combined into one analysis (e.g., if 20 people participated in the study and each person were measured at three different speeds on a treadmill, then the statistical analysis might be conducted with 60 observations). This approach violates the assumption of independence of observations. The results of violating this assumption are that the correlation between the measured and predicted criterion scores may be wrong and the associated standard errors are probably wrong. A possible solution to this problem would be to randomly select only one observation for each individual such that even though all participants were measured at multiple observation points, only one data point for a given individual would be included in the analysis. This, of course, requires that an adequate number of people be included in the sample for validation studies.

A popular approach for examining concurrent evidence of validity in activity monitoring devices is to correlate output from an activity monitor (e.g., activity counts, vector magnitude) with the metabolic cost of the activity performed (e.g., Bouten et al., 1994; Fehling et al., 1999; Freedson et al., 1998; Hendelman et al., 2000; Leaf & MacRae, 1995; Meijer et al., 1989; Melanson & Freedson, 1995; Nichols et al., 2000; Pambianco et al., 1990; Sallis et al., 1990; Strath et al., 2000; Trost et al., 1998; Welk et al., 2000). The metabolic cost is often measured with a metabolic measurement system, and oxygen consumption is transformed to MET levels or calories expended during the activity.

Evidence of predictive validity is desired when one wishes to predict the criterion measure some time in the future from the surrogate measure (called the *predictor*). Similar analyses are used for concurrent and predictive evidence of validity. The difference between concurrent and predictive evidence of validity is a function of when the criterion is measured. For concurrent validity, the predictor and criterion scores are measured at about the same point in time.

The use of predictive evidence of validity has a long history. In the Air Force Aviation Psychology Program (Flanagan, 1948), scores on pencil-and-paper performance tests were used to predict success as a bombardier, navigator, or pilot. The criterion measures of performance were administered several weeks after the pencil-and-paper performance tests. Colleges and universities also use predictive evidence of validity when they decide whether to accept an applicant into the institution. The acceptance is usually based on several predictor variables (e.g., Scholastic Assessment Test score, high school grade-point average) and the criterion variable of success in college is measured some time in the future (e.g., grade-point average at the end of the freshman year).

In physical activity research, predictive evidence of validity is important for identifying measures of physical activity and other variables during childhood or adolescence that might predict activity levels during adulthood. In addition, prediction of future events, such as heart attacks, from cardiovascular disease risk factors (one of which is level of physical activity) uses predictive evidence of validity. The calculation of a correlation between the measured criterion (Y) and predicted criterion (Y'), along with regression analysis statistics, are used to provide evidence of predictive validity.

In summary, concurrent evidence of validity can be thought of as a shortcut method for conferring validity on an instrument. Rather than go through the complete validation process, we demonstrate that a surrogate measure of physical activity correlates highly with a criterion measure for which we already have strong validity evidence. If scores obtained with the criterion (gold-standard) measure can be estimated accurately from scores obtained on the surrogate measure, the confidence we already have in the criterion measure is conferred on the surrogate measure. That is, we accept the surrogate measure as a suitable substitute for the criterion measure in situations where the criterion measure is impractical. Predictive evidence of validity is used to examine whether the surrogate measure under investigation (and therein the construct) is a reliable predictor of a future event (e.g., heart attack), behavior (e.g., adult physical activity), or other future indicators of health status (e.g., cholesterol level at age 50).

Known Difference Evidence

This type of evidence is similar to what previously has been labeled the *known groups* method, originating from a paper by Cronbach and Meehl (1955). The term *known difference evidence* is preferred here because two slightly different designs, though similar in purpose, can be used. First, if two or more populations differ on a construct (according to theory or strong rationale), this should be reflected in significant (and meaningful) mean differences on a measure of that construct. A study could therefore be set up in which a representative sample from each of the two (or more) different populations is selected and the measurement instrument under investigation is administered. An alternative is to use an experimental intervention on a single sample that should theoretically change levels of the construct in the population they were drawn from. This can be tested by setting up an experiment and measuring the construct before and after the intervention to see if the measurement instrument detects any significant changes. If results based on the scores fail to support the researchers' hypotheses (in either type of design), this would lead to a lack of confidence in the validity of the measurement instrument, or possibly in the quality of the intervention in the study. It is obviously important to ensure that the intervention is administered correctly.

An example of a known difference study is that of Kochersberger et al. (1996). In this study, nurses classified 30 nursing home residents as sedentary, moderately active, or active, based on their everyday observation of the residents over a prolonged period. Each of the residents was asked to wear a Tritrac accelerometer for several days. Tritrac counts/min differed significantly among the three groups (i.e., 20.0 versus 46.6 versus 102.4), providing adequate known difference evidence of validity. Additionally, 10 participants wore Tritracs during eating, walking, and a treadmill stress test. Tritrac counts/min differed among these three activities as well (mean counts/min were 7.0 versus 272.5 versus 636.9). This latter part of the study is an example of experimental manipulation to obtain the second type of known difference evidence, in which data are collected from only one group but under different conditions that correspond to different types of activity.

The temptation in these types of studies is to choose maximally differing populations or maximally differing experimental conditions. Although this might ensure a successful study, in terms of obtaining the desired results, the data may have limited practical application. The experimental known differences evidence in the sample study of Kochersberger et al. (1996) was probably not representative of the range of physical activity levels in typical older adults (i.e., most older adults do not jog or participate in activities of a similarly high intensity). It would be more useful to know whether the Tritrac can differentiate between finer differences of lower-intensity physical activity (e.g., walking at a variety of lower speeds, various household chores). In practice, this is the level of sensitivity that many researchers would wish to detect in older adults. In the designing of studies to obtain known difference evidence of validity, selection of populations and experimental manipulations should be guided by the intended use of the instrument to answer important applied research questions.

The second stage of construct validation involves the gathering of data to confirm (or disconfirm) hypotheses set up regarding the nature of physical activity and how well instruments measure the construct as described. We have described four different types of designs for this purpose and recommend that evidence from all four designs be gathered. After extensive research of the different designs described has given us confidence in our conceptualization of the construct dimensionality, and in the instruments we intend to use, only then are we ready to move on to the final stage of construct validity: theory testing.

Theory-Testing Stage

The final stage of construct validation research involves the testing of our theories surrounding the construct of interest (physical activity) using the scientific method. Messick (1989), in a landmark paper, first introduced the notion that theory testing should be at the heart of a strong construct validation program. Messick's view is perhaps most apparent in the final stage of construct validation. In real life, constructs do not occur in isolation. Human attributes, whether they are behavioral, psychological, or physical, exist in a dynamic and complex framework of interrelationships. Many factors (constructs) influence physical activity, sometimes directly and at other times indirectly. We typically refer to these as *determinants* or *correlates* of physical activity. Some determinants have a strong influence on physical activity, while others have a less powerful influence. Some have a direct influence, while the influence of others is exerted via a mediating factor. An example hypothesis might be that in children, parental support (in the form of positive feedback) influences physical activity participation indirectly by increasing physical self-efficacy (the mediating factor).

Physical activity, in turn, influences other constructs. We usually refer to these as *outcomes*. Determinants and outcomes of physical activity may be human attributes of the person whose physical activity we are investigating (e.g., age, self-esteem, body image, disposable income, lipid profile), or characteristics of the person's environment (e.g., social support, availability of transport, role modeling of peers, existence of community facilities). An hour spent browsing through the physical activity literature would expose the reader to a seeming abundance of proposed determinants and outcomes of physical activity.

When one considers the countless possible determinants and outcomes of physical activity, the final stage of construct validation appears to pose an overwhelming task. Even considering that this book has been written for researchers interested primarily in the health outcomes of physical activity (which narrows the research area somewhat), the job of testing complex theoretical models seems intimidating. We have already alluded to the complexity of the task of measuring physical activity in isolation. The task of investigating a host of determinants and outcomes simultaneously increases the complexity of the task. The relationship between physical activity and health is even different for different health outcomes. For example, physical activity such as yoga may have limited benefit for cardiovascular fitness and obesity but may have particular benefit for psychological health outcomes and flexibility. Similarly, total physical activity energy expenditure is more relevant to management of body weight, while higher-intensity physical activity is more relevant to improvement of fitness.

There is no denying that the establishment of validity is, as Shepard (1993) described it, "a never-ending process" (p. 407). Little comfort can be gained from Cronbach's (1989) observation that the development of a solid theoretical framework for even a single construct is a long-term project. He illustrated this point with the observation that we have arrived at our present understanding of intelligence more than a century after Binet began his work in this area. We prefer to focus on the more comforting view, proposed in the same paper, of validation as a "community process," necessitating "widespread support in the relevant community"

and entailing a process that is "social as much as rational" (Cronbach, 1989, p. 164). A sharing of the responsibilities involved not only ensures a lightening of the workload but also encourages different viewpoints applied to the validation process.

In this context, physical activity researchers seem to be in a good position. There is currently a wealth of ongoing physical activity research, and it appears that the trend will continue. Consider that a keyword search for the word *activity* among the abstract titles from the most recent (2001) Annual Meeting of the American College of Sports Medicine produced a list of over 10% of the research presentations. We certainly appear to have plenty of the pieces that would go toward completing the theory-testing puzzle. What we apparently have not done (yet) is make use of a procedure for putting the pieces together. The nomological network and structural equation modeling can be used to help us complete the puzzle by organizing knowledge from previous studies into a coherent theory and by testing the proposed theory.

The nomological network was introduced in an influential article by Cronbach and Meehl (1955). They described the nomological network as an "interlocking system of laws which constitute a theory" (p. 290). This definition is probably less than helpful to most researchers. We include it here because its use marks an important departure from the traditional test-centered understanding of validity as a property of a test that can be demonstrated via a single study involving procurement of any one of a variety of types of evidence (e.g., content validity, concurrent validity, predictive validity).

In practical terms, the nomological network is a representation of what was described earlier: the patterns of association and influence between physical activity and other constructs in which we are interested. We acknowledge that it is neither reasonable nor practical to propose and test a model that contains all possible determinants and outcomes of physical activity that fits all sections of the population. Cronbach and Meehl recognized that, especially at the early stages of theory development, the network would be limited and the construct would have few connections. We have several pieces of the puzzle in the existing body of physical activity research. Before using any of these pieces, however, it is extremely important that we determine the trustworthiness of the information in these studies. This should be determined both in terms of the research design and the way in which the constructs in the study were operationalized, using criteria and guidelines described earlier in this chapter and in the literature we have referenced.

As with a jigsaw puzzle, we should start small, by trying to complete small sections. (The corners always make a good start in a jigsaw puzzle; choosing the starting point for the nomological network takes a little more planning.) A couple of examples might help here. One area of the nomological network might involve the interaction between physical activity and a particular health outcome, such as cardiovascular health. A review of the various studies linking different aspects and types of physical activity to different aspects of cardiovascular health (e.g., vascular tone, coronary blood flow, heart rate variability) could be used to guide a proposed model combining the knowledge gained in these previous studies. A second example would be part of the nomological network that describes the pattern of association between some of the determinants and physical activity. Fortunately, Welk (1999) has already proposed the Youth Physical Activity Promotion Model for explaining the physical activity of children and youth. The model was developed from a review of the existing literature, utilizing aspects of Green and Kreuter's (1991) Precede-Proceed health promotion planning model. For an in-depth description of the model, see the original article. However, a simplistic representation of the model is presented in figure 3.2.

The circles in the diagram represent constructs. A straight line with an arrowhead indicates a direct relationship between one construct and another. The arrowhead indicates the direction of influence. For example, in this model, reinforcing factors (e.g., family influence and peer influence) are proposed as directly influencing physical activity. The double-headed arrow linking enabling factors (e.g., fitness and motor skills) and physical activity indicate a reciprocal influence. The proposal here is that being fitter and having higher levels of motor skill would increase the likelihood of engaging in physical activity. In return, engaging in more physical activity will improve fitness and motor skills. As we stated, this is a simplification of the types of diagrams conventionally used to represent a nomological network. A complete diagram would also include squares representing the items (or instru-

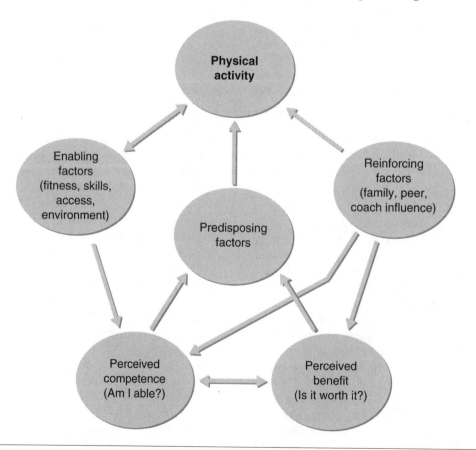

Figure 3.2 Example of a nomological network.
Adapted from Welk, 1999.

ments) used to operationalize each construct, similar to figure 3.1 shown earlier (p. 58).

To our knowledge, the article by Welk (1999) is the first in the physical activity research area in which an explicit theoretical model has been proposed and represented diagrammatically in this way. The field of research into physical activity has fallen behind many other fields of study in this regard. The next step is to use an appropriate research design and statistical method to test the model. This would involve a cross-sectional design conducted with the population of interest (which would be children, in the example). Every child in the sample would be measured on every construct in the model, using instruments for which acceptable previous evidence of validity already exists (if not, this must be done *before* embarking on the complete model-testing study).

The outcome of this data collection would be a large covariance matrix (a covariance is simply a nonstandardized correlation), containing all of the covariances between each pair of measures. The number of covariances would be $(v \times [v - 1]) \div 2$, where v = the number of measurement instruments used in the study. If each child were measured with 10 different instruments, for example (which is a fairly conservative estimate), there would be 45 covariances in the covariance matrix. The covariance matrix would then be used as input to a data analysis procedure generally described as *structural equation modeling*. A description of the complex procedures involved in performing and interpreting these procedures correctly requires more space than is available here. For further reading, refer to books on the subject by Bollen (1989), Bollen and Long (1993), Hoyle (1995), and Schumacker and Lomax (1996). Several excellent overview articles have been written in the journal *Sociological Methods and Research*. The relatively new (since 1996) journal *Structural Equation Modeling* has been devoted to methods and applications of the procedure, usually reported using a quite practical approach that is accessible to most people with a good grasp of regression techniques.

What does structural equation modeling enable us to do? The following list summarizes the major potential outcomes of this method of analysis:

1. Summary statistical indices, used to evaluate the overall fit, or suitability, of the proposed model to explain the data set.
2. Statistical indices that enable us to compare two or more competing models (this is helpful if there are separate schools of thought on the exact specifications of the model).
3. Parameter estimates, used to draw inferences about the size and direction of relationships between the different constructs in the model.
4. Factor loadings that enable us to determine how well our instruments are measuring the intended construct.
5. Modification indices that lead us, via empirical (data-centered) evidence, to possible modifications of the model. An example might be that two constructs previously thought to be unrelated might truly be related. Another example might be that an instrument believed to measure one construct also seems to be measuring another construct in the model.

Structural equation modeling clearly has powerful possibilities for improving our ability to test theory. It is not without its problems, however, and should be used only by researchers who understand how to use the procedure appropriately. New software packages such as AMOS (Smallwaters Corp., Chicago, IL) enable the analysis to be conducted increasingly simply, even by drawing the model diagram on the computer. By making it easier to run the analysis, the software also increases the risk that it will be used by researchers with insufficient understanding of some of the fairly restrictive assumptions, without adequate criteria for selecting different options for fit indices, and so on. The use of structural equation modeling as described in point 5 also contradicts the central tenet of construct validation: that it should be driven by theory. Data should be used primarily to help us confirm (or disconfirm) theory-driven models. The use of empirical evidence as the primary basis for driving new theory should be viewed with great caution by all physical activity researchers.

The final stage of the construct validation process is theory testing. The nomological network should be proposed based on previous work in the area and should be tested with an appropriate statistical technique. Correct use and interpretation of these statistical techniques require expertise and experience, and we recommend that physical activity researchers collaborate with statistics or measurement specialists with knowledge of structural equation modeling techniques.

Summary

In this chapter, we presented recommendations for construct validation in physical activity research, using a three-stage paradigm. Much of the information presented is not new; it can be found in varying forms throughout the psychometric literature. The field of human movement science is different in many respects from the field of psychology. Consequently, we have organized this information into a new structure that is better suited to the nature of research into physical activity (and other areas of human movement research). We have also provided guidance on novel applications of existing techniques to match the kinds of data that are collected in our field. In conclusion, we emphasize the following three thoughts, which apply to every aspect of the procedures described in this chapter:

1. Validity is an ongoing process. Construct validation cannot be concluded in a single study. Validity is a property not of a measurement instrument but of uses and interpretations of people's scores on a test. Validity is ascertained through the understanding of a construct that differs in different populations, in different situations, and at different times. We should seek to extend our understanding of the construct by extending the nomological network. As society changes with time, we should reevaluate previously established theories to determine whether they are still relevant. We should never believe that we have finished the job of explaining a construct.

2. Validity is a community process. The scope of construct validation is too great for a single researcher (or even a single team of researchers) to complete the task alone. Within physical activity research, we are fortunate to have an extensive community of highly competent researchers and thinkers, and so the concept of research as a community process is not unfamiliar. Many of our suggestions for new research designs in construct validation differ from previous validity research in this area, in two important ways. First, designs such as the MTMM and multivariate analyses such as factor analysis and structural equation modeling require extremely large sample sizes. Interlaboratory or interuniversity collaboration is essential to much of this type of research because many of our measurement procedures (with the exception of questionnaires) are costly, time consuming, and require technical expertise. Second, execution and interpretation of confirmatory factor analysis and structural equation modeling are particularly challenging and require expertise if they are to be used correctly. We encourage research teams of physical activity specialists to extend the community by recruiting measurement or statistics specialists from the earliest stages of the research process. To paraphrase Charles Babbage, a statistician brought in at the closing stages of a study can often only tell you what the experiment died of. Past experience also leads us to recommend that such a person should have some competency in an allied area so that they understand the variables being measured.

3. Validity involves consequences. This issue is gaining increasing recognition in the psychometric world (it first appeared in the 1985 version of the APA standards) and has had legal ramifications via numerous lawsuits stemming from important decisions made on the basis of test scores (e.g., from job aptitude tests). The use and interpretation of measurement instruments entail consequences, often far-reaching ones. Practitioners may conclude from self-report questionnaires that clients are sufficiently active, not considering the possibility that they may have answered in a socially desirable way by exaggerating responses. A government agency may conclude from a handful of items at the end of a lengthy and tedious survey that we are a nation of couch potatoes, and change public policy as a result. Similarly, researchers disseminate important conclusions from scores on measurement instruments that may change current thinking in their field. We hope that before doing so, physical activity researchers are confident that sufficient rigor was applied to the design and conduct of the study, that they have confidence in the trustworthiness of the scores obtained, and that they have considered all plausible alternative explanations for the findings of the study.

References

Ainsworth, B.E., Haskell, W.L., Leon, A.S., Jacobs, D.R., Jr., Montoye, H.J., Sallis, J.F., & Paffenbarger, R.S., Jr. (1993). Compendium of physical activities: Classification of energy costs of human physical activities. *Medicine and Science in Sports and Exercise, 25,* 71-80.

American Psychological Association, American Educational Research Association, & National Council on Measurement in Education. (1954). Technical recommendations for psychological tests and diagnostic techniques. *Psychological Bulletin, 51,* 1-38.

American Psychological Association, American Educational Research Association, & National Council on Measurement in Education. (1966). *Standards for educational and psychological tests and manuals.* Washington, DC: American Psychological Association.

American Psychological Association, American Educational Research Association, & National Council on Measurement in Education. (1974). *Standards for educational and psychological tests.* Washington, DC: American Psychological Association.

American Psychological Association, American Educational Research Association, & National Council on Measurement in Education. (1985). *Standards for educational and psychological testing.* Washington, DC: American Psychological Association.

American Psychological Association, American Educational Research Association, & National Council on Measurement in Education. (1999). *Standards for educational and psychological testing.* Washington, DC: American Psychological Association.

Bassett, D.R., Jr., Cureton, A.L., & Ainsworth, B.E. (2000). Measurement of daily walking distance—Questionnaire versus pedometer. *Medicine and Science in Sports and Exercise, 32,* 1018-1023.

Baumgartner, T.A., & Jackson, A.S. (1999). *Measurement for evaluation in physical education and exercise science* (6th ed.). Dubuque, IA: McGraw-Hill.

Bollen, K.A. (1989). *Structural equations with latent variables.* New York: Wiley.

Bollen, K.A., & Long, J.S. (1993). *Testing structural equation models.* Newbury Park, CA: Sage.

Bouten, C.V., Westerterp, K.R., Verduin, M., & Janssen, J.D. (1994). Assessment of energy expenditure for physical activity using a triaxial accelerometer. *Medicine and Science in Sports and Exercise, 26,* 1516-1523.

Campbell, D.T., & Fiske, D.W. (1959). Convergent and discriminant validation by the multitrait-multimethod matrix. *Psychological Bulletin, 56,* 81-105.

Caspersen, C.J. (1989). Physical activity epidemiology: Concepts, methods, and applications to exercise science. *Exercise and Sport Science Reviews, 17,* 423-473.

Caspersen, C.J., Bloemberg, B.P.M., Saris, W.H.M., Merritt, R.K., & Kromhout, D. (1991). The prevalence of selected physical activities and their relation with coronary heart disease risk factors in elderly men: The Zutphen Study. *American Journal of Epidemiology, 133,* 1078-1092.

Comrey, A.L. (1978). Common methodological problems in factor analytic studies. *Journal of Consulting and Clinical Psychology, 46,* 648-659.

Comrey, A.L., & Lee, H.B. (1992). *A first course in factor analysis* (2nd ed.). Hillsdale, NJ: Lawrence Erlbaum.

Cronbach, L.J. (1971). Test validation. In R.L. Thorndike (Ed.), *Educational measurement* (2nd ed., pp. 443-507). Washington, DC: American Council on Education.

Cronbach, L.J. (1989). Construct validation after thirty years. In R. Linn (Ed.), *Intelligence: Measurement, theory, and public policy (Proceedings of a symposium in honor of Lloyd Humphreys)* (pp. 147-171). Urbana, IL: University of Illinois.

Cronbach, L.J., & Meehl, P.E. (1955). Construct validity in psychological tests. *Psychological Bulletin, 52,* 281-302.

DiPietro, L., Caspersen, C.J., Ostfeld, A.M., & Nadel, E.R. (1993). A survey for assessing physical activity among older adults. *Medicine and Science in Sports and Exercise, 25,* 628-642.

Dishman, R.K., Darracott, C.R., & Lambert, L.T. (1992). Failure to generalize determinants of self-reported physical activity to a motion sensor. *Medicine and Science in Sports and Exercise, 24,* 904-910.

Eston, R.G., Rowlands, A.V., & Ingledew, D.K. (1998). Validity of heart rate, pedometry, and accelerometry for predicting the energy cost of children's activities. *Journal of Applied Physiology, 84,* 362-371.

Fehling, P.C., Smith, D.L., Warner, S.E., & Dalsky, G.P. (1999). Comparison of accelerometers with oxygen consumption in older adults during exercise. *Medicine and Science in Sports and Exercise, 31,* 171-175.

Flanagan, J.C. (1948). *The aviation psychology program in the Army Air Forces.* Washington, DC: U.S. Government Printing Office.

Freedson, P.S., Melanson, E., & Sirard, J. (1998). Calibration of the Computer Science and Applications, Inc. accelerometer. *Medicine and Science in Sports and Exercise, 30,* 777-781.

Freedson, P.S., & Miller, K. (2000). Objective monitoring of physical activity using motion sensors and heart rate. *Research Quarterly for Exercise and Sport, 71,* S21-S29.

Garcia, A.W., George, T.R., Coviak, C., Antonakos, C., & Pender, N.J. (1997). Development of the Child/Adolescent Activity Log: A comprehensive and feasible measure of leisure-time physical activity. *International Journal of Behavioral Medicine, 4,* 323-338.

Garrett, H.E. (1937). *Statistics in psychology and education.* New York: Longmans, Green.

Green, L.W., & Kreuter, M.W. (1991). *Health promotion planning: An educational and environmental approach.* Toronto, ON: Mayfield.

Gorsuch, R.L. (1983). *Factor analysis* (2nd ed.). Hillsdale, NJ: Lawrence Erlbaum.

Hendelman, D., Miller, K., Baggett, C., Debold, E., & Freedson, P. (2000). Validity of accelerometry for the assessment of moderate intensity physical activity in the field. *Medicine and Science in Sports and Exercise, 32,* S442-S449.

Hoyle, R.H. (1995). *Structural equation modeling: Concepts, issues, and applications.* Thousand Oaks, CA: Sage.

Jackson, A.S., & Pollock, M.L. (1978). Generalized equations for predicting body density of men. *British Journal of Nutrition, 40,* 497-504.

Jackson, A.S., Pollock, M.L., & Ward, A. (1980). Generalized equations for predicting body density of women. *Medicine and Science in Sports and Exercise, 12,* 175-182.

Jacobs, D.R., Ainsworth, B.E., Hartman, T.J., & Leon, A.S. (1993). A simultaneous evaluation of 10 commonly used physical activity questionnaires. *Medicine and Science in Sports and Exercise, 25,* 81-91.

Janz, K.F., Witt, J., & Mahoney, L.T. (1995). The stability of children's physical activity as measured by accelerometry and self-report. *Medicine and Science in Sports and Exercise, 27,* 1326-1332.

Jöreskog, K.G. (1979). Basic ideas of factor and component analysis. In K.G. Jöreskog & D. Sörbom, *Advances in factor analysis and structural equation models* (pp. 5-20). Cambridge, MA: Abt.

Kelley, T.L. (1927). *Interpretation of educational measurements.* Yonkers-on-Hudson, NY: World Book.

Kenny, D.A., & Kashy, D.A. (1992). Analysis of the multitrait-multimethod matrix by confirmatory factor analysis. *Psychological Bulletin, 112,* 165-172.

Kochersberger, G., McConnell, E., Kuchibhatla, M.N., & Pieper, C. (1996). The reliability, validity, and stability of a measure of physical activity in the elderly. *Archives of Physical Medicine and Rehabilitation, 77,* 793-795.

Kowalski, K.C., Crocker, P.R.E., & Faulkner, R.A. (1997). Validation of the Physical Activity Questionnaire for Older Children. *Pediatric Exercise Science, 9,* 174-186.

Leaf, D.A., & MacRae, H. (1995). Validity of two indirect measures of energy expenditure during walking in the elderly. *Journal of Aging and Physical Activity, 3,* 97-106.

Loevinger, J. (1957). Objective tests as instruments of psychological theory. *Psychological Reports, 3,* 635-694.

Lowe, N.K., & Ryan-Wenger, N.M. (1992). Beyond Campbell and Fiske: Assessment of convergent and discriminant validity. *Research in Nursing and Health, 15,* 67-75.

Lowther, M., Mutrie, N., Loughlan, C., & McFarlane, C. (1999). Development of a Scottish physical activity questionnaire: A tool for use in physical activity interventions. *British Journal of Sports Medicine, 33,* 244-249.

Marsh, H.W., & Hocevar, D. (1983). Confirmatory factor analysis of multitrait-multimethod matrices. *Journal of Educational Measurement, 20,* 231-248.

Marsh, H.W., & Hocevar, D. (1988). A new, more powerful approach to multitrait-multimethod analyses: Application of second-order confirmatory factor analysis. *Journal of Applied Psychology, 73,* 107-117.

McKenzie, T.L., Sallis, J.F., & Nader, P.R. (1991). SOFIT: System for observing fitness instruction time. *Journal of Teaching in Physical Education, 11,* 195-205.

McMurray, R.G., Harrell, J.S., Bradley, C.B., Webb, J.P., & Goodman, E.M. (1998). Comparison of a computerized physical activity recall with a triaxial motion sensor in middle-school youth. *Medicine and Science in Sports and Exercise, 30,* 1238-1245.

Meijer, G.A., Westerterp, K.R., Koper, H., & Ten Hoor, F. (1989). Assessment of energy expenditure by recording heart rate and body acceleration. *Medicine and Science in Sports and Exercise, 21,* 343-347.

Melanson, E.L., Jr., & Freedson, P.S. (1995). Validity of the Computer Science and Applications, Inc. (CSA) activity monitor. *Medicine and Science in Sports and Exercise, 27,* 934-940.

Messick, S. (1980). Test validity and ethics of assessment. *American Psychologist, 35,* 1012-1027.

Messick, S. (1989). Validity. In R. Linn (Ed.), *Educational measurement* (3rd ed., pp. 13-104). Washington, DC: American Council on Education.

Nichols, J.F., Morgan, C.G., Chabot, L.E., Sallis, J.F., & Calfas, K.J. (2000). Assessment of physical activity with the Computer Science and Applications, Inc. accelerometer: Laboratory versus field validation. *Research Quarterly for Exercise and Sport, 71,* 36-43.

Nunnally, J.C. (1978). *Psychometric theory* (2nd ed.). New York: McGraw-Hill.

Pambianco, G., Wing, R.R., & Robertson, R. (1990). Accuracy and reliability of the Caltrac accelerometer for estimating energy expenditure. *Medicine and Science in Sports and Exercise, 22,* 858-862.

Rowlands, A.V., Eston, R.G., & Ingledew, D.K. (1999). Relationship between activity levels, aerobic fitness, and body fat in 8- to 10-yr-old children. *Journal of Applied Physiology, 86,* 1428-1435.

Safrit, M.J., & Wood, T.A. (1995). *Introduction to measurement in physical education and exercise science* (3rd ed.). St. Louis: Mosby.

Sallis, J.F., Buono, M.J., Roby, J.J., Carlson, D., & Nelson, J.A. (1990). The Caltrac accelerometer as a physical activity monitor for school-age children. *Medicine and Science in Sports and Exercise, 22,* 698-703.

Sallis, J.F., Condon, S.A., Goggin, K.J., Roby, J.J., Kolody, B., & Alcaraz, J.E. (1993). The development of self-administered physical activity surveys for 4th grade students. *Research Quarterly for Exercise and Sport, 64,* 25-31.

Sallis, J.F., Strikmiller, P.K., Harsha, D.W., Feldman, H.A., Ehlinger, S., Stone, E.J., Williston, J., & Woods, S. (1996). Validation of interviewer- and self-administered physical activity checklists for fifth grade students. *Medicine and Science in Sports and Exercise, 28,* 840-851.

Schmitt, N., & Stults, D.M. (1986). Methodology review: Analysis of multitrait-multimethod matrices. *Applied Psychological Measurement, 10,* 1-22.

Schumacker, R.E., & Lomax, R.G. (1996). *A beginner's guide to structural equation modeling.* Hillsdale, NJ: Lawrence Erlbaum.

Shepard, L.A. (1993). Evaluating test validity. *Review of Research in Education, 19,* 405-450.

Simon, J.A., & Smoll, F.L. (1974). An instrument for assessing children's attitude toward physical activity. *Research Quarterly, 45,* 407-415.

Sirard, J.R., Melanson, E.L., Li, L., & Freedson, P.S. (2000). Field evaluation of the Computer Science and Applications, Inc. physical activity monitor. *Medicine and Science in Sports and Exercise, 32,* 695-700.

Strath, S.J., Swartz, A.M., Bassett, D.R., Jr., O'Brien, W.L., King, G.A., & Ainsworth, B.E. (2000). Evaluation of heart rate as a method for assessing moderate intensity physical activity. *Medicine and Science in Sports and Exercise, 32,* S465-S470.

Trost, S.G., Ward, D.S., McGraw, B., & Pate, R.R. (1999). Validity of the Previous Day Physical Activity Recall (PDPAR) in fifth-grade children. *Pediatric Exercise Science, 11,* 341-348.

Trost, S.G., Ward, D.S., Moorehead, S.M., Watson, P.D., Riner, W., & Burke, J.R. (1998). Validity of the Computer Science and Applications (CSA) activity monitor in children. *Medicine and Science in Sports and Exercise, 30,* 629-633.

Warnecke, R.B., Johnson, T.P., Chavez, N., Sudman, S., O'Rourke, D.P., Lacey, L., & Horm, J. (1997). Improving question wording in surveys of culturally diverse populations. *Annals of Epidemiology, 7,* 334-342.

Welk, G.J. (1999). The Youth Physical Activity Promotion Model: A conceptual bridge between theory and practice. *Quest, 51,* 5-23.

Welk, G.J., Blair, S.N., Wood, K., Jones, S., & Thompson, R.W. (2000). A comparative evaluation of three accelerometry-based physical activity monitors. *Medicine and Science in Sports and Exercise, 32,* S489-S497.

Weston, A.T., Petosa, R., & Pate, R.R. (1997). Validation of an instrument for measurement of physical activity in youth. *Medicine and Science in Sports and Exercise, 29,* 138-143.

Widaman, K.F. (1985). Hierarchically nested covariance structure models for multitrait-multimethod data. *Applied Psychological Measurement, 9,* 1-26.

Youth Risk Behavior Surveillance System (YRBSS). (2001). 2001 Youth Risk Behavior Survey [On-line]. Available: **www.cdc.gov/nccdphp/dash/yrbs/2001survey.htm.**

4

Physical Activity Data: Odd Distributions Yield Strange Answers

Jerry R. Thomas, EdD
Katherine T. Thomas, PhD
Iowa State University

There are several approaches used to measure physical activity, including direct observation, written instruments (e.g., questionnaires, diaries), direct measures of movement (e.g., motion sensors, accelerometer), physiological instruments (e.g., heart rate monitors), and unobtrusive measures (e.g., indicators of policies on physical activity). Most parametric statistical techniques—analysis of variance (ANOVA), multivariate analysis of variance (MANOVA), regression—that are used to analyze this type of data are based on the assumption that the data are normally distributed. We have shown that data from exercise science and other related fields often go unevaluated for the assumption of normality and, in fact, often have nonnormal distributions (Thomas et al., 1999). Yet scholars in our field continue to believe either one or both of two premises about their data:

1. Most data in the study of physical activity are normally distributed.

2. If data are not normally distributed, violation of that assumption is not a serious statistical problem.

Both of these premises are likely to be false (Bradley, 1980; Micceri, 1989; Thomas et al., 1999). As stated over 50 years ago by Geary (1947, p. 241), "Normality is a myth; there never was, and never will be, a normal distribution." In the words of Thomas et al. (1999), "God may not love the normal curve" (p. 11). Most researchers continue to rely on parametric approaches in their data analysis because they are more familiar with these methods and the methods are available on most statistical programs. It is also generally assumed that these procedures are more powerful than nonparametric methods. Alternative procedures are now available for non-normal data, and these calculations can be performed as easily as they are performed with traditional methods and with little or no loss of power.

In this chapter, we describe the use of a general linear model rank order approach, developed by Puri and Sen (1969, 1985), which we have previously advocated (Thomas et al., 1999). First we provide an overview of the general linear model (GLM) and contrast it with the rank order approach. Then we examine typical data collected in physical activity studies and evaluate it for normality. To illustrate the application of the rank order procedures, we directly compare results and interpretations from the GLM and the rank order approaches using data sets from physical activity.

Overview of the General Linear Model and Rank-Order Procedures

Scholars in the study of physical activity will be familiar with the general linear model (GLM) and the fact that it is the underlying model for all of the standard parametric statistical techniques (e.g., ANOVA, MANOVA, regression). The GLM assumes that a straight line is the best fit for data plotted on an X versus Y graph and can be written as follows (Timm, 1975, p. 185):

$$Yp = B[M] * Xq + Ep \qquad (4.1)$$

where Y = vector of scores on p (j = 1, 2, ...p) dependent (or criterion) variables, X = vector of scores on q (k = 1, 2, ...q) independent (or predictor) variables, B = p × q matrix of regression coefficients (based on original scores), and E = vector of errors.

This equation applies to both univariate and multivariate procedures, depending on whether X and Y represent a single or weighted composite of variables and whether the variables are coded vectors or continuous variables. For example, the traditional univariate procedures to test differences among groups (t and ANOVA procedures using F) all represent a single dependent variable with coded vectors for independent variables (group membership). A coded vector is simply a system for defining group membership using 1s and 0s; for example, in a two-group case (experimental versus control), participants from the experimental group would all be coded "1" while the participants in the control group would be coded "0." Extending this to multivariate procedures (discriminant analysis, MANOVA) involves the use of multiple dependent variables—the independent variables are still coded vectors for group membership. For testing relationships among variables, r is simply the relationship between two variables, R (multiple regression) uses a linear composite of predictors and one criterion, and Rc (canonical correlation) is two linear composites—one of predictors and one of criteria (Campbell & Taylor, 1995; Pedhazur, 1982; Thompson, 1984).

The GLM also underlies the Puri and Sen (1969, 1985) procedures for rank-order data; that is, when the original data are changed to ranks, the assumption is that a straight line is the best fit. Thus, the standard parametric techniques of differences among groups (e.g., t test, ANOVA, discriminant analysis, MANOVA) and relationships among variables (simple correlation, multiple regression, and canonical correlation) can all be performed in a similar way with ranked variables. However, two items are different when using rank-order versions of the typical statistical procedures:

1. Original data must be changed to ranks.
2. Comparisons are made to the chi-square (χ^2) table instead of to the tables for t or F distributions.

Comparisons to the chi-square table do not require the assumption of normal distribution of data that are required for the use of parametric tables involving t and F. Instead of the t or F ratio provided in the GLM procedures, a chi-square approximation, L, is calculated with its associated degrees of freedom:

$$L = (N - 1)\, r^2 \qquad (4.2)$$

where N = number of participants, r^2 = proportion of true variance (SSregression/SStotal), with pq degrees of freedom where p = df for the number of independent (predictor) variables and q = df for the number of dependent variables.[1]

Using equation 4.2, the L statistic may be calculated as the significance test for any regression (r, R, Rc) or difference among groups technique (t, ANOVA, MANOVA). Appropriate dfs are determined and the L statistic is compared to the chi-square table for the selected level of alpha.

Determining Whether Data Are Normally Distributed

The first step before any statistical analysis is to evaluate the distribution of the data (Micceri, 1989; Thomas et al., 1999). To use the GLM, a straight line must be the best fit if the data are plotted on an X versus Y graph. If the GLM is the choice, then the distribution of the data must be evaluated for normality. Of course, as pointed out previously (Geary, 1947; Micceri, 1989), the perfect normal curve is like a unicorn: nonexistent. So the question becomes this: How non-normal do data have to be (Thomas et al., 1999)? Scholars might consider the following questions in evaluating the data distribution:

1. Are the mean and median about the same?
2. How large is the standard deviation compared to the group mean?
3. What does the distribution look like?
   a. Is the hump of the curve shifted left or right (skewed)?
   b. Is the curve more peaked or flat than normal (kurtosis)?

If data are plotted as a frequency bar graft or a stem-and-leaf graph, skewness and kurtosis values are typically provided. Table 4.1 is a stem-and-leaf graph that can be readily compared to figure 4.3 (frequency distribution of the same data; p. 78) by rotating one or the other. In the left-hand column of the stem-and-leaf graph is the frequency a score occurs; next (on the right) is the whole number of the score (the stem). The leaf represents both the frequency of that score and the decimal portion of the score. So, 812 subjects had a score of 0.0; 166 had a score of 0.33; 171 had a score of 0.66; 144 had a score

of 1.0; and so forth. Although the two plots (stem-and-leaf in table 4.1 and histogram in figure 4.3) provide essentially the same information, there are two differences. First, the stem-and-leaf graph provides an exact frequency for each score, whereas one has to estimate these from the histogram. Second, the histogram provides all scores (e.g., extremes are represented such as the scores of 7 and 8), whereas in the stem-and-leaf graph the extreme scores are grouped and reported as extreme. Most statistical and graphics programs allow the normal curve to be superimposed over the bars to provide a picture of the normality of the data. Skewness and kurtosis values are usually presented as z-scores (mean of 0 and standard deviation of 1). Values that exceed +1 suggest that the data do not look normal; in addition, the skewness and kurtosis values can be tested against their standard errors (Thomas et al., 1999), and values larger than 1.96 are considered significant ($p < .05$). However, the decision about whether data are non-normal enough to fail the assumption for parametric procedures is an arbitrary choice made solely by the researcher.

Sometimes non-normal data can be made more normal by evaluating for outliers (data points that are not representative). Identifying outliers is also a difficult decision. First, the researcher has to identify the data point as a potential outlier. Then, standard outlier tests contrast the data point against either the overall variability (with that point included) or the next most extreme data point. Both decisions—what data point should be examined as an outlier and which standard of comparison to test for outliers—are relatively arbitrary decisions. Sometimes data are trimmed, meaning a certain percentage of the extreme scores are discarded. All these procedures result in data loss. If the data points are "real" and the researcher wants to keep them in the data set, the rank-order procedures reported here have the advantage of simply changing the most extreme value(s) to the first (or last) rank rather than discarding it based on some arbitrary decision. Further, the assumptions of the GLM are neither ignored nor violated using these techniques. For example, one or two extreme data points may cause a straight line (that is the GLM) to not fit the data; however, when

TABLE 4.1   Stem-and-Leaf Drawing: Average of Physical Activity for 3 Days

Freq	Stem	Leaf
812	0	.000000000000000000000000000000000000000000000000000
166	0	.3333333333
171	0	.6666666666
144	1	.00000000
130	1	.3333333
97	1	.666666
97	2	.000000
73	2	.3333
51	2	.666
50	3	.000
41	3	.33
19	3	.6
42	4	.00

Note: Compare to figure 4.3.
125.00 extremes (> = 4.3)
Stem width: 1.00
Each leaf: 17 case(s)

changed to ranks, a straight line may be more likely to fit.

Application of Rank-Order Procedures

In the following sections, we use two sets of data to illustrate the application of rank-order procedures as an alternative way to analyze data (we use the Statistical Programs for the Social Sciences [SPSS], but any standard statistical program provides the procedures). The data used in these examples represent typical data that are used in physical activity research. One data set (PA1) is from a recall-based, self-report instrument that is commonly used with children—the Previous Day Physical Activity Recall. (Dr. Greg Welk provided this data.) This instrument provides information about the number of bouts of moderate and vigorous activity children report each day. The data used in the present application were collected from more than 2,000 children from 16 different elementary schools over 3 consecutive days. The other data set (PA2) represents estimated minutes of moderate and vigorous activity obtained from an accelerometry-based activity monitor. (This data set was provided by Drs. Russ Pate, Patty Freedson, Jim Sallis, and Wendell Taylor.) In this case, an algorithm was applied to the raw data counts to determine the number of minutes the children spent in moderate or vigorous activity.

The examples are not used to suggest that previous analyses with these data are inappropriate. We are simply showing that these two sets of data (PA1 is active time during periods of the day; PA2 is counts during varying periods of the day) in their original forms have some characteristics that suggest non-normality. We then demonstrate the use of the rank-order procedures on these data and compare the outcomes to standard parametric analyses.

First, we use the procedures described previously to present the data distributions and evaluate the data sets for normality. We compare the parametric and rank-order techniques using the correlation procedures of r and R. Then we compare the rank-order procedures and the L statistic for tests of differences among groups using the t and F tests. Finally, we discuss the extension of the procedures to multivariate models and the overall utility of the rank-order techniques.

Data Distributions and Correlation

In figure 4.1a the original data (PA2) for 4 hours of moderate physical activity are plotted against 4 hours of vigorous physical activity. In figure 4.1b, the same data are plotted after they are changed to ranks. Note that the ranked data in figure 4.1b are considerably more linear going from lower left to upper right than the original data were. This produces a stronger relationship between the variables.

Figure 4.2a is a histogram of 4 hours of moderate physical activity for the 199 children in the sample. The curve is slightly shifted left (skewness = 0.72) but is relatively normal in height (kurtosis = –0.05). The median (33 minutes) is relatively far from

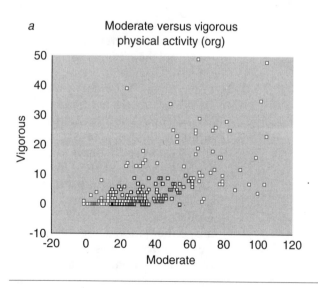
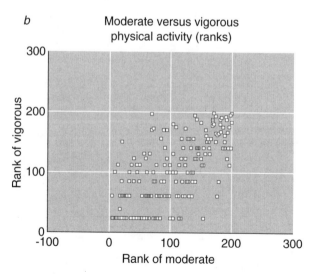

Figure 4.1 Moderate versus vigorous physical activity: *(a)* original data, *(b)* ranked.

the mean (38.31 minutes), and the standard deviation (24.43) is large when compared to the group mean. Figure 4.2b is a histogram of the same 4 hours for the same children but for vigorous physical activity. Note the pronounced shift left of the curve (skewness = 2.37) and the high level of peakedness (kurtosis = 6.49). The median is 3 minutes and the mean is 6.26 minutes with a large standard deviation (8.72). The moderate data are slightly non-normal with the vigorous data more seriously non-normal, based on the five criteria. If one expected children who are moderately active to be vigorously active also, correlation would test this prediction. The Pearson r is significant ($r = .598$, $p < .0001$) between moderate and vigorous physical activity. However, when both moderate and vigorous physical activity data are changed to ranks and correlated, the $r = .706$ and the proportion of variance accounted for is 14% greater (from 35.8% to 49.8%). The $L(1) = 98.60$ for the rank-order correlation and is significant at $p < .0001$ when compared to the chi-square table.

Can vigorous physical activity be predicted from moderate physical activity plus individual characteristics? In this multiple regression example, we use weight, height, and moderate physical activity to predict vigorous physical activity. Using the original GLM measures, the three predictors in a linear composite have an $R = .617$ (38.1% variance accounted for) and is significant, $F(3,195) = 39.85$, $p < .0001$. If all data are changed to ranks, the linear composite of the three predictors has an $R = .725$ (52.6% variance accounted for) and the $L(2) = 104.15$, $p < .0001$. Thus, a 14.5% increase in variance ac-

counted for is obtained by changing the original data to ranks before applying multiple regression. It is also acceptable to mix the ranked variables with original data as long as the L test rather than the F test is used. In this instance, if we keep height and weight in their original metric form but use ranks for moderate and vigorous physical activity, the multiple $R = .720$, a value slightly less than the completely ranked model but greater than the original model. In this example, the rank technique produces clearer and more compelling results and does not violate the assumption of normality.

In looking at both simple and multiple regression, it is clear that when the data are non-normal (see figures 4.2a and b) and changed to ranks, additional power is obtained as the r^2 and R^2 account for considerably greater proportions of the variance and significance levels remain similar. Also, comparisons of figure 4.1a and b make it clear that the data better meet the GLM assumption when changed to ranks.

Data Distributions and Differences Among Groups

In figure 4.3 the average number of bouts of vigorous physical activity reported per day is presented in a histogram (PA1). The median (0.67) and the mean (1.16) are somewhat different and the standard deviation is large (1.54) relative to the mean. Also the curve is shifted left (skewness = 1.71) and is very peaked (kurtosis = 3.11). Once again, four of the five indices of non-normal data are present.

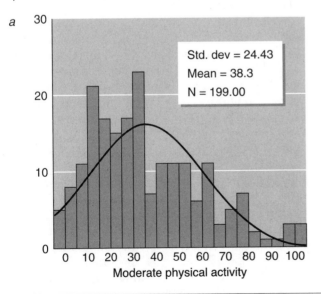

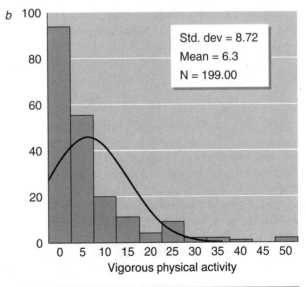

Figure 4.2 *(a)* Moderate physical activity for 4 hours; *(b)* vigorous physical activity for 4 hours.

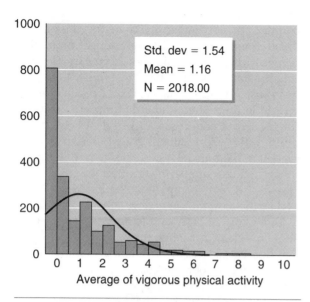

Figure 4.3 Average of vigorous physical activity for 3 days.

A common question in physical activity epidemiology is whether differences exist in activity patterns for different demographic groups. If the hypothesis is made that boys are more likely to be physically active than girls, this can be tested for the 3-day average of vigorous physical activity with a t test on the original data displayed in figure 4.3. The results were 1.42 ± 1.73 bouts per day for the boys and 0.93 ± 1.31 bouts per day for the girls. The t test was significant [$t(2{,}016) = 7.20$, $p < .0001$]. However, because the data are rather non-normal, the results from parametric analyses may not be correct. Changing the data to ranks allows the t test to be changed to an L statistic. Now comparisons can be made to the chi-square table instead of a table of t values. The r^2 can be obtained from the t for the ranked data using equation 4.3:

$$r^2 = t^2/(t^2 + df), \qquad (4.3)$$

where $df = (n_1 + n_2)$.

For the ranked data, $t = 6.54$. In solving equation 4.3, $r^2 = .021$ and $L(1) = 42.36$, $p < .0001$. In this case, the interpretation is the same but the L test is the better choice given the non-normal distribution of the data.

Returning to the accelerometer data (PA2), we evaluated an ANOVA model to determine whether there was any influence of race or body size on the activity levels in the sample. We used two characteristics as independent variables (race = black or white; weight = 100 pounds or less or greater than 100 pounds) and the 4-hour sum of vigorous physi-

cal activity (figure 4.2b) as the dependent variable in a 2 × 2 ANOVA. The ANOVA tables for these comparisons are provided in table 4.2. The only significant effect was for weight—that is, the lighter participants had more vigorous physical activity than the heavier participants; $F(1{,}183) = 7.09$, $p = .008$. When the data were changed to ranks and run as a 2 × 2 ANOVA, $L(1) = 4.65$, $p < .05$. Using the sums of squares for weight and sums of squares for total from the ANOVA table [$r^2 = $ (SSregression/ SStotal) = (15271.76 / 614275.20) = .025] and the total number of participants for degrees of freedom [$df = (N - 1) = (187 - 1) = 186$], one can calculate r^2 and thereby L ($r^2 = .025$; $.025 \times 186 = L$). In this example, the p value for the chi-square test of L was slightly lower than the corresponding test of F, but both comparisons resulted in the same outcome.

Thus, in the t and factorial ANOVA examples, power was maintained and significance was declared using the GLM rank-order procedures on the non-normal data. These approaches can be extended into any factorial ANOVA design. For example, both the number of levels of the independent variable can be expanded (e.g., 3 × 4 ANOVA), as can the number of independent variables (e.g., 2 × 3 × 4 ANOVA) using the rank-order procedures.

Extensions of GLM Rank-Order Procedures Into Multivariate Analyses

Correlation and differences among group procedures can be extended into the multivariate cases (Thomas et al., 1999). For rank-order canonical correlation, the procedure involves ranking the multiple predictor and criterion variables, running the standard canonical analysis, and calculating L using the R_c^2 ($N - 1$). Degrees of freedom are the number of predictors (p) multiplied by the number of criterion (q), or $df = pq$. For discriminant analysis and MANOVA, the multiple dependent variables are changed to ranks, Pillai's Trace is used as R^2, then L is calculated in the normal way (see equation 4.2).

Pillai's Trace is one of several ways to estimate effects in discriminant analysis and MANOVA. It is the best choice to use with rank-order multivariate techniques because it can be interpreted as an estimate of R^2 for use in calculating the L statistic (see Thomas et al., 1999 for more information). In repeated measures designs, Pillai's Trace is also used as the R^2 for each factor, including the repeated measures. One additional step is required

TABLE 4.2   Source Tables for 2 by 2 ANOVA Using Original Data and Ranked Data As Dependent Variables

Source	ANOVA source table for original data				
	Sums of squares	*df*	Mean square	F	Sig. (*p*)
Main effects					
Weight	522.18	1	522.18	7.09	.008
Race	103.66	1	103.66	1.41	.237
Interaction	154.39	1	154.39	2.09	1.49
Residual	13483.341	183	73.68		
Total	14217.68	186	76.44		

Source	ANOVA source table for ranked data				
	Sums of squares	*df*	Mean square	L	Sig. (*p*)
Main effects					
Weight	15271.76	1	15271.76	4.65	.05
Race	360.61	1	360.61	.112	.739
Interaction	6230.48	1	6230.48	1.93	.167
Residual	591124.20	183	3230.19		
Total	614275.20	186	3302.56		

$r^2 = (\text{SSregression}/\text{SStotal}) = (15271.76/614275.20) = .025$

$df = N - 1 = 187 - 1 = 186$

$L(1) = .025\,(186) = 4.65$

in repeated measures designs: The repeated measures must all be ranked as a single variable and then copied into the columns for repeated measures in SPSS or any other statistical package.

Value of GLM Rank-Order Statistical Procedures

The techniques developed by Puri and Sen (1969, 1985) have considerable value and application to our field, particularly to physical activity data because typical variables are expected to have nonnormal distributions (e.g., the data used here as well as other physical activity data). We have advocated these procedures for consideration by exercise science professionals (Thomas et al., 1999), and others have advocated the procedures for educational research (Harwell, 1990; Harwell & Serlin, 1989).

One historical drawback to nonparametric analyses has been the lack of standard computer techniques for data analysis. These are now more available, but the advantage of the GLM rank-order procedures described here is that they are used with standard parametric models and pro-

grams well understood by researchers in physical activity. By simply changing the data to ranks (a standard option on statistical programs), all of the typically used regression and differences among groups routines are available to the researcher in parallel form to their parametric counterparts. All that is required is one simple calculation for the L statistic (equation 4.2).

While nonparametric procedures are often thought of as lacking the power of parametric ones, that does not appear to be the case if the data are non-normal. In each example in this chapter or previously cited (Thomas et al., 1999) the results are either equivalent or stronger. While these procedures may not have great power with small sample sizes, neither do parametric ones. These procedures do maintain Type 1 error rates effectively in moderate sample sizes when data are non-normal (Harwell, 1990). As Thomas et al. (1999, p. 22) summarize for these rank-order nonparametric statistical techniques:

the researcher has a cohesive and flexible set of procedures to choose from, depending on the characteristics of the data distribution. Consequently, with the ease and flexibility of the L test, the software (e.g., SPSS, 1997) available, and the high degree of power for nonnormal distributions, the procedures reported here have much to recommend them.

Summary

In this chapter we have shown that with nonnormal physical activity data, rank-order procedures using the GLM have considerable value for statistical analyses.

Five strengths of this technique are as follows:

1. The assumption of normality is not required for ranked data.
2. The statistical techniques for the ranked data are familiar to scientists (and more familiar than typical nonparametric techniques).
3. The parametric statistics used for ranked data are available in computer statistical packages.
4. The results of the rank order procedures are similar to or stronger than parametric analyses performed on the original data.
5. The rank-order procedure maintains Type I error rates.

These rank-order techniques are exact parallels for the standard parametric techniques that most researchers in physical activity use on a regular basis. All interpretation of significance and meaningfulness remains constant except the use of the chi-square table for evaluating the L statistic rather than the use of the parametric evaluation tables for t and F.

Endnote

[1]To compare values to the chi-square table using regression procedures (r or R), calculate $p \times q$ where p is the number of predictor variables and q is the number of criterion variables; for r, $p = 1$ and $q = 1$, thus $p \times q = 1$. In ANOVA, p = number of groups – 1 and q = number of dependent variables; thus in a simple ANOVA with 3 groups, $p = 3 - 1 = 2$ and q = number of dependent variables (1 in ANOVA). Then, $p \times q = 2 \times 1 = 2$. The pq degrees of freedom are used to enter the chi-square table to determine whether the estimated chi-square is significant.

References

Bradley, J.V. (1980). Nonrobustness in *Z, t,* and *F* tests at large sample sizes. *Bulletin of the Psychonomic Society, 16,* 333-336.

Campbell, K.T., & Taylor, D.L. (1995). Canonical correlation analysis as a general linear model: A heuristic lesson for teachers and students. *Journal of Experimental Education, 64,* 157-171.

Geary, R.C. (1947). Testing for normality. *Biometrika, 34,* 209-242.

Harwell, M.R. (1990). A general approach to hypothesis testing for nonparametric tests. *Journal of Experimental Education, 58,* 143-156.

Harwell, M.R., & Serlin, R.C. (1989). A nonparametric test statistic for the general linear model. *Journal of Educational Statistics, 14,* 351-371.

Micceri, T. (1989). The unicorn, the normal curve, and other improbable creatures. *Psychological Bulletin, 105,* 156-166.

Pedhazur, E.J. (1982). *Multiple regression in behavioral research* (2nd ed.). New York: Holt, Rinehart and Winston.

Puri, M.L., & Sen, P.K. (1969). A class of rank order tests for a general linear hypothesis. *Annals of Mathematical Statistics, 40,* 1325-1343.

Puri, M.L., & Sen, P.K. (1985). *Nonparametric methods in general linear models.* New York: Wiley.

Thomas, J.R., Nelson, J.K., & Thomas, K.T. (1999). A generalized rank-order method for nonparametric analysis of data from exercise science: A tutorial. *Research Quarterly for Exercise and Sport, 70,* 11-23.

Thompson, B. (1984). *Canonical correlation analysis: Uses and interpretations.* Newbury Park, CA: Sage.

Timm, N.H. (1975). *Multivariate analysis with applications in education and psychology.* Monterey, CA: Brooks/Cole.

5

Equating and Linking of Physical Activity Questionnaires

Weimo Zhu, PhD

University of Illinois at Urbana-Champaign

Surveys with self- or interviewer-administered questionnaires remain the most commonly used approach in measuring physical activity, although more than 30 other methods are now available. The major reasons for the popularity of the survey methods are that they can be administered quickly to a large number of respondents, they are inexpensive, they have a minimal respondent burden, and they can provide information on the patterns of physical activity participation over time (Sallis & Saelens, 2000). To date, more than 40 different physical activity questionnaires have been developed for use in adult populations (Lamb & Brodie, 1990; Kriska & Caspersen, 1997; Montoye et al., 1996) and new questionnaires targeting a particular subpopulation or serving a special purpose are regularly being developed.

The results and interpretations from physical activity research are greatly influenced by the measurement properties of the instruments that are used. Therefore, a major research challenge has been to try to compare results from studies using differ-

ent self-report assessments. In general, scores from different questionnaires are not compatible and, therefore, not exchangeable. The reason for this is that questionnaires are typically set on different scales. No study, according to a literature search, has examined the comparability of the questionnaires, although a few attempts were made to evaluate the validity of several questionnaires simultaneously (e.g., Jacobs et al., 1993). Fortunately, this important, yet ignored, issue can be addressed using test equating: A set of statistical methods is developed in educational and psychological measurement research to set two or more measures on the same scale.

The purpose of this chapter is to provide an overview of test equating, including its definition and essentials, major equating methods, related conditions, and several major practical issues. In addition, the remaining challenges and future research directions in this area are described. To foster understanding of the issues in the context of assessing physical activity using questionnaires, the term *scale equating*, rather than *test equating*, is used in the rest of this chapter.

What Is Scale Equating?

To better understand the scale equating of physical activity questionnaires, the nature of physical activity surveys should be examined. The process of measuring physical activity using questionnaires is, essentially, to take a sample of a respondent's physical activity behavior from a predefined, overall behavioral universe. This universe usually includes several categories, such as leisure-time, household, and occupational activities. Based on the sample taken, the respondent's individual physical activity behavioral universe can be estimated or predicted.

Typically, there are two ways to take physical activity behavioral samples using physical activity questionnaires, and both try to collect information on intensity, frequency, and duration of physical activity that respondents participated in. The first way is to present a number of specific activities from selected physical activity categories (e.g., running from a "conditioning exercise" category) and ask respondents if they have participated in that activity during a past period of time (e.g., one month or one year; see questionnaire A in figure 5.1). The intensity in this case is controlled by the nature of the physical activity. If the respondents say yes, they are asked to provide more detailed information (e.g., frequency and duration). The well-known Minnesota Leisure-Time Physical Activity Questionnaire (Taylor et al., 1978) is a good example of this kind of sampling method.

The second way to take physical activity behavioral samples is to present just a sample of physical activity categories at different intensity levels (e.g., moderate, hard, very hard) and ask respondents if they engaged in any activity in these categories during a past period of time (see questionnaire B in figure 5.1). Again, if they did, they will be asked to provide more detailed information (e.g., frequency and duration). The Godin Leisure-Time Exercise Questionnaire (Godin & Shephard, 1985) and the

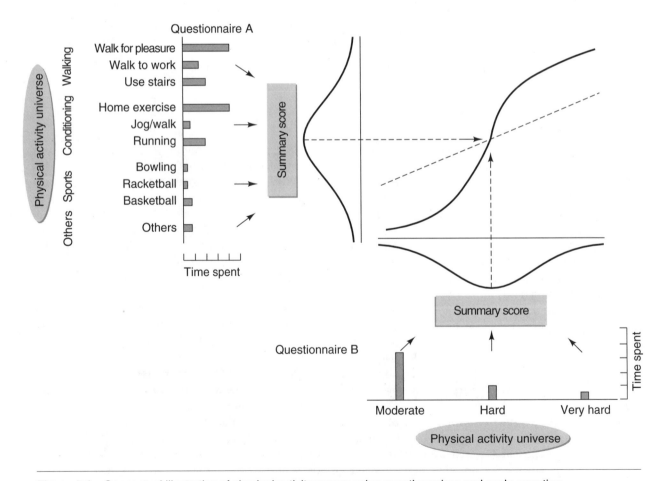

Figure 5.1   Conceptual illustration of physical activity survey using questionnaires and scale equating.

Seven-Day Physical Activity Recall (Sallis et al., 1985) are two good examples of this kind of sample method.

Either way, a summary score is generated, usually by timing respondents' response scores with some kind of energy expenditure weights (e.g., a MET value) and summing them together. The summary score is then used as an estimate of respondents' physical activity participation level at a predetermined scale. Since questionnaires are usually set on different scales (e.g., "MET-min/day" in the Minnesota Leisure-Time Physical Activity Questionnaire and "weighted times per week" in the Godin Leisure-Time Exercise Questionnaire), absolute values and units of the summary scores, even if they were generated from the same respondent, are usually different. As a result, these scores cannot be directly compared with each other.

The task of scale equating is to establish statistically a conversion relationship among summary scores from two or more questionnaires. The relationship could be linear (the straight line in figure 5.1) or nonlinear (the curved line in figure 5.1), depending on the scale equating method employed.

Equating Methods

Scale equating methods, according to the testing theory on which they are based, can be generally classified into two categories: traditional equating (Kolen, 1988) and item response theory (IRT) equating (Cook & Eignor, 1991).

Traditional Equating

Traditional equating methods are based on classical test theory (CTT), in which observed scores are believed to consist of true scores and errors (Hambleton & Jones, 1993). In traditional equating methods, score correspondence of questionnaires is established by setting the characteristics of the score distributions equal for a specified group of examinees (Kolen, 1988). This means that although the absolute values of respondents' summary scores may be different from one another, their relative positions in a sample should be approximately the same, as long as both questionnaires take behavioral samples from a similarly defined activity universe. Linear and equipercentile equating are the two most commonly used traditional equating methods.

Linear Equating

In the linear equating method, the means and standard deviations of the two questionnaires for a particular group of examinees are set equal, thus:

$$\frac{X_1 - \overline{X}_1}{S_1} = \frac{X_2 - \overline{X}_2}{S_2} \tag{5.1}$$

where X_1 and X_2 are summary scores, $\overline{X}_1$ and $\overline{X}_2$ are means, and S_1 and S_2 are standard deviations of questionnaires 1 and 2, respectively. One may immediately realize that the elements on each side of equation 5.1 are, indeed, the familiar z-score equations. Linear equating, therefore, can be conceptually considered as the establishment of equivalent z-scores for two different questionnaires. Based on equation 5.1, the conversion constants *A* and *B*, known as the *slope* and *intercept* of linear equating, can be derived (Kolen, 1988):

$$X_1 = \frac{S_1}{S_2}X_2 + \left[\overline{X}_1 - \frac{S_1}{S_2}\overline{X}_2\right] = AX_2 + B \tag{5.2}$$

Therefore,

$$A = \frac{S_1}{S_2} \tag{5.3}$$

and

$$B = \overline{X}_1 - \frac{S_1}{S_2}\overline{X}_2. \tag{5.4}$$

If, for example, $\overline{X}_1 = 400$, $S_1 = 150$, $\overline{X}_2 = 35$, and $S_2 = 2$, then $A = 75$ and $B = -2225$. A score of 37 from questionnaire 2, according to the conversion based on equation 5.2, is equivalent to a score of 550 in questionnaire 1. In this way, questionnaire 2 was equated to questionnaire 1. But there is an important assumption in linear equating that the score-distribution shapes of questionnaires 1 and 2 should be the same, or at least approximately the same, which is often not the case in reality.

Equipercentile Equating

In equipercentile equating, score distributions are set to be equal so that the same percentile ranks from different questionnaires are considered to indicate the same level of performance. The first step in equipercentile equating is to determine the percentile ranks for the score distributions on each of the two questionnaires to be equated. Percentile ranks are then plotted against the raw scores for each of the two questionnaires and so-called rank-raw score curves can be constructed. Figure 5.2 illustrates plots of the rank-raw score curves of two hypothetical questionnaires on different scales, in which the relative cumulative frequency distributions (i.e., percentile ranks/100) were first plotted for each questionnaire. As long as the percentile rank-raw score curves are constructed, equivalent scores can easily be converted from them. For example, to convert a score in questionnaire 2 to questionnaire 1, one can simply first draw a vertical line up from the scale of questionnaire 2 to its own rank-raw score curve, a horizontal line to the curve of questionnaire 1, and another vertical line down to the scale of questionnaire 1 (see also figure 5.2). To reduce sampling error, especially when the sample size is small, these curves are usually smoothed, which is often completed using analytical smoothing procedures (Kolen, 1984).

Analytical smoothing procedures, depending on what is smoothed, can be classified as pre- and post-smoothing. In pre-smoothing, the score distributions are smoothed. Commonly used pre-smoothing procedures include the polynomial log-linear and the strong true score methods (Kolen & Brennan, 1995). In post-smoothing, the equipercentile equivalents are smoothed. The commonly used post-smoothing procedure is the cubic smoothing

splines method developed by Kolen (1984). In this method, a set of smoothing parameters is usually set and the equating results are compared by examining the smoothed and observed equipercentile relationship. For more information on smoothing strategies, refer to Kolen and Brennan (1995).

Item Response Theory (IRT)

IRT is also known as the *latent trait theory* or *item characteristic curve theory*. At the heart of the theory is a mathematical model describing how examinees or respondents, at different levels for the ability or trait to be measured, should respond to an item or question (Hambleton & Jones, 1993). IRT has several measurement advantages over CTT (Hambleton & Swaminathan, 1985). First, item parameters are independent of the ability level of the respondents responding to the item and, at the same time, ability is also independent of the performance of other respondents and the items used in questionnaires, if model assumptions are met. This is known as the *invariance* feature of IRT. Because of this feature, the interpretation of item characteristics and respondent ability is consistent in IRT. Second, the precision of measurement can be determined at any ability level, while the measurement error in CTT is usually determined at the group level (e.g., the commonly used standard error is an overall precision index). Finally, item and respondent ability are set on the same scale, which makes it

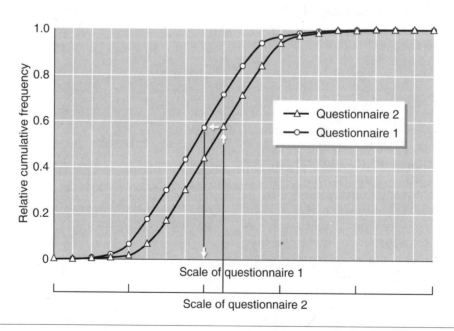

Figure 5.2 Equipercentile equating of two hypothetical physical activity surveys.

much easier to determine the appropriateness of an item and to interpret test scores. Very little information can be obtained when an item or question administered is too far from a respondent's ability level.

IRT was introduced to our field some time ago (Spray, 1987) and a number of successful applications have been reported (see e.g., Looney, 1997; Safrit et al., 1989; Zhu, 1996; Zhu & Cole, 1996). IRT, however, is more restricted in model assumptions. Satisfactory model-data fit is usually required before applying IRT results. Also, a larger sample size is usually required to implement IRT calibrations. For more detail on IRT and its differences from CTT, refer to Hambleton and Jones (1993), Lord (1980), and Spray (1989).

IRT Equating. If parameters in IRT are invariant (i.e., similar estimations should be obtained regardless of which questionnaire respondents are to take, or which group of respondents will be used to calibrate questionnaires), there seems little need to equate questionnaires. While the IRT invariance feature is true in theory, it is not completely true in real measurement practice. The invariance feature will hold if there is only a single calibration in which a scale is set up. This, however, is often not the case in practice (e.g., two calibrations are needed when linking one questionnaire to another). When there is an additional calibration, the IRT invariance feature will no longer hold because in each separate IRT calibration, a score of zero has to be arbitrarily assigned, either to the mean of examinee ability or to the mean of an item parameter. Therefore, to set two questionnaires on the same scale, differences caused by arbitrary assignments must be adjusted. The process to adjust the difference is called *equating*, which indeed is the scaling. Thus, in the framework of IRT, scaling (rather than equating) is necessary, but the terms are often used interchangeably in the literature (Hambleton et al., 1991).

Steps of IRT Equating. Major steps of IRT equating include data collection, selection of an IRT model, estimation of parameters, determination of scaling constants, and scale transformation. These steps are briefly described in the following paragraphs. See Kolen and Brennan (1995) for more information about IRT equating, and Zhu (2001) for an application example in motor function assessment.

The data collection starts from selecting a data collection design. Although all three data collection designs described subsequently have been used in IRT equating, the anchor-test design is the one most commonly used. This is because group abilities involved in the anchor-test design are often different from each other; this is when traditional equating methods often do not work well (Cook & Eignor, 1983).

After the data are collected, the next step is to select an appropriate IRT model. The commonly used IRT models for dichotomous scores are one-, two-, and three-parameter logistic models (Hambleton & Swaminathan, 1985). The one-parameter model is also known as the *Rasch model* (Rasch, 1960; Wright & Masters, 1982). Other multiple response category and multiple parameter IRT models have also become available. See Baker (1992, 1993b), Hambleton and Swaminathan (1985), and van der Linden and Hambleton (1997) for more detail.

With the selected IRT model, item and respondent parameters can be estimated, which is usually accomplished by employing certain computer programs (Hambleton et al., 1991). Model-data fit is also examined statistically at the same time. If the model and data do not fit, either a new model should be considered or new data should be collected. If the model and data do fit, the equating can then move to the next step, in which parameter estimates from separate calibrations are placed on a common scale.

Determining scaling constants is the key process in the step to place parameter estimates on a common scale. A number of methods to determine scaling constants have been developed, and they can generally be classified into four categories: regression, mean and sigma, robust means and sigma, and characteristics curve methods (Hambleton & Swaminathan, 1985). These methods differ from each other in which parameter (e.g., item difficulty or discrimination) or what kind of information (e.g., mean and standard deviation) is included in the computations (Hambleton et al., 1991; Stocking & Lord, 1983).

IRT equating has basically been completed after parameters from two separate estimations have been set on the same scale. In practice, however, it may be necessary to further scale ability estimates into other predetermined scales. There are two reasons for this. First, IRT computer programs usually provide a scale called the *logit*, which is very similar to *z*-scores, with zero representing the mean and negative values representing scores below the mean. The general public, however, is often uncomfortable with, and sometimes even confused by, these zero and negative values. Therefore, instrument developers need to transfer ability estimates into scales having only positive values. Second, many assessment programs have established their own scales (e.g., the Graduate Record Examination [GRE] is calibrated on a scale with a mean of 500 and a standard deviation of 100). To report IRT estimates into an established scale, further transformation becomes necessary.

If instrument developers want merely to transfer the ability estimates to a scale having only positive values, the transformation is indeed very straightforward because ability estimates are already set on an interval scale, which means that the scale can be linearly transferred into any other scale. If, however, instrument developers want to transfer ability estimates into established scales, a more complex two-step approach is often needed: (1) transfer ability estimates into true scores, and (2) further transfer the true scores into reported scores (Lord, 1980; Cook & Eignor, 1991; Kolen & Brennan, 1995).

Practical Issues of Scale Equating

Scale equating, like other aspects of instrument development, involves a rather complex process, and many practical decisions have to be made. Several key practical issues are reviewed briefly in this chapter, including data collection design, selection of an equating method, equating errors, sample size, and computer program. For more detailed information, refer to Kolen and Brennan (1995) and Zhu (1998b).

Data Collection Designs

An equating starts from data collection. Three commonly used data collection designs are single-group, equivalent-groups, and common item nonequivalent groups designs. In the single-group design, two or more questionnaires are administered to the same group of respondents. The advantage of this design is that the measurement error is relatively small. This is because only one group of respondents is involved in the equating process. Therefore, differences among questionnaires are not confounded by differences among groups. This design, however, requires that respondents answer many questions; therefore, fatigue is a major concern. Practice effect (i.e., respondents may change their responses due to their responses on previous questionnaires) is another concern. To avoid fatigue and practice effects, some sort of spiraling process should be applied. For example, the order of administration of the testing forms can be counterbalanced (Kolen, 1988; Kolen & Brennan, 1995).

In the equivalent-groups design, two questionnaires to be equated are administered to two equivalent groups of respondents. The groups may be chosen randomly, which is why this design is sometimes called the *random-groups* design. The advantage of this design is that the problems related to the single-group design, such as fatigue and practice effects, can be eliminated. Furthermore, administration time is minimized and the questionnaire can be finished in a single administration. The disadvantage of this design is that an unknown degree of error is introduced in the equating process because groups are often not exactly the same in their ability distributions. To control sample-related error, larger samples generally are required for this design (Kolen, 1988; Kolen & Brennan, 1995).

Finally, in the common item nonequivalent groups design (Kolen, 1988), questionnaires are also administered to two different groups of respondents. In contrast with the equivalent-groups design, however, groups can be different from each other in their ability distributions. Furthermore, a set of common, or anchor, items or questions is included in both questionnaires so that differences between the two questionnaires can be adjusted based on common-item statistics. Because the two groups do not have to be equivalent, this design is extremely useful when measuring growth or change in which two groups are known to be not equivalent, or when it is impossible to administer more than one questionnaire at a time per date due to practical concerns. This design is also necessary when developing an item bank, which will be described later. Strong statistical assumptions are required to remove the confounding effects of group and test differences and, quite often, statistical procedures can provide only limited adjustments (Petersen et al., 1989).

Practicality is often the major concern in selecting a data collection design. The common item nonequivalent groups design is often more practical because both single-group and equivalent-groups designs require the administration of two questionnaires with little or no time intervening between test administrations, which is often difficult to implement in practice. Many equating procedures may not be applicable when distributions of common items from two subpopulations are markedly different (Angoff, 1984). In addition, there should be enough common items with representative content to be measured. A rule of thumb for the minimum length of a common item nonequivalent group is 20 to 25% of the number of items in either of the questionnaires.

Selecting an Equating Method

Measurement advantages and disadvantages, as well as the tenability of assumptions, accuracy, and practicality (Crocker & Algina, 1986), are the major considerations for choosing an equating procedure.

IRT Equating Versus Traditional Equating. Besides its usefulness for the nonequivalent group design, IRT equating has five other advantages over traditional equating (Cook & Eignor, 1983). First, the quality of equating is improved because better equating at the upper end of the score scale is usually provided, which is extremely important in the case of vertical equating. Second, greater flexibility is provided in choosing previous versions of a questionnaire because all previous versions were calibrated on the same scale. Third, re-equating becomes easier because if an item is dropped, the shortened test can be easily reconstructed based on the item information from the remaining test items. Fourth, the impact of sampling error is decreased because of the invariance feature of IRT equating. And fifth, pre-equating becomes possible when applying IRT equating. Pre-equating is an attempt to prepare raw-to-scale score conversion tables even before a question is administered so that quick score reporting can be accomplished. IRT equating makes it possible because items in the "new" questionnaire have been previously administered, and they will be invariant when applied to new groups. (For more detail on these advantages, refer to a comprehensive review by Skaggs and Lissitz [1986].)

IRT equating, however, is not always superior to traditional equating. In fact, a number of researchers (e.g., Lord, 1975; Rentz & Bashaw, 1977) have compared the performance of these equating methods and found that traditional equating often worked as well as IRT equating methods, except for the common item nonequivalent groups design (Lord, 1975). Furthermore, IRT equating is often more complex technically than traditional equating approaches. Therefore, the general rule is to apply IRT equating when

- a questionnaire is already calibrated using an IRT model,
- traditional equating is not appropriate (e.g., anchor-test design with nonequivalent groups), and
- other obvious benefits can be achieved (e.g., pre-equating) (Cook & Eignor, 1983; Kolen & Brennan, 1995).

Traditional equating should be employed when IRT equating is not applicable.

Linear Equating Versus Equipercentile Equating. Then which equating method, linear or equipercentile, should be chosen when traditional equating is employed? Again, the tenability of assumptions, practicality, and accuracy should be considered. When compared to linear equating, which assumes that the only differences between the distributions of questionnaires 1 and 2 are the mean and the variance, equipercentile equating makes fewer restricted assumptions.

The equipercentile methods, however, have larger equating errors than linear equating. Thus, linear equating is generally preferable if the distributions of z-scores from questionnaires to be equated are approximately the same. On the other hand, if the distribution shapes of questionnaires to be equated are not the same, linear equating may not be appropriate and equipercentile equating should be employed.

In the past, equipercentile equating has been considered more complicated to carry out and, therefore, was less practical. Due mainly to today's computing power, practicality is no longer a major constraint in selecting an equating method because commonly used equating computer programs often include several equating methods. The RAGE equating computer program (Zeng et al., 1995), for example, provided equating results by the linear, unsmoothed equipercentile method, and several smoothed equipercentile methods, simultaneously.

Equating Errors and Evaluating Criteria

Two kinds of errors, random and systematic, often occur in the process of test equating (Kolen, 1988). Random equating errors occur when samples are used to estimate parameters (e.g., population means, percentile ranks, item difficulties). Generally, random error is a major concern when the sample size is small. Therefore, random error can be reduced by increasing the sample size. Employing appropriate data collection designs may also help to reduce random error. Systematic errors occur when assumptions, or conditions, of a particular data collection design or an equating method are violated. In the single-group design, for example, failure to take fatigue and practice effects into consideration may lead to systematic errors. Systematic errors can also occur whenever the new and old questionnaires differ in content, difficulty, or reliability. Thus, every effort should be made to control and monitor random and systematic errors in the process of test equating.

To control or at least monitor equating errors, a number of evaluation criteria and procedures have been proposed: for example, weak equity (Yen, 1983), standard errors (Angoff, 1984), standardized difference score (Wright, 1968), and cross-validation (Kolen, 1981). Unfortunately, according to an extensive review by Harris and Crouse (1993), none of these criteria or procedures will apply to all equating situations, and much more research is needed in this area. Even so, to ensure the quality of test equating, some sort of evaluation plan is strongly urged whenever test equating is conducted or a new equating procedure is introduced. Kolen and Brennan (1995) have provided a thorough description of the characteristics of equating situations and related evaluation plans and methods that should be employed.

Sample Size

Sample size, as mentioned earlier, has a direct impact on random equating errors. Thus, larger samples generally lead to better equating results. Kolen and Brennan (1995) have proposed two rules of thumb for sample size. The first is based on standard deviation units; the second is based on comparisons with identity equating. For traditional equating, a sample size of 400 per questionnaire is generally needed for linear equating and slightly over 1,500 for equipercentile equating. For IRT equating, a sample size of 400 is generally needed for the Rasch model and 1,500 for a three-parameter model (Kolen & Brennan, 1995). When the sample size is small, other equating approaches, such as log-linear smoothing (Livingston, 1993) and the collateral information method (Mislevy et al., 1993), may be considered.

Computer Programs

Scale equating, like other areas of test development, also takes advantage of the power of modern computers. A set of computer programs for test equating, including both traditional and IRT equating methods, has been developed along with an excellent test equating book by Kolen and Brennan (1995). In addition, a computer program for IRT nominal and grade response models has been developed by Baker (1993a). These computer programs can be obtained at no cost by contacting the program authors.

Remaining Challenges and Future Research Directions

While equating procedures have been well developed and described in educational and psychological measurement research literature, and successfully applied to physical fitness (Zhu, 1998a; 1998b) and motor function tests (Zhu, 2001), many challenges remain to apply these procedures to physical activity questionnaire equating. These challenges, along with future research directions, are discussed briefly in this section.

The major challenge related to classical test equating is that most existing computer software has not been developed to handle continuous data with different response formats. In the practice of educational measurement, forms or tests to be equated usually have the same number of items and similar response formats. Existing test equating computer programs were, therefore, developed for these kinds of forms or tests. For example, the RAGE computer program developed by Zeng et al. (1995) is limited to a total of 80 items, or score categories, in each form. The number of questions (or the range of the scores) in physical activity questionnaires and their response formats, however, vary greatly from each other. As a result, the data from these questionnaires may not be analyzed by that software. An effort to modify existing software according to the nature of continuous data in physical activity questionnaires and evaluation of the appropriateness of their utility is greatly needed.

When such appropriate software becomes available, the next immediate task is to determine the equivalence of commonly used physical activity questionnaires. More important, along with other psychometric techniques (e.g., factor analysis), systematic efforts should be made to determine which set of questionnaires can be equated to each other. If one questionnaire is designed to measure leisure activities while the other measures occupational activities, for example, the two questionnaires may not be equated to the same scale because they measure different things.

The continuous nature of physical activity questionnaire data is also a challenge to IRT equating. Most existing IRT models are developed for educational or psychosocial measures, where commonly used response categories are dichotomous or polyotomous. While a few continuous IRT models (e.g., Samejima, 1973) have been developed, few computer programs are available for their applications. To be able to take the measurement advantages of IRT models, efforts to develop computer software programs that implement these models should be made.

The unidimensionality assumption (i.e., only one major attribute or latent trait is measured) related to the commonly used IRT model could be another constraint to apply IRT equating. This is because most physical activity questionnaires may measure more than one dimension of physical activity participation (e.g., leisure activities, sports, occupational activities). Fortunately, some progress has recently been made to develop and apply multidimensional IRT models (Reckase, 1997). Exploring these new measurement models for physical activity questionnaires should be an exciting research area.

Developing an item bank for physical activity assessment and computerized adaptive administering of a survey based on such a bank are the future research directions. An item bank is a collection of items or questions organized and catalogued to take into account the content of each item, as well as other measurement characteristics (e.g., validity, reliability, difficulty) (Umar, 1997). More important, all of the items in the bank share a common scale. With an item bank, several measurement advantages can be achieved. First, since all of the items are set on the same scale, the problem of score nonequivalence among different questionnaires is automatically eliminated. Thus, scores can be directly compared with each other. Through some sort of validation process, the connection of the scale to other, more objective physical activity measures (e.g., energy expenditure measured by doubly labeled water or movement assessment devices) can be established, and scale scores become more meaningful.

Second, a stable scale can be developed even if new versions or items are later added to the bank. That is, the scale in a physical activity item bank will function like the scales developed in standardized tests (e.g., the GRE), where the mean and standard deviation of the scale are consistent across different times even if new items are used. A stable scale is essential to measure physical activity across occasions and to communicate the results of physical activity research studies.

Third, since the characteristics of the items are already known, constructing new questionnaires for different purposes of a survey becomes much easier. For example, if a researcher is interested in constructing a survey to screen a population's physical activity participation, items that cover a broad range of activity participation can be selected. As a result, a broad-ranged questionnaire (shown in figure 5.3a) can be easily constructed. If, on the other hand, the interest is to construct a survey to determine if a respondent belongs to a certain subpopulation (e.g., sedentary versus nonsedentary), only items around a theoretical cutoff point will be selected. As a result, a shorter survey, but one with peaked information at the cutoff point, can be constructed (see the peaked questionnaire in figure 5.3b). As a result, a more accurate classification can be achieved.

More important, with a well-developed item bank, a new and tailored way to administer a survey using computers, known as *computerized adaptive testing* (CAT) (Wainer, 1990), becomes possible. CAT, originally developed in educational measurement practice, is very similar to a high-jump competition, in which competitors choose their jumping heights depending on their perceived abilities. In CAT, a question is first asked in the middle of an ability range that an item bank has defined. If it is answered correctly, the next question is more difficult; if it is incorrectly answered, the next question is easier. This continues until the examinee's proficiency is established to within a predetermined level of accuracy. Because of IRT invariance, examinees' abilities can be estimated even if they respond to different items in CAT. In fact, a more accurate estimation is expected because the items selected are close to an examinee's ability. Because responses have been entered into a computer during the test administration, the final estimation of an examinee's ability can be reported as soon as the test is finished. With a physical activity item bank, physical activity surveys can be administered in a similar manner (i.e., select a survey question based on a respondent's previous responses; see the computerized-tailored questionnaire in figure 5.3c).

Developing and maintaining such an item bank is not an easy task, however. It takes long and continuous efforts to link items measuring the same construct onto the same scale and the equating process described earlier is just a part of that effort. To develop an item bank, one must continuously develop, pilot test, and add items to the bank (i.e., link them on the same scale). Usually, an IRT-based calibration is a requirement for developing such a bank (Hambleton et al., 1991). Highly specialized professionals in both measurement and computer technology are also needed on the development team. It is, therefore, more expensive to develop an item bank for general measurement purposes, especially at the beginning. Finally, the requirements of advanced measurement theory are sometimes difficult to satisfy.

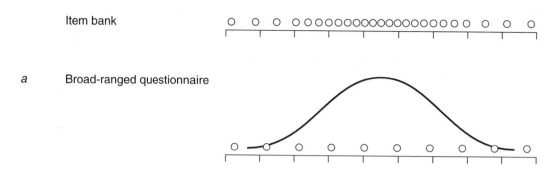

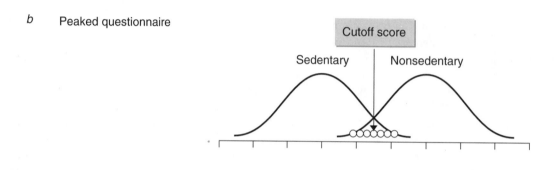

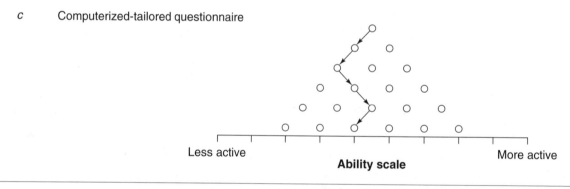

Figure 5.3 Questionnaire construction based on an item bank: (*a*) broad-ranged, (*b*) peaked, and (*c*) computerized-tailored questionnaires.

Considering the measurement benefits already described, however, efforts to overcome these difficulties are clearly needed and worthwhile. It is also expected that some of these difficulties will be overcome in the very near future with newly developed measurement models and more user-friendly computer or Web-based software. For the area of physical activity measurement, realizing measurement benefits of the item bank and making an initial exploration of such a system should be promoted and encouraged.

Summary

Physical activity questionnaires must be set on the same scale before their scores can be compared. A set of statistical procedures, equating, has been developed and is available for such a purpose. To apply them to physical activity questionnaires, however, considerable effort is needed to develop new software to implement both classical and IRT equating methods and to evaluate the utility of these methods in equating physical activity questionnaires, which are usually multidimensional and have continuous data and different response formats. Based on IRT equating, developing item banks for future physical activity assessment becomes possible. With such a bank, computerized, individually tailored physical activity assessment could become practical, less time consuming, and more accurate.

References

Angoff, W.H. (1984). *Scales, norms, and equivalent scores.* Princeton, NJ: Educational Testing Service.

Baker, F.B. (1992). Equating tests under the graded response model. *Applied Psychological Measurement, 16,* 87-96.

Baker, F.B. (1993a). EQUATE 2.0: A computer program for the characteristic curve method of IRT equating. *Applied Psychological Measurement, 17,* 20.

Baker, F.B. (1993b). Equating tests under the nominal response model. *Applied Psychological Measurement, 17,* 239-251.

Cook, L.L., & Eignor, D.R. (1983). Practical considerations regarding the use of item response theory to equate tests. In R.K. Hambleton (Ed.), *Applications of item response theory* (pp. 175-195). Vancouver, BC: Educational Research Institute of British Columbia.

Cook, L.L., & Eignor, D.R. (1991). An NCME instructional module on IRT equating methods. *Educational Measurement: Issues and Practice, 10,* 37-45.

Crocker, L., & Algina, J. (1986). *Introduction to classical and modern test theory.* New York: Holt, Rinehart and Winston.

Godin, G., & Shephard, R.J. (1985). A simple method to assess exercise behavior in the community. *Canadian Journal of Applied Physiology, 10,* 141-146.

Hambleton, R.K., & Jones, R.W. (1993). An NCME instructional module on comparison of classical test theory and item response theory and their applications to test development. *Educational Measurement: Issues and Practice, 12,* 38-47.

Hambleton, R.K., & Swaminathan, H. (1985). *Item response theory: Principles and applications.* Boston: Kluwer-Nijhoff.

Hambleton, R.K., Swaminathan, H., & Rogers, H.J. (1991). *Fundamentals of item response theory.* Newbury Park, CA: Sage.

Harris, D.J., & Crouse, J.D. (1993). A study of criteria used in equating. *Applied Measurement in Education, 6,* 195-240.

Jacobs, D.R., Ainsworth, B.E., Hartman, T.J., & Leon, A.S. (1993). A simultaneous evaluation of 10 commonly used physical activity questionnaires. *Medicine and Science in Sports and Exercise, 25,* 81-91.

Kolen, M.J. (1981). Comparison of traditional and item response theory methods for equating tests. *Journal of Educational Measurement, 18,* 1-11.

Kolen, M.J. (1984). Effectiveness of analytic smoothing in equipercentile equating. *Journal of Educational Statistics, 9,* 25-44.

Kolen, M.J. (1988). An NCME instructional module on traditional equating methodology. *Educational Measurement: Issues and Practice, 7,* 29-36.

Kolen, M.J., & Brennan, R.L. (1995). *Test equating: Methods and practices.* New York: Springer.

Kriska, A.M., & Caspersen, C.J. (1997). Introduction to a collection of physical activity questionnaires. *Medicine and Science in Sports and Exercise, 29,* 5-9.

Lamb, K.L., & Brodie, D.A. (1990). The assessment of physical activity by leisure-time physical activity questionnaires. *Sport Medicine, 10,* 159-180.

Livingston, S.A. (1993). Small-sample equating with log-linear smoothing. *Journal of Educational Measurement, 30,* 23-29.

Looney, M.A. (1997). Objective measurement of figure skating performance. *Journal of Outcome Measurement, 2,* 143-163.

Lord, F.M. (1975). A survey of equating methods based on item characteristic theory. *Research Bulletin No. 75-13.* Princeton, NJ: Educational Testing Service.

Lord, F.M. (1980). *Applications of item response theory to practical testing problems.* Hillsdale, NJ: Lawrence Erlbaum.

Mislevy, R.J., Sheehan, K.M., & Wingersky, M. (1993). How to equate tests with little or no data. *Journal of Educational Measurement, 30,* 55-78.

Montoye, H.J., Kemper, H.C.G., Saris, W.H.M., & Washburn, R.A. (1996). *Measuring physical activity and energy expenditure.* Champaign, IL: Human Kinetics.

Petersen, N.S., Kolen, M.J., & Hoover, H.D. (1989). Scaling, norming, and equating. In R.L. Linn (Ed.), *Educational measurement* (3rd ed., pp. 221-262). New York: Macmillan.

Rasch, G. (1960). *Probabilistic models for some intelligence and attainment tests.* Copenhagen: Danish Institute for Educational Research. [Reprinted, Chicago, IL: University of Chicago Press, 1980].

Reckase, M.D. (1997). The past and future of multidimensional item response theory. *Applied Psychological Measurement, 21,* 25-36.

Rentz, R.R., & Bashaw, W.L. (1977). The national reference scale for reading: An application of the Rasch model. *Journal of Educational Measurement, 14,* 161-179.

Safrit, M.J., Cohen, A.S., & Costa, M.G. (1989). Item response theory and the measurement of motor behavior. *Research Quarterly for Exercise and Sport, 60,* 325-335.

Sallis, J.F., Haskell, W.L., Wood, P.D., Fortmann, S.P., Rogers, T., Blair, S.N., & Paffenbarger, R.S.J. (1985). Physical activity assessment methodology in the five-city project. *American Journal of Epidemiology, 121,* 91-106.

Sallis, J.F., & Saelens, B.E. (2000). Assessment of physical activity by self-report: Status, limitations, and future directions. *Research Quarterly for Exercise and Sport, 71,* 1-14.

Samejima, F. (1973). Homogeneous case of the continuous response model. *Psychometrika, 38,* 203-219.

Skaggs, G., & Lissitz, R.W. (1986). IRT test equating: Relevant issues and a review of recent research. *Review of Educational Research, 56,* 495-529.

Spray, J.A. (1987). Recent developments in measurement and possible applications to the measurement of psychomotor behavior. *Research Quarterly for Exercise and Sport, 58,* 203-209.

Spray, J.A. (1989). New approaches to solving measurement problems. In M.J. Safrit & T.M. Wood (Eds.), *Measurement concepts in physical education and exercise science* (pp. 229-248). Champaign, IL: Human Kinetics.

Stocking, M.L., & Lord, F.M. (1983). Developing a common metric in item response theory. *Applied Psychological Measurement, 7,* 201-210.

Taylor, H.L., Jacobs, D.R., Shucker, B., Knudsen, J., Leon, A.S., & DeBacker, G. (1978). A questionnaire for the assessment of leisure-time physical activities. *Journal of Chronic Diseases, 31,* 741-755.

Umar, J. (1997). Item banking. In J.P. Keeves (Ed.), *Educational research, methodology, and measurement: An international handbook* (2nd ed., pp. 923-930). New York: Elsevier Science.

van der Linden, W.J., & Hambleton, R.K. (Eds.) (1997). *Handbook of modern item response theory.* New York: Springer.

Wainer, H. (1990). *Computerized adaptive testing: A primer.* Hillsdale, NJ: Lawrence Erlbaum.

Wright, B.D. (1968). Sample-free test calibration and person measurement. *Proceedings of the 1967 Invitational Conference on Testing Problems* (pp. 85-101). Princeton, NJ: Educational Testing Service.

Wright, B.D., & Masters, G.N. (1982). *Rating scale analysis.* Chicago, IL: MESA Press.

Yen, W.M. (1983). Tau-equivalence and equipercentile equating. *Psychometrika, 48,* 353-369.

Zeng, L., Kolen, M.J., & Hanson, B.A. (1995). Random groups equating program (RAGE, Version 2.0) [Computer software]. Iowa City: American College Testing.

Zhu, W. (1996). Should total scores from a rating scale be used directly? *Research Quarterly for Exercise and Sport, 67,* 363-372.

Zhu, W. (1998a). Test equating of commonly used physical fitness tests. *Research Quarterly for Exercise and Sport, 69* (Supplement 1), A-56 [Abstract].

Zhu, W. (1998b). Test equating: What, why, how. *Research Quarterly for Exercise and Sport, 69,* 11-23.

Zhu, W. (2001). An empirical investigation of Rasch equating of motor function tasks. *Adapted Physical Activity Quarterly, 18,* 72-89.

Zhu, W., & Cole, E. (1996). Many-faceted Rasch calibration of a gross motor instrument. *Research Quarterly for Exercise and Sport, 67,* 24-34.

CHAPTER

6

Sample Size and Power Considerations in Physical Activity Research

Diane J. Catellier, DrPH
Keith E. Muller, PhD
University of North Carolina at Chapel Hill

The guiding principle for decisions regarding study design and analysis is accuracy in estimation of the *parameter of interest*, such as the mean of a population characteristic or the effect of an intervention or exposure on the probability of a successful response. Accuracy is quantified as the difference between the sample estimate and its corresponding true value in the population, also known as the error in estimation. The two main sources of error in estimation are random error and systematic error. The distinction is important because the strategies for minimizing them are different.

The *precision* of an estimator is directly related to the magnitude of random error. Random error in estimation of a parameter arises purely by chance and is similar to randomly selecting a number and adding it to or subtracting it from the true value over repeated samples. This type of error is equally likely to underestimate or overestimate the true value. The average error introduced over repeated

samples of the same sample size is equal to the standard error (SE) of the estimator = standard deviation (SD) of individual observations/$\sqrt{n}$ (n = sample size). Thus, to reduce random error, or increase precision, one must reduce the standard error of the estimator. This can be done in two ways: increase the sample size or choose a measurement with lower variability.

The *validity* of inferences drawn from a study sample to the source population (i.e., internal validity) requires a lack of systematic error in the estimation of effects. Systematic error, or bias, occurs when the estimated effect tends to vary consistently in one direction (higher or lower) from the true value. For example, in tests of human performance, subjects might perform better on a second trial than on the first trial because they are motivated to show an improvement or are more skilled at the task. Such systematic change is less of a concern for a researcher performing a controlled study because only the relative change in means for both groups provides evidence of an effect. Other common types of systematic error that detract from internal validity include the following:

1. Confounding. The observed effect of the intervention or exposure is not real (some third factor is associated with the outcome and the study group).
2. Observer bias. This involves conscious or unconscious distortion of the measurement by an observer.
3. Nonresponse bias. Measurements are different for respondents and nonrespondents.
4. Measurement instrument bias (e.g., bias resulting from leading questions).

The size of the sample does not affect systematic error. Increasing the sample size will give a more precise estimate, but the estimate will continue to systematically overestimate or underestimate the true value. In addition to the strategies set in "Improving Validity (Reducing Systematic Error)" (see page 99) for eliminating or reducing systematic error (choosing objective measurement instruments, administering enough practice trials before applying the intervention so that leaning effect is minimized, standardizing training of interviewers), statistical procedures can be used to reduce certain types of systematic error (e.g., confounding).

Concepts and Language Associated With Power Analysis

Statistical power is the ability to detect a specified effect if there truly is one. Testing for a significant effect involves assessing how the variation explained by the controlled factor under study (e.g, treatment condition) differs from that associated with random error. If variation caused by the controlled factor is much larger than the error variation, we attribute the difference in variation to the factor and conclude that it has an effect on the outcome. If controlled variation does not exceed random error, we conclude that the differences are likely to be random and not due to the controlled factor. The problem arises when a real, nonrandom pattern exists but our test does not find it to be statistically significant. In this case, the random error is so large as to have obscured the real nonrandom pattern of variation. The chance of making such an error (failing to detect a real effect) can be decreased by decreasing the random error. One way this can

be done is to enlarge the size of the study. The aim of power analysis is to determine the appropriate sample size to allow a high probability of detecting a real effect while keeping the probability that a significant effect results purely from random chance suitably low.

It is worthwhile to define more formally the concepts covered in the previous paragraph. The reader interested in additional references and discussion on power may wish to consult Muller and Benignus (1992) and O'Brien and Muller (1993). Suppose the objective of a study is to estimate the effect of an intervention or exposure on an outcome variable. All statistical conclusions involve deciding between two mutually exclusive hypotheses, termed the null (H_0) and the alternative (H_1) hypothesis. For example, in a study to determine whether an intervention increases physical activity, a researcher might randomly assign a sample of subjects to the intervention (I) or control (C) condition. It is typical for the null hypothesis to refer to the case in which there is no difference in mean level of physical activity between the groups, or equivalently that there is no intervention effect (δ)

$$H_0: \delta = \mu_I - \mu_C = 0$$

while the alternative might be

$$H_1: \delta = \mu_I - \mu_C \ne 0,$$

where μ_I and μ_C represent the population mean values for intervention and control conditions, respectively. If one rejects the null hypothesis and concludes the intervention is effective when it is not, then one has committed a Type 1 error. On the other hand, if one fails to reject the null hypothesis and concludes that the intervention is not effective when it is, then one has committed a Type 2 error. The probabilities of Type 1 or Type 2 errors are usually denoted by α and β, respectively. The power of a test, Power = $1 - \beta$, is the probability of rejecting the null hypothesis, if indeed an intervention effect exists.

If researchers believe that the intervention will result in an effect, then they must design the study to have adequate statistical power to detect this effect. A study with 80% power (recommended by Cohen [1988] for behavioral science research) means that the researcher has given himself an 80% chance of confirming a true effect. The Type 2 error is 20%, meaning that the researcher is prepared to accept a 20% chance of being wrong (in this case meaning the researcher would report nonsignificant results when in fact the treatment was actually effective).

Thus, statistical power provides the researcher with the probability of getting the data that will lead to a correct rejection of the null hypothesis. Historically, research in the social sciences has been woefully underpowered.

Factors That Control Power

In designing a study to test the effect of a particular intervention or exposure factor, there are four factors that influence power, or the likelihood that one will find an effect when it occurs: the level of significance to be achieved, the precision of the effect estimator, the hypothesized effect, and the sample size.

Significance Level

The significance level, α, represents the degree of certainty required to claim that an intervention is effective. Reducing α (e.g., from .05 to .01) makes it more difficult to reject the null hypothesis, and thus reduces the chances of committing a Type 1 error. If all other factors are held constant, increasing α increases the power.

An informal rule that many researchers follow to determine the appropriate sample size for a study is to set the significance level $\alpha = .05$ and $\beta = .20$. This rule implies that the risk of falsely rejecting the null hypothesis is four times more serious than failing to detect a true significant effect. In the context of drug trials, this rule is well supported because it is more important to reduce the risk of falsely claiming a new drug's effectiveness in the treatment of a disease (especially when there might be other treatments that truly are effective) than to reduce the risk of not detecting its true effectiveness. However, it is questionable whether this notion applies in the context of physical activity research. It is unlikely that falsely claiming the benefits of an intervention on physical activity will have any harmful effects. On the other hand, it can be argued that a more conservative Type 2 error than $\beta = .20$ should be used because physical activity intervention studies are susceptible to loss in power caused by one or more of the following: insensitivity or unreliability of measurement method, inadequate control of extraneous variables, or a weak implementation of the intervention. In other words, failure to show a significant effect may not have been due to the theoretical basis on which the treatment was based, but rather the weaknesses of the study in several aspects of the design and imple-

mentation of the treatment. Thus, to reduce the risk of unfairly "shelving" an effective intervention, researchers may well want to pursue higher power (i.e., reduce the Type 2 error risk) at the expense of inflating the Type 1 error risk to something larger than .05.

The historical "$\alpha = .05$" rule was devised for a single test or comparison, referring to a 1 in 20 chance of claiming a significant effect when there is none. However, if you perform multiple tests of hypotheses, the probability that you might get at least one false positive can be much larger than .05. For example, suppose we perform five independent tests at the usual $\alpha = .05$ level and that all null hypotheses are true. The probability of finding at least one false positive is equal to 1 – the probability of no false positives (that is, do not reject any null hypotheses) = $1 - (.95)^5 = .23$.

As we have shown, it is important to be aware that multiple testing or multiple comparisons can increase α. Depending on your study objectives, several approaches can be taken to control the Type 1 error rate at or below a desired level, say $\alpha = .05$. If the goal of the study is to decide whether the intervention affects *any* of a set of outcomes, regardless of which outcome is affected, there is no need for adjustment of α. In this case, the researcher tests the global, or *composite*, null hypothesis of no effect on any outcome and only tests the individual null hypotheses if the global hypothesis is rejected. If the goal of the study is to evaluate the effect of the intervention on k independent outcomes, one should use the Bonferroni procedure to adjust the significance level for each individual test to α/k. Power calculations in the context of multiple testing are complex because of the multiple null and alternative hypotheses specified. A complete treatment of this topic is found in Westfall et al. (1999).

Precision of the Estimator

Reducing the random error, or "noise," or increasing precision of the effect estimator will increase power. The precision of an estimator is represented by its standard error (SE), which is equal to the standard deviation (SD) of the outcome variable divided by the square root of the sample size: $SE = SD/\sqrt{n}$. Thus, to increase the precision of an estimator, one must study more subjects or choose a more precise outcome variable. For example, in using the t test to estimate a sample size, a 20% decrease in the standard deviation of the outcome variable results in a 36% decrease in the sample size. Techniques for increasing the precision of a variable, such as using the average of repeated measurements, are presented in "Improving Precision (Reducing Random Error)."

Hypothesized Effect

The independent groups to test for the difference in mean response between two treatments in a clinical trial will be used to correctly illustrate some of the important concepts in power analysis.

We can think of the null hypothesis as the case in which there is no effect (H_0: $\delta = \mu_I - \mu_C = 0$). Acceptance of the null hypothesis implies an effect of zero, whereas rejection implies a non-zero effect (H_1: $\delta = \mu_I - \mu_C \neq 0$). The larger the departure from a null effect, the larger the value of δ. A large δ is easier to detect, leading to increased power.

Whenever possible, subject-matter experts should decide what effect size is most appropriate given the scientific context of the study. However, when the choice remains unclear, two principal strategies for defining a meaningful effect are employed. Cohen (1988, 1992) popularized the idea of translating a qualitative description of an effect ("small," "medium," or "large") to *standardized effect*, defined as the effect (δ) divided by the standard deviation of the outcome variable, which could then be used to compute power. For example, the definition of a medium effect for the *t* test is a standardized effect of 0.5, or a difference in sample means that is half of one standard deviation (Cohen, 1988). The utility of this approach is that power can be determined using only general qualitative terms for the effect size and does not require an estimate of the SD of the outcome variable. However, this approach is not recommended for two reasons. First, the approach will necessarily result in an imprecise estimate of power because it relies on interpretation of such imprecise concepts as a "small effect," which in turn is converted to standardized effect based on "conventional" definitions that are arbitrary. Second, there is a danger in focusing on a standardized effect rather than its components. For example, consider using this approach to design a study to detect a *medium*-sized effect of an intervention to increase physical activity. Two designs that have identical sample sizes, power, significance levels, and standardized effect sizes but use different methods of physical activity measurement are proposed. However, because the SD of the outcome variable in the first design is half of that in the second design, the first design will allow us to detect intervention effect half that of the second design. Despite the clear design advantage of the first design, focusing only on the identical stan-

dardized effect may result in selecting the second design because of lower measurement costs. In the words of Lenth (2000), "We get no credits (or demerits) for using good (or bad) measurement procedures, and we actually get negative credit for using a good design."

A second strategy that is often used to specify an important effect is to get an idea of the natural variation of the response in various subgroups and environmental settings. What are differences in average physical activity between genders, in athletes versus non-athletes, during the summer versus fall, and so on? Knowledge of the size of these effects helps one to define a plausible range of differences that might be expected (Muller & Benignus, 1992).

Sample Size

As a rough rule of thumb, it helps to think of small, medium, or large studies as involving 10s, 100s, or 1000s of subjects. This type of scaling is appropriate for evaluating not only practical aspects of a study but also the performance of statistical methods and methods for power analysis. In some cases, changing the sample size by 5, 10, or 20% can have a substantial impact on power. More commonly, changing sample size by a factor of two or more is needed to make a large change in power. The same idea also applies in evaluating the accuracy of power approximations. Any use of a power approximation, as distinct from an exact method, should be justified by appeal to research on the accuracy of the method for the size of the study being considered. Fortunately, commercial and freeware programs for power analysis increasingly use exact methods when they exist, at least if requested by the user. The limitations of approximations in small samples provide an important motivation for using commercial software in lieu of tables of power values or calculator-friendly formulas.

Improving Precision (Reducing Random Error)

As described in the introduction to this chapter, the most straightforward way to improve precision is to increase the size of a study. The following are other features of the study protocol affecting precision.

Measurement Scale and Schedule

The scale of the outcome variable greatly affects the sample size needed. Most importantly, replacing

Example 1. Use of Continuous Versus Dichotomous Variables

Consider a clinical trial to determine the effect of an intervention (e.g., educational materials promoting activity) on physical activity in adolescent girls. Suppose that national guidelines recommend a minimum of 30 minutes of moderate-to-vigorous physical activity (MVPA) per day. A previous study in adolescent girls found that the number of minutes per day of MVPA, determined using counts from the Computer Science and Applications (CSA) accelerometer, is normally distributed with a mean of 30 minutes and a standard deviation of 13 minutes and that about 50% of girls meet the national guidelines. No change is expected in the average daily minutes of MVPA for the control group, and the intervention group is expected to have a 5-minute increase in minutes of MVPA. This change is estimated, based on the distributional assumptions, to correspond to an increase of 65% in the proportion of girls meeting the national guidelines.

One design might treat MVPA as a dichotomous variable: meeting versus not meeting the national guidelines. Another might treat the number of minutes per day of MVPA as continuous. How many girls would each design require at $\alpha = .05$ (two-sided) and $\beta = .20$?

Solution: Interest in comparing two binomial probabilities leads to using the exact "unconditional" test. The sample size required for the exact one-sided (unconditional) test so that the two proportions are equal (dichotomous outcome) is 136 per group. The approximate sample size based on the F test or the chi-squared test is 134 per group (see "Comparing Proportions for Two Independent Samples," p. 101). Using a one-sided t test to detect a difference in means (continuous outcome) would require 85 in each group (see "Comparing Means for Two Independent Samples," p. 101).

a categorical outcome, such as obesity indicated as yes or no, with a continuous outcome, such as body mass index measured as weight over height squared, often greatly reduces the sample size needed to reliably distinguish between groups. For example, with a t test analysis, cutting the residual variance of the outcome in half has essentially the same impact on power as doubling the sample size. See example 1 for further insight.

The measurement of physical activity may be made on a continuous, ordinal, or nominal scale. The scale of the variable constrains the choice of analysis. In turn, different analyses with the same number of participants can have dramatically different powers. For example, a repeated measures analysis of longitudinal data may have much greater power than a collection of univariate tests. This is especially true when an appropriate Bonferroni correction is used to control the multiple testing bias.

Under all circumstances, it is important to ensure that the outcome is aligned with the goals of the study. For example, the choice between the mean number of minutes of daily activity or the proportion meeting guideline levels of activity should reflect which is being targeted by the intervention. The choice should not reflect which outcome yields greater power.

In exercise science studies with a continuous outcome, taking repeated measures and evaluating trends over time usually provides both more information and also more power than available with simple cross-sectional designs. Such longitudinal designs control for differences among initial levels of the outcome measure, among the study participants. This typically decreases bias in the estimates of response, and also typically increases precision and therefore power.

Design Strategies

Most often, an exercise science design with approximately equal numbers of subjects in each group is best in terms of precision and power. However, for some outcome measures, achieving the most precise estimate of a non-zero effect (rather than testing it) may require using unequally sized groups (Walter, 1977) for large effects.

Precision can be improved by separating subjects into strata (subgroups), according to one or more characteristics known to be correlated with the outcome, and randomizing subjects to study groups within strata. See example 2 for a concrete illustration. Stratification increases precision by creating greater homogeneity within a stratum than exists in the total, unstratified population. Taking advantage of the increased homogeneity requires accounting for the stratification variable in the analysis, such as having the stratification variable serve

Example 2. Use of Stratification

Goran et al. (1998) used doubly labeled water to measure activity energy expenditure of 9-year-old boys. Data showed activity energy expenditure averaged 660 kcal/day (SD = 286). An investigator wants to study the effect of adding a brisk 10- to 15-minute walk to baseline daily activity in a similar group of boys. If the intervention is expected to increase activity energy expenditure by 50 kcal/day (i.e., standardized effect size of .17), 515 participants per group are required to achieve 80% power, at α (two-sided) = .05 (see "Comparing Means for Two Independent Samples" on page 101 for appropriate sample size formula). Given the rarity and cost of doubly labeled water, a study of this size would not be feasible. Doubling the duration of the brisk walk, thus doubling the expected increase in energy expenditure to 100 kcal/day, would require 130 boys per group; again, this would be too cost prohibitive. The issue, of course, is the large standard deviation in the activity energy expenditure of boys, resulting in a low standardized effect and high sample size. An effective strategy in this setting is to stratify children according to baseline physical activity behaviors to minimize random error variation. If this strategy results in an average within-strata SD of 200, the sample size would be reduced by half (n = 64 per group).

as a covariate. This increase in precision corresponds to a decrease in uncertainty around estimates of effect, and hence increases power for any significance test.

Reliability

The issue of reliability of a measurement refers to its "repeatability," or "consistency." A measure is considered *reliable* if it would give us the same result in repeated testing, assuming that what we are measuring isn't changing.

Where possible, the reliability in measurement of key study variables should be quantified and described in pilot studies. *Retest* reliability is one of the most common ways of estimating reliability. Subjects are measured on two or more occasions separated by a time interval sufficiently short that we can assume the underlying characteristic is unlikely to have changed. In the case of recall measurements, one must be careful that the interval not be so short that individuals could remember answers from the previous occasion. Suppose σ_y^2 denotes the total variance of a continuous outcome, and σ_b^2 and σ_w^2 denote the between-subject and within-subject variances, respectively. The classical definition of reliability is the fraction of the total variance $\sigma_y^2 = \sigma_b^2 + \sigma_w^2$ attributable to between-subject variation.

$$r = \frac{\sigma_b^2}{\sigma_b^2 + \sigma_w^2}$$

One of the best ways to increase reliability is to take multiple measurements and average the scores. The result of this strategy is to divide the within-

subject variance by the number of occasions the response is measured. For example, if the average of four measurements is used as the outcome variable, and all four measurements may be assumed to be comparable, the Spearman-Brown formula (Traub, 1994) can be used to estimate the reliability coefficient:

$$r = \frac{\sigma_b^2}{\sigma_b^2 + \sigma_w^2/4}$$

Several factors make subjectively determined measures less reliable than objectively determined measures: vague or ambiguous wording of questionnaires, inadequate interviewer instructions leading to variability between and within interviews, varying context of the data collection (interviewer or setting may influence response), etc. Where possible, directly measured data are preferred. If subjective reporting cannot be avoided, reliability can be increased by (1) using multiple items (and analyzing the sum of scores), (2) improving the research instrument as a consequence of a pilot study, and (3) carefully recruiting and training interviewers to ensure they execute their role in a neutral and consistent manner. See example 3 for a comparison of multiple versus single measures.

Analytic Strategies

Inferential procedures are techniques for drawing objective conclusions about whether the data agree or disagree with specified hypotheses. They can be

Example 3. Use of a Response Measured on One Occasion Versus Multiple Occasions

Recall example 1, in which the investigator studies the effect of an intervention on physical activity in adolescent girls. A pilot study is performed to determine the minimum number of days of monitoring activity to be averaged to ensure a reliability of approximately .80. The random sample of girls recruited for the pilot study wore the CSA monitors for 7 consecutive days. Analysis of variance results were used to calculate a between-person variance of 144 and a within-person error variance of 220. Using a single day of monitoring results in a reliability of .4. Using a 7-day monitoring protocol would increase reliability to .82.

divided into two types: parametric procedures and nonparametric procedures, which differ in their underlying assumptions, the type of data to which they are best suited, and their associated power. Parametric tests typically use continuous data and are based on the assumption that the random error component is normally distributed. Nonparametric tests use rank order or categorical information and have less restrictive assumptions. This results in nonparametric tests generally having lower power than corresponding parametric tests. On the other hand, nonparametric tests are simple to use and distribution-free, and their efficiency can be fairly close to corresponding parametric tests.

Violation of assumptions can affect both the power of the inferential tests and the probability of making a Type 1 error. Although many parametric tests are "robust" to minor shortfalls in the normality assumption, transformation of the dependent variable in the case of extreme violation and small sample size will lead to improvement in power. More serious consequences result from the use of a test that assumes independence of observations, such as the t test or analysis of variance (ANOVA), with correlated data. The higher the correlations, the more deviant the actual significance level will be from the nominal level (i.e., the more likely we are to falsely reject the null hypothesis). Appropriate adjustment for the correlation is necessary.

A further aspect to consider, particularly in small sample studies, is whether the question to be assessed concerns a general difference (two-tail test) or a specific directional difference (one-tail test). The latter is more powerful than the former.

Decreasing variability of dependent measure that is attributable to sampling error makes *smaller* the estimate of sampling error against which treatment differences are tested. One way this can be done is to identify one or more factors expected to be related to the dependent variable. Consider a

repeated-measures design in which physical activity levels are measured on participants at the beginning of sixth grade and the end of eighth grade. The question to be answered is this: Is there a difference in the average change in physical activity levels of girls versus boys? The baseline is the fitness level of participants before their assignment to an experimental condition. This information could be used to increase the efficiency of the experiment by blocking participants with respect to fitness levels and by employing a matched or randomized block design or by retaining the simple randomized design and using the baseline measure as a covariable in an analysis of covariance (ANCOVA) of the outcome measure. ANCOVA typically delivers a greater gain in efficiency than blocking does because, first, some heterogeneity of factors remains within the blocks after blocking and, second, estimation of the covariable effect (assuming a linear or low-order polynomial relationship with the outcome) will consume fewer degrees of freedom than estimation of block effects. The major caution that should be observed, however, is that the measurements for the covariates are obtained before the assignment of individuals to experimental conditions. This ensures that the treatment effect will not be correlated with the covariates and helps ensure that the error will be homogenous across treatment groups.

Improving Validity (Reducing Systematic Error)

There are several design and analytic strategies for increasing the likelihood that our measuring instruments (e.g., questionnaires, scales) measure what they are intended to measure and that the results are attributable to the intervention or exposure alone (i.e., reduce sources of bias).

Design Strategies

Select a sample that is representative of the population of interest and randomly allocate subjects to treatment groups. Random assignment creates treatment groups that are initially comparable on most, if not all, participant characteristics. It can then be concluded that any final outcome differences are due to treatment effects alone. Research staff and participants should be blinded to treatment assignment to eliminate the risk of responses being influenced by the subjects' or the experimenter's expectations. In studies where participants are measured at more than one point in time, a control group should be included to enable comparison with the experimental group. This design eliminates the threats to validity arising from learning effects, maturation, or other changes that would have occurred even in the absence of the intervention.

To rule out confounding caused by an important factor, one can design a study to guarantee balance across levels of every factor, either by stratified random assignment in the case of a randomized study or by matching exposed subjects to unexposed subjects who share the same profile with respect to the stratification variables.

In general, objectively determined responses have better validity than subjective responses. Self-reported data can often be inaccurate because of faulty memory and other psychological factors. Cook and Campbell (1979) have pointed out that subjects tend to report what they believe the researcher expects to see or what reflects positively on their own abilities, knowledge, or beliefs. For example, epidemiologic studies have shown that men and women tend to overreport their height and underreport their weight. As indicated in "Improving Precision (Reducing Random Error)" on page 96, directly measured data are preferred to self-reported data. If self-reported data cannot be avoided, strategies for increasing validity include improving the self-report instrument based on feedback on ease of understanding and item sensitivity from pilot respondents, exploring patterns of nonresponse, and training interviewers in a standardized fashion to ensure they are not inadvertently leading the respondent to a specific answer.

Analytic Strategies

Even with careful planning and execution of a study, investigators are often faced with one or more biases after the data are collected (e.g., missing data, confounding). Suppose, for example, that the study groups are not comparable with respect to one or more important risk factors (e.g., race) known to be related to the outcome of interest. This can cause serious confounding. One can maintain validity by using statistical methods of adjustment to simulate the effect of designing a study to guarantee balance across levels of every factor.

Use of an intent-to-treat analysis minimizes bias resulting from differences in the number of subjects who cross over, are lost to follow-up, or die across study groups.

Common Formulas for Calculating Sample Size

It should be noted that these formulas *always* provide approximate sample sizes. Kupper and Hafner (1989) and Gatsonis and Sampson (1989) highlight the limitations of using rough approximations. In general, it is always best to use special purpose power software for calculations. Hand calculation has a risk of human error. Given valid input, commercial computer software will typically produce numerically accurate results. Furthermore, power software makes it very easy to consider many "what if" scenarios and examine a range of possible designs. Such explorations may help greatly in choosing and refining the design.

Many simple formulas were developed before computers were widely available in order to allow hand calculation. The approximations vary greatly in accuracy, especially for small samples. The competition of the marketplace tends to lead software vendors to have the best available methods, even when the methods require extensive programming. In contrast, formulas for hand calculation are never used if they are too complicated.

There are many useful study designs for which *only* rough approximations are available for computing hypothesis tests and confidence intervals. Naturally only equally rough power approximations are available. There are two broad classes of data in activity research for which this is likely true: categorical data and continuous data with complex models, especially those involving missing data, mistimed data, and covariate values that change across time. Such data are often analyzed with what statisticians refer to as "mixed models."

In the following section, we mention some commonly needed tests in activity research. We then briefly review power methods for each, giving special attention to known limitations.

Comparing Means for Two Independent Samples

When the response is approximately normally distributed, the t test is most often used to compare the means of two independent groups, such as when subjects are assigned at random to an experimental group and a control group. The null hypothesis is that the population mean scores are equal for the two groups, H_0: $\delta = \mu_1 - \mu_2 = 0$. Any commercial power software has special features to handle this situation conveniently. Alternately, many of the textbooks on power listed in the references give tables.

Comparing Proportions for Two Independent Samples

When members of two independent samples are classified on a common dichotomous outcome, the null hypothesis of interest is that the proportion of participants classified at one level of the outcome in each group is equal. Suissa and Shuster (1985, table 3) tabulated sample sizes for exact one-sided tests of the difference between proportions. For applications not considered in their tabulations, one should rely on commercial software, or if large sample tests apply, approximate sample size formulae may suffice. Dozier and Muller (1993, tables 5 and 6) described a convenient F approximation that works well even with relatively small sample sizes. Fleiss (1981) described a more traditional chi-squared approximation that is less accurate in small sample sizes.

Testing the Linear Relationship Between Two Continuous Variables

One of the most commonly used statistical methods for quantifying the linear relationship between two continuous variables, X and Y, is the Pearson correlation coefficient, ρ. Researchers might have more experience with ρ^2, the proportion of variance in either X or Y that may be predicted by the variance of the other. If the true relationship is linear, and X can be measured without appreciable error, then the slope of the regression line for predicting Y from X is

$$\beta_{YX} = \rho \frac{\sigma_Y}{\sigma_X}.$$

Hence testing whether the slope is equal to zero is equivalent to testing whether the correlation is equal to zero.

Gatsonis and Sampson (1989) provide tabulations of exact sample sizes necessary to test H_0: $\rho = 0$. A more general case involves testing H_0: $\rho = \rho_0$, with ρ_0 the minimal acceptable level of correlation between X and Y. Approximate sample sizes can be calculated for the more general case using Fisher's transformation

$$Z(\rho) = \frac{1}{2}\ln\left[(1+\rho)/(1-\rho)\right].$$

The number of subjects required to detect a difference between ρ_0 and ρ_1 is given by

$$n = \frac{\left(z_{1-\alpha/2} + z_\beta\right)^2}{\left[Z(\rho_1) - Z(\rho_0)\right]^2} + 3.$$

Reliability Studies for Continuous Measures

Reliability studies involve taking a series of k measurements on the same subject to estimate the reliability of a measurement method. For the special case of a test-retest design ($k = 2$), reliability can be assessed by Pearson's correlation coefficient.

Example 4. Comparing Two Correlations

Consider an investigator who wishes to study the effect of race on the correlation between fat-free body mass and a bioelectrical impedance measurement from a new machine. The minimal difference of interest is H_1: $\rho_1 - \rho_2 = .05$ between Caucasians and African-American children. If the two correlations are $\rho_1 = .97$ and $\rho_2 = .92$, a test of the difference in correlations will have 90% power with 45 children per race group, at α (two-sided) = .05 (see "Testing the Linear Relationship Between Two Continuous Variables," this page). Note that considering other values of ρ_1 and ρ_2, which also differ by .05, will typically give a different answer.

Example 5. Sample Size for Reliability Studies

Trost et al. (2000) calculated daily time spent in moderate-to-vigorous physical activity (MVPA) over 7 days recorded by the Computer Science and Applications (CSA) activity monitor in children. The reliability coefficient for 7 days of monitoring among 97 students in grades seven through nine was .8. An investigator wishes to test whether increasing the number of days of monitoring to 10 can increase reliability to .9. If the null and alternative hypotheses are H_0: $\rho_0 = .8$ and H_1: $\rho_1 = .9$, the exact sample size (using $k = 10$) required to ensure 80% power is 22, assuming α (two-sided) = .05. The sample size from the approximate formula is 23.

However, for $k > 2$, the intraclass correlation coefficient is a reasonable estimate of the reliability of the measurement method, assuming all measurements are comparable (Bravo & Potvin, 1991). Donner and Eliasziw (1987) provide tabulations of exact sample sizes (for a wide range of values of k) necessary to test H_0: $\rho = \rho_0$, where ρ_0 is the minimally acceptable reliability. Walter et el. (1998) provide a formula for calculation of approximate sample sizes.

Summary

Considering strategies that will minimize the sample size is a critical component of designing a good study. These strategies include using approximately equal numbers of subjects in each study group, having a continuous outcome, having an average of replicate measurements as the outcome, and adjusting for one or more covariates in the analysis.

The formulas for calculating power and choosing an appropriate sample size in this chapter were necessarily limited to designs that are most pertinent to research on physical activity. An obvious exception is power analysis for multivariate and repeated measures power, to which the reader is referred to Muller et al. (1992). There are many commercial and freeware software packages available to help researchers perform power analysis for a broader range of applications: nQuery Advisor (Elashoff, 2000), StatXact (CYTEL Software Corporation, 1998), Power and Precision (Borenstein et al., 2001), PASS (Hintze, 2000), and UnifyPow (O'Brien, 1986). Readers interested in a general text that covers the general concepts of power are referred to Kraemer and Thiemann (1987), Cohen (1988), Lipsey (1990), and O'Brien and Muller (1993).

References

Borenstein, M., Rothstein, H., Cohen, J., Schoefeld, D., Berlin, J., & Lakatos, E. (2001). *Power and Precision*. Englewood, NJ: Biostat Inc.

Bravo, G., & Potvin, L. (1991). Estimating the reliability of continuous measures with cron bach's alpha and intraclass correlation coefficient: Toward the integration of two traditions. *Journal of Clinical Epidemiology, 44*, 381-390.

Cohen, J. (1988). *Statistical power analysis for the behavioral sciences* (2nd ed.). Hillsdale, NJ: Lawrence Erlbaum.

Cohen, J. (1992). A power primer. *Psychological Bulletin, 112*, 155-159.

Cook, T.D., & Campbell, D.T. (1979). *Quasi-experimentation: Design and analysis issues for field settings*. Boston: Houghton Mifflin.

CYTEL Software Corporation. (1998). *StatXact4 for Windows®: Statistical software for the exact nonparametric inference user manual*. Cambridge, MA: CYTEL.

Donner, A., & Eliasziw, M. (1987). Sample size requirements for reliability studies. *Statistics and Medicine, 6*, 441-448.

Dozier, W.G., & Muller, K.E. (1993). Small-sample power of uncorrected and Satterthwaite corrected t tests for comparing binomial proportions. *Communications in Statistics: Simulation and Computation, 22*, 245-264.

Elashoff, J.D. (2000). *nQuery Advisor release 4.0 users guide*. Boston: Statistical Solutions.

Fleiss, J.L. (1981). *Statistical methods for rates and proportions* (2nd ed.). New York: Wiley.

Gatsonis, C., & Sampson, A.R. (1989). Multiple correlation: Exact power and sample size calculations. *Psychological Bulletin, 106*, 516-524.

Goran, M.I., Gower, B.A., Nagy, T.R., & Johnson, R.K. (1998). Developmental changes in energy expenditure and physical activity in children: Evidence for a decline in physical activity in girls before puberty. *Pediatrics, 101*, 887-891.

Hintze, J.L. (2000). *PASS user's guide: PASS 2000 power analysis and sample size for Windows®*. Kaysville, NC: NCSS Inc.

Kraemer, H.C., & Thiemann, S. (1987). *How many subjects? Statistical power analysis in research*. Newbury Park, CA: Sage.

Kupper, L.L., & Hafner, K.B. (1989). How appropriate are popular sample size formulas? *The American Statistician, 43*, 101-105.

Lenth, R.V. (2000). Two sample-size practices I don't recommend. In *Proceedings of the section on physical and engineering sciences*, 8-11. American Statistical Association.

Lipsey, M.W. (1990). *Design sensitivity: Statistical power for experimental research*. Newbury Park, CA: Sage.

Muller, K.E., & Benignus, V.A. (1992). Increasing scientific power with statistical power. *Neurotoxicology and Teratology, 14*, 211-219.

Muller, K.E., LaVange, L.M., Ramey, S.L., & Ramey, C.T. (1992). Power calculations for general linear multivariate models including repeated measures applications. *Journal of the American Statistical Association, 87*, 1209-1226.

O'Brien, R.G. (1986). UnifyPow: A SAS Macro for sample size analysis. In *Proceedings of the 22nd SAS users group international conference* (pp. 1353-1358). Cary, NC: SAS Institute.

O'Brien, R.G., & Muller, K.E. (1993). Unified power analysis for t-tests through multivariate hypotheses. In L.K. Edwards (Ed.), *Applied analysis of variance in behavioral science* (pp. 297-344). New York: Marcel Dekker.

Suissa, S., & Shuster, J.J. (1985). Exact unconditional sample sizes for the 2 x 2 binomial trial. *Journal of the Royal Statistical Society, Series A, 145*, 317-327.

Traub, R.E. (1994). *Reliability for the social sciences: Theory and applications*. London: Sage.

Trost, S.G., Pate, R.P., Freedson, P.S., Sallis, J.F., & Taylor, W.C. (2000). Using objective physical activity measures with youth: How many days of monitoring are necessary. *Medicine and Science in Sports and Exercise, 32*, 426-431.

Walter, S.D. (1977). Determination of significant relative risks and optimal sampling procedures in prospective and retrospective comparative studies of various sizes. *American Journal of Epidemiology, 105*, 387-397.

Walter, S.D., Eliasziw, M., & Donner, A. (1998). Sample size and optimal designs for reliability studies. *Statistics in Medicine, 17*, 101-110.

Westfall, P.H., Tobias, R.D., Rom, D., Wolfinger, R.D., & Hochberg, Y. (1999). *Multiple comparisons and multiple tests using the SAS system*. Cary, NC: SAS Institute.

Techniques for Physical Activity Assessment

7

Use of Self-Report Instruments to Assess Physical Activity

Charles E. Matthews, PhD
University of South Carolina

The goal of this chapter is to provide practical information about the most effective ways to collect and interpret self-reported physical activity data for research purposes. Physical activity behaviors present a complex challenge for measurement because they occur frequently during one's waking hours, can occur in a variety of social domains (e.g., household, occupational, leisure), and can be characterized in several different dimensions (e.g., total amount, type, frequency, intensity, or duration of activity) (Casperson et al., 1985). Self-reported physical activity instruments have become an invaluable tool for physical activity research because of their minimal expense and scoring flexibility. The only equipment needed are a data collection strategy, a pencil, and a piece of paper. Moreover, information derived from these methods allow, with appropriate data collection and reduction methods, the greatest flexibility in providing either a behavioral description of physical activity patterns or a more quantitative summary estimate of physical activity energy expenditure, or both.

Physical activity has traditionally been defined as "any bodily movement produced by skeletal muscles that results in energy expenditure" (Casperson et al., 1985). While this definition broadly describes one outcome of physical activity behavior, it does not describe what self-reported physical activity instruments actually measure. Self-report methods gather information about the characteristics of physical activity behaviors performed in specific periods of time and in specific activity domains. The behavioral characteristics of physical activity are the type (e.g., aerobic, anaerobic, occupational, household), intensity (i.e., energy cost), frequency (e.g., times/week), and duration (e.g., minutes/occasion). These characteristics are often translated into an estimate of physical activity energy expenditure using standardized methods (Ainsworth, Haskell et al., 1993; Ainsworth, Haskell et al., 2000).

Rather than act as a catalog of self-reported physical activity methods, with implementation details and reliability and validity information about specific instruments, this chapter seeks to provide a conceptual framework for collecting and interpreting self-reported activity data. A number of comprehensive reviews have been published in the last 20 years describing the implementation and reliability and validity information for many of the individual self-report instruments (Baranowski, 1988; Kohl et al., 2000; Laporte et al., 1985; Sallis & Saelens, 2000; Washburn & Montoye, 1986). In addition, the most frequently employed instruments have individual methodological papers that provide a detailed description of the original implementation and scoring procedures for the method. Two exceptional references containing collections of self-reported physical activity instruments and their scoring methods, as well as listings of published reliability and validity studies, can be found in Montoye et al. (1996) and in the 1997 supplement to *Medicine and Science in Sports and Exercise, A Collection of Physical Activity Questionnaires* (Pereira et al., 1997).

In an effort to provide a conceptual framework to guide individuals in the selection of instruments and in interpreting self-reported physical activity data, this chapter reviews the cognitive aspects of the self-report process as well as models describing the likely sources of variation that exist in self-reported physical activity measures. This approach attempts to isolate and understand the complex set of factors that are involved in measuring physical activity behaviors by self-report. In addition, this chapter reviews the four broad categories of available instruments, the basic outcome measures derived from these measures, and some general measurement properties that should be considered when implementing a particular instrument.

Principals of Method: Cognitive Issues in Self-Reports of Physical Activity

Self-reported physical activity instruments, unlike all of the other physical activity assessment methods outlined in this book, rely completely on a respondent's ability to provide good information about his or her own behaviors. For the respondent, this is pure cognitive work that involves the manner in which memories are stored at the time the physical activity events occur as well as the challenge of retrieving memories from past autobiographical events. In the past 50 years, tens of thousands of questionnaires and interviews have

produced important data that have been used to identify links between physical activity and many chronic diseases, and to elucidate many determinants of physical activity behaviors. Still, we know very little about the cognitive processes utilized in providing self-reports of physical activity. There are two commonly cited papers that have specifically addressed the cognitive issues involved in self-reported physical activity.

Baranowski (1988) was the first to carefully consider the self-reporting of physical activity from an information-processing perspective. He presented a basic model to explain the self-report process in cognitive terms (i.e., encoding, storage, and retrieval of information from memory) and used these theoretical concepts to design experiments demonstrating that revision of the methods for the original 7-day physical activity recall (7DPAR) (Sallis et al., 1985) was warranted. This work resulted in important modifications of the 7DPAR interview methods that have now become common practice (Sallis, 1997). Durante and Ainsworth (1996) continued this work by incorporating many of the findings of the fruitful collaboration between cognitive psychologists and survey researchers that has continued our understanding of the general processes involved in self-reporting (Jobe & Mingay, 1991). Durante and Ainsworth (1996) presciently noted that self-reported physical activity measures typically explain only about half of the variance in validation studies, and they contended that such modest validity coefficients were "attributable to error in the cognitive operations employed in recalling and reporting physical activity" (p. 1282).

The first section of this chapter seeks to build upon these important works by considering additional information about storage of physical activity events in memory as well as insight from cognitive scientists that have accumulated in recent years. It should be noted that very little direct research exists that addresses the cognitive issues related to self-reported physical activity. However, research by cognitive scientists has sought to understand the fundamental processes involved in answering survey questions that should apply to reports of activity.

Storage of Memories

It is tempting to think of memories of past behavior as being immediately retrievable on demand as a complete record of past events. Unfortunately, memory does not appear to work this way, in most cases. Therefore, it is important to consider what type of physical activity information may be stored in memory and how these items are organized because this information forms the basic substrate from which individuals provide physical activity self-reports. A significant issue to consider when evaluating self-reports of physical activity, particularly as our instruments have shifted from the assessment of a single domain of physical activity (i.e., leisure) to the assessment of the full range of activities encountered in daily life (e.g., household chores, occupational activity), is how information about physical activity behavior gets stored in memory. Obviously, if physical activity events are never stored in memory, there is no chance that a specific episode of activity can be recalled with any accuracy. It should also be noted that a lack of storage of information about specific activity events does not mean that respondents will fail to report these activities when they are asked to do so on a survey. Remarkably, this may be acceptable in certain situations.

The cognitive model of Baranowski and Domel (1994) has been adapted here to illustrate the basic cognitive processes thought to be involved in the storage of autobiographical memories. This model, by carefully describing the individual processes involved in the storage of memories, provides a useful tool for identifying potential weaknesses in the storage process that are germane to the assessment of physical activity. The basic cognitive structures in the model are the sensory register, short-term memory, and long-term memory (see figure 7.1).

The *sensory register* works to screen the hundreds of bits of possible information present at any moment in our immediate environment (e.g., sights and sounds), and it can focus attention on specific sensory items in the environment or on our personal behaviors. For information about physical activity to be perceived by the sensory register, attention must be given to the behavior. Information that is attended to is perceived (or comprehended) and enters short-term memory where it is evaluated by the experience monitor.

Short-term memory is where conscious intellectual activity occurs and, perhaps, up to five items can be held here at any one time (Baranowski & Domel, 1994). Certain information is transferred from short- to long-term memory for storage. The process of labeling and storing items in long-term memory is called *encoding* (Sudman et al., 1996). Autobiographical information appears to be encoded in long-term memory in a hierarchical manner with specific individual events being embedded within meaningful clusters of information sequences. These sequences often appear to be chronologically arranged (Sudman et al., 1996). That is, individual episodes of physical activity—unless they are extremely meaningful to the individual—may not be stored as individual items in long-term memory. Most physical activity events are likely to be embedded within memories of larger contextual events, such as locations, people, or social occasions. We use the term *contextual information* to refer to these memories because they often serve as the basis from which physical activity information is derived. In reviewing the function of each of the cognitive structures (i.e., sensory register, short-term memory, and long-term memory), it becomes clear that if certain physical activity behaviors are not perceived by the sensory register, and thus fail to enter short-term memory, the behaviors are unlikely to be transferred to and stored as an individual item

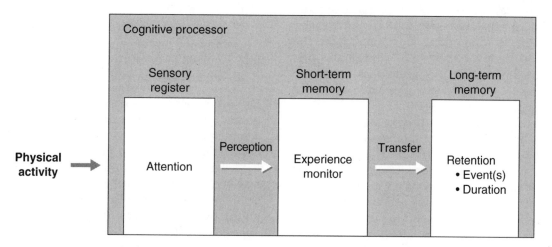

Figure 7.1 Cognitive structures and memory storage.

Adapted from Baranowski and Domel, 1994. © Am. J. Clin. Nutr. American Society for Clinical Nutrition.

in long-term memory. Thus, it is unlikely that they will become available for recall at some point in the future. If this is the case, how are we able to provide useful information about experiences that we may not be able to recall directly?

Autobiographical information is thought to be stored as two conceptually distinct types of memory: episodic and generic. Episodic memories are specific recollections of individual and innumerable autobiographical events. This type of memory is generally employed in a recall task when one seeks to enumerate individual episodes of less-frequent occurrences, or a small number of events that occur in a short time frame (e.g., the number of days of exercise last week). Episodic memories are thought to contain contextual information about the actual events that occurred (e.g., locations, people, actions) but may not contain specific information about absolute time (Brewer, 1994). Reconstruction of time from episodic memories, and for our purpose physical activity duration, appears to be derived from the contextual information that is available on recall. In other words, in response to a question about the number of minutes spent walking yesterday morning, a value for the specific duration of walking in this period is unlikely to reside in memory, but an estimate of duration could be reconstructed using the contextual information that is available.

In contrast to episodic memories, generic memories are recollections of the general events or patterns of events (or activities) that occur in one's life. This form of memory is employed to provide self-reports of physical activity when specific and innumerable memories are not stored. For example, one may be able to provide a generic report of the frequency of playing golf last summer (once or twice a week) but would be unable to recall and enumerate each individual outing. In general, more frequent and familiar events in one's life appear to be stored as generic memories rather than as discrete memories of the individual events that make up the generic whole (Brewer, 1994).

With the exception of self-reported physical activity methods that require a short recall period (e.g., the last 24 hours), or specific behaviors that occur over a short period of time (e.g., exercise bouts in last 7 days), most self-reported physical activity instruments rely on information derived from generic memory. This cognitive aspect of physical activity reporting may be one of the factors contributing to the generally lower reproducibility coefficients for light- to moderate-intensity physical activities on various physical activity assessments that seek to quantify these common and frequently encountered behaviors (Dipietro et al., 1993; Jacobs et al., 1993; Sallis, 1997). For example, evaluation of the 14-day repeatability data from the exemplary and detailed validation work for the Yale Physical Activity Survey (YPAS) of DiPietro et al. (1993) suggests that more prevalent household activities, which are likely to be reported using generic memory, generally have lower reliability coefficients compared to more specific exercise activities that may be reported using episodic memory (see figure 7.2). Given that the YPAS time frame for respondents to report is stated as a "usual week in the last month," the 14-day recall interval from which these analyses were derived should largely reflect variation in reporting the same physical activity behaviors rather than variation in the underlying behaviors themselves.

Careful consideration of the type of memory used in self-reports of physical activity on surveys or in interviews is likely to suggest ways to more effectively merge our assessment strategies with the cognitive processes employed and, in the process, facilitate better reporting of the behavioral characteristics of the physical activity on these instruments. Consideration of these cognitive factors will also provide investigators with an additional framework from which to interpret the information provided by the respondent. Now that we've outlined some of the basic aspects of memory storage, and the substrate from which self-reports of physical activity are based, let us examine the process of retrieving information from memory in response to a question.

Retrieval From Memory

The four primary stages a respondent goes through in answering a question are

1. question comprehension,
2. decision/judgment,
3. retrieval from memory, and
4. response generation.

Each of these stages is generally completed in the question-answering process, but the order of the completion of some stages may be quite flexible, depending on the type of question that is asked and the level of effort employed by the respondent. A number of models describing these processes have been described (Jobe & Hermann, 1996), and most entail the four stages originally outlined by Tourangeau (1984). We present the model of Willis et al. (1991), the Flexible Processing Model (Jobe & Hermann, 1996), because it accounts for many of the variations that are likely to occur in the cogni-

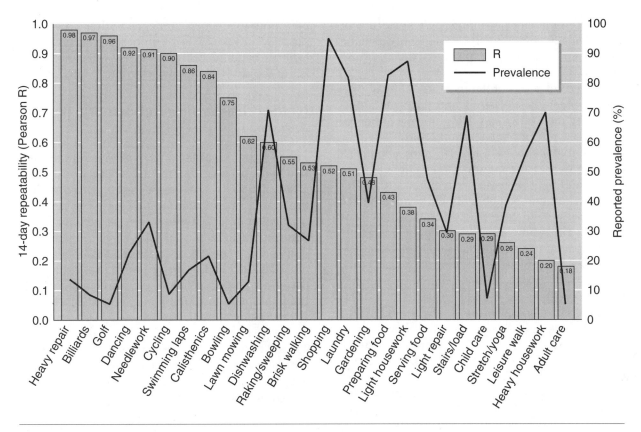

Figure 7.2 Repeatability and prevalence of reporting of common physical activities among elderly adults over 14 days.
Adapted from DiPietro et al., 1993.

tive processing that takes place in answering survey questions (see figure 7.3).

The first stage of the response process, question comprehension, is obviously important because comprehension determines whether or not the respondent understands what is being asked. Once a question is asked, the respondent will form his or her own interpretation of the question. This interpretation guides all further efforts in the reporting process. Consider this question about walking behaviors from a recent version of the Behavioral Risk Factor Surveillance Survey physical activity module: "In a usual week, do you walk for at least 10 minutes at a time while at work, for recreation, exercise, to get to and from places, or for any other reason?" To respond in a manner that is consistent with the intent of the question, a respondent must comprehend at least three things:

1. The time frame of the question (i.e., a usual week)
2. How to define a bout of walking (10 minutes at a time)
3. That all of the walking done in a day should be reported

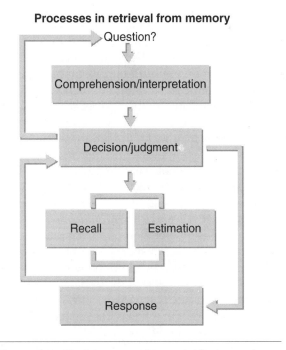

Figure 7.3 Flexible processing model for answering survey questions.
Adapted from Willis et al., 1991.

Incorrect comprehension or interpretation of any of these elements could send the respondent down the wrong path in subsequent phases of the reporting process.

The second phase of the memory retrieval process involves the decisions and judgments employed by the respondent that guide his or her retrieval and response generation efforts. The first decision may be to ask for clarification about certain elements of the question to ensure adequate understanding. If they do not require clarification about the question, respondents will decide whether or not they need to retrieve information from memory. For example, if a respondent was asked about an activity he or she had not done in the prescribed recall interval (e.g., running in the last year), no retrieval work would be required to respond. If the respondent did need to retrieve information from memory, however, he or she would either work to *recall* episodic information pertaining to specific events that were asked about in the question or would use generic memories and *estimation* strategies to answer the question.

The retrieval phase appears to be quite fluid because the process of gathering information from long-term memory is thought to be iterative, or "reconstructive" (Sudman et al., 1996), in nature. That is, during the retrieval process, one reconstructs memories first by gathering basic contextual information about the event(s) of interest (e.g., locations, times of day) and, from these initial recollections, works to build a report. Reconstruction of the memories that go into the final report may have gone through several different memory searches and a series of decisions and judgments about the usefulness of the information retrieved. In general, respondents will work to retrieve information from memory until they make the judgment that they have gathered adequate information to answer the question posed. As noted in figure 7.3, there is likely to be considerable interaction between decision/judgment processes and the work done during retrieval (i.e., recall and/or estimation). The final phase, response generation, is the time in which respondents edit their responses to fit the structure of the question that was posed to them and can be thought of as the inverse of the initial comprehension stage. That is, respondents retranslate their answers to fit what they feel is the intent and meaning of the question originally posed.

In summary, careful consideration of the discrete processes involved in the storage of memories and the processing employed in retrieving information from memory may suggest ways to more effectively merge our assessment strategies with the cognitive processes employed. In the process, this merging will facilitate better reporting of the relevant characteristics of the physical activity behaviors of interest.

Implications for Physical Activity Assessment

The key cognitive elements in most self-report instruments to be considered are whether or not the physical activity event of interest occurred within the time frame of the question (i.e., yes/no) and the frequency and duration of the activity. Recollection of participation in specific activities over short recall periods could be a simple or a challenging cognitive task, depending on the time frame of the question and the type of activities that are being assessed. For example, questions that can logically be answered using episodic memory (e.g., Did you walk for exercise yesterday? or Did you run or jog for at least 20 minutes at a time last week?) are relatively simple to comprehend and recall. In contrast, questions that rely more heavily on generic memories of a broad range of activities, or that ask about longer time frames, are cognitively more challenging.

Individuals appear to be able to accurately report never having participated in a particular behavior. For example, cognitive scientists have noted that respondents can accurately report not eating specific food items in the last month and not having a cancer screening test in the past 5 years (Jobe et al., 1993). Although this phenomenon has not been carefully studied by physical activity researchers, these findings suggest that individuals may be able to accurately report that they did not exercise regularly or did not have a physically demanding occupation. This notion is consistent with findings from long-term recall of leisure-time activities (Blair et al., 1991; Slattery & Jacobs, 1995) and lends support to the ability to separate exercisers from nonexercisers in relatively simple epidemiological questionnaires that have sought to evaluate recent habitual activity (Ainsworth, Sternfeld et al., 2000; Washburn et al., 1993) or lifelong activity behaviors (Friedenreich et al., 1998; Kriska et al., 1988).

Providing reports of activity behavior for the time period of a particular assessment instrument, or "event dating" (Sudman et al., 1996), has been noted to be cognitively difficult. A common error introduced when asking about specific time periods is a phenomenon called *telescoping*. Telescoping occurs in the retrieval phase when events from outside the time frame of the question are reported, resulting in overreporting. The effects of telescoping can be minimized by using repeated interviews that employ a "bounded reference period" (Durante & Ainsworth, 1996), using landmark or life events (Friedenreich et al., 1998; Sudman et al., 1996), and providing more contextual cues to aid in recall (Durante & Ainsworth, 1996). Detailed descriptions of this phenomenon and potential remedies have been described elsewhere (Durante & Ainsworth, 1996; Sudman et al., 1996).

Recollection of the frequency of activity events is a critical component in many self-reported physical activity measures. Understanding how individuals provide these values in response to a question is useful for understanding the strengths or limitations of the responses provided. When asked to report a frequency value to a question, respondents are likely to either recall and count each episode of the activity in the time period (i.e., enumeration) or provide an answer that is based on a base rate of the behavior that has been multiplied to meet the time frame of the question (i.e., estimation) (Sudman et al., 1996). For example, in response to the question "How many times did you exercise last month?" respondents may think of their base rate (i.e., how many times they usually exercised each week in the last month) and multiply this value by 4 to provide an appropriate response to the question. Cognitive researchers have determined that, in general, individuals use enumeration to report unusual events and tend not to attempt to count events that occur more than 10 times (Sudman et al., 1996). Information about activities that are similar to one another or that occurred more frequently (e.g., daily) appear to be estimated rather than enumerated (Sudman et al., 1996). Unfortunately, the factors influencing the use of enumeration (episodic memory) or estimation (generic memory) of physical activity events can only be inferred from the findings of cognitive scientists. More work focusing specifically on recalling physical activity behavior is needed to better understand the processes. These studies could suggest ways to improve the assessment of many common moderate-intensity lifestyle activities.

As noted in the previous section on memory storage, explicit memories of absolute time may not reside in memory (Brewer, 1994). This notion would appear to have important implications for providing duration estimates on physical activity instruments. While structured and repetitive activities such as exercise, transportation, or occupational activity may provide good contextual cues and allow for reasonable estimates of activity duration to be provided, less-structured (i.e., spontaneous) activities done during the course of the day (e.g.,

daily walking) may have far fewer contextual cues associated with the activity of interest. This may be an important factor in the ability of respondents to provide accurate duration information for many common lifestyle activities that are considered to be of light to moderate intensity.

How can we enhance the accuracy of information provided in physical activity assessments? Contextual cues are useful in helping respondents to identify the larger clusters of meaningful information stored in memory that are likely to contain memories of the activity behaviors that are relevant to the question being asked (e.g., frequency, intensity, duration). Effort on the part of the respondent is required to do the cognitive work needed for accurate recall or self-assessment. Time is necessary for respondents to comprehend the questions posed of them and then to search memory, reconstruct relevant events, and make decisions about the appropriateness of the information retrieved. While providing respondents adequate time to form their response makes intuitive sense, this element of the process is often at odds with time constraints of ambitious assessment protocols. As noted previously, little research has been done by physical activity researchers to examine the cognitive aspects of self-report instruments. It seems logical that an extension of our knowledge in this area could lead to more effective ways to assess physical activity behaviors using the simplest, most flexible, and inexpensive method currently available (i.e., self-reported measures).

One technique that would clearly be useful is that of cognitive interviews. This method has not been employed by physical activity researchers until relatively recently (Willis, 1994). Cognitive interviews provide qualitative information about the question-answering process, particularly with respect to question comprehension and the decisions and judgments employed in providing responses to survey questions. Individual in-depth interviews of small groups of individuals (n = 10 to 30) are geared for obtaining qualitative rather than quantitative information about response processes. Greater use of these methods may lead to a better understanding of some of the basic features of the cognitive processes involved in providing self-reports of physical activity.

Instrumentation

A variety of instruments have been developed and validated to assess physical activity behaviors over short (e.g., 1 to 30 days) and longer periods (e.g., 30 days, 12 months). Although beyond the scope of this chapter, a number of comprehensive reviews describing many of the individual instruments and their measurement properties (i.e., reliability and validity) are available and should be consulted for details on specific instruments (Laporte et al., 1985; Montoye et al., 1996; Pereira et al., 1997; Sallis & Saelens, 2000). Many exceptional instruments may not be cited in this section of the chapter. Inclusion of certain instruments here does not imply that they are in any way preferable to those not cited.

The mode of administration, interviewer or self-administration, is an important consideration when selecting an instrument for use in a particular situation. Self-administered instruments simply require that the respondent complete a structured questionnaire. Responses to the questionnaire can then be scored manually or entered into a computer using an optical scanning device, or by hand, preferably using a form of double entry to minimize transcription errors. Relatively sophisticated data entry systems can be created using free software (EpiInfo) that is available from the Centers for Disease Control and Prevention (**www.cdc.gov/epiinfo**). After the responses have been entered into digital form, data reduction and scoring algorithms, preferably containing quality assurance checks (e.g., range checks, missing values), can be written to summarize the physical activity information obtained. Self-administered instruments are favored in larger population-based research where it is not feasible to expend a great deal of resources on a per subject basis. Instruments in this category tend to be less complex to maximize comprehension of the questions by the respondent and often focus on the specific activity domain(s) of interest in the project in which they were originally developed (Ainsworth, Sternfeld et al., 2000; Baecke et al., 1982; Paffenbarger et al., 1981; Washburn et al., 1993; Wolf et al., 1994).

Administration of instruments by an interviewer is often the preferred mode of administration for more detailed assessments of activity behavior (Dipietro et al., 1993; Friedenreich et al., 1998; Kriska et al., 1988; Matthews et al., 2000; Sallis et al., 1985; Taylor et al., 1978). Interviewers perform the important function of facilitating comprehension of the questions posed to the respondent and of ensuring that the responses provided make logical sense and have been refined appropriately. For example, reports of activity duration should be refined during the interview to exclude periods of physical inac-

tivity (e.g., rest periods, water breaks, and so on) prior to the duration information entering the data stream. Sallis (1997) provides an exceptionally detailed summary of the many issues involved in conducting 7-day physical activity interviews that are applicable to many different interview situations. Because the interviewer acts as a facilitator of the recall process and also as a quality assurance watchdog, interviewer-administered instruments have greater credibility in terms of validity, although this comes with an increase in cost. Physical activity interviews can take from 10 minutes to one hour, depending on the length and complexity of the instrument. Interviewers must be hired and trained. Projects utilizing multiple interviewers and those that involve interviews being conducted at more than one point in time need to limit systematic variation both between and within interviewers during the course of the project.

Table 7.1 summarizes the four major categories of self-reported physical activity instruments. This summary indicates the utility of the instrument types in large-scale studies, implementation costs, administrative and respondent burden incurred, whether or not the measures are likely to influence behavior, acceptability of the method, and whether or not specific activity domains can be evaluated. The following sections briefly review each of the four broad categories of instruments with particular emphasis on cognitive and sampling issues for each type of instrument.

Physical Activity Records and Diaries

Diary-based self-report instruments provide, with good participant compliance, a detailed record of the many individual physical activities undertaken in a day, more or less as the activities are completed. The difference between physical activity records and logs is that with records, participants are instructed to record the individual bouts of activity as they occur during the day (Ainsworth et al., 1999). In contrast, logs capture the time participants spend in broad categories of activity (e.g., sitting, standing, walking) (Bouchard et al., 1983). In general, records provide more detail about the types, intensity, and patterns of activity completed during the day but require more effort on the part of the participant and a higher level of resources from the study team in terms of data entry and reduction. The recent development of palm-top personal digital assistants has enabled participants to record detailed information directly into an electronic medium.

TABLE 7.1   Summary of the Four Major Categories of Self-Reported Physical Activity Assessments

Measurement	Use in large-scale studies	Low cost	Low administration burden	Low subject burden	Likely to influence behavior	Acceptable to persons	Socially acceptable	Activity specific
Records/logs	Yes	Yes	No	No	Yes	Not always	Yes	Yes
Recall questionnaire	Yes	Yes	Yes	Yes	No	Yes	Yes	Yes
Quantitative history	Yes	Yes	No	No	No	Yes	Yes	Yes
Global self-report	Yes	Yes	Yes	Yes	No	Yes	Yes	Not always

Because of the intensive effort required of the participants, records and diaries are typically implemented for short periods of time, usually 1 to 7 days. This short sampling interval has a number of implications for potential systematic and random variation in the physical activity information obtained that are germane to this type of data in an analysis. The primary advantages of physical activity records and logs are that (1) they provide rich detail about the diverse range of physical activity individuals encounter in their daily lives, (2) bouts of physical activity during a day can be quantified, and (3) the cognitive difficulties of recalling past physical activity behaviors are minimized, at least as long as the participants record their activities as instructed. Unfortunately, the act of recording physical activity may influence a participant's behavior. This phenomenon has been termed *reactivity* and may result in physical activity levels that are higher than average for the participant during the assessment period. The potential impact of reactivity on the internal validity of the instrument in a particular application should be carefully considered when using these instruments, as this phenomenon may be more problematic in some applications (e.g., intervention research) than others (e.g., self-monitoring).

Global Self-Report

Global self-report measures are brief one- to four-item instruments that ask the respondents to provide a generic classification of their usual activity patterns in a specific time period, usually for a specific activity domain (e.g., leisure or occupation). Short and simple assessments such as these have frequently been used in epidemiological studies where the objective is to stratify a population into high and low physical activity exposure categories so that rates of disease in each activity category can be compared (Ainsworth, 1997). Global assessment

may also be useful as a screening tool in the clinical setting to identify individuals who are at increased risk for a number of chronic diseases as a consequence of their low physical activity levels. Results from such screening can then be used to tailor intervention efforts to each individual's current activity level. Because of their brevity, global assessments only attempt to characterize the broad patterns of physical activity habitual behavior, usually over a relatively long time period (e.g., one year). Clearly, the reports of physical activity offered on these instruments are derived from generic memory. Systematic variation derived from seasonal and day-of-the-week effects should be minimal because of the long time frame of the questions. On the other hand, random error in these measures will be primarily due to analytic variation, or variation in reporting by the respondent from one administration to the next.

Recall Questionnaires

Recall-based questionnaires are usually relatively short and simple instruments (e.g., 5 to 15 items), take 5 to 15 minutes to complete, and seek to quantify physical activity patterns in the recent past (e.g., past week or month). The objective of these instruments is to classify individuals into broad categories of activity and provide some basic quantification of the major behavioral characteristics of the activity patterns reported (Ainsworth, Sternfeld et al., 2000; Baecke et al., 1982; Casperson, 1997; Voorips et al., 1991). They are particularly useful for descriptive epidemiological studies designed to assess the prevalence of various activity patterns in large populations. This category of instrument forms the cornerstone of surveillance efforts by public health agencies. For example, the physical activity module on the current Behavioral Risk Factor Surveillance Survey (Macera et al., 2000) uses 12 questions to evaluate the population prevalence for the following categories of physical activity behaviors: physical inactivity (i.e., no moderate-vigorous activity), insufficiently active (i.e., not inactive but does not meet guidelines), and meets current physical activity guidelines (i.e., at least 30 min/day of moderate activity on 5 to 7 days/week, or at least 20 min/day of vigorous activity on at least 3 days/week) (U.S. Department of Health and Human Services, 1996, 2001).

Quantitative History Questionnaires

Quantitative history questionnaires are detailed instruments that evaluate physical activity behaviors using 15 to 60 items that seek to capture the intensity, frequency, and duration of activity patterns in various categories, such as occupation, household, sports and conditioning, transportation, caring for others, and recreational activities. They allow the researchers to obtain detailed information about physical activity energy expenditure and patterns of activity from the past day (Matthews et al., 2000), week (Sallis et al., 1985), month (Dipietro et al., 1993), year (Ainsworth, Jacobs et al., 1993; Taylor et al., 1978), and over one's lifetime (Friedenreich et al., 1998; Kriska et al., 1988). However, because of their length and complexity, these instruments may take from 15 to 60 minutes to complete and are usually interviewer administered.

Quantitative history instruments are appropriate for studies designed to examine issues of dose response and historical activity patterns. These instruments generally account for seasonal variation in activity levels by allowing respondents to report the number of months per year they participated in an activity. In terms of random error in these instruments, analytic variation is the primary concern given the cognitive challenges of consistently recalling some physical activities (e.g., daily walking). Behavioral variation in behavior (e.g., daily or monthly) should be minimized by the relatively long time frame of the instruments (e.g., past year). The cognitive challenges of these types of instruments can vary considerably, depending on the length of the recall period and the activity domains evaluated.

Assessment Procedures and Outcome Variables

Information derived from self-reported instruments about the characteristics of physical activity behaviors can be used directly as a summary measure, or these data can be employed to estimate physical activity energy expenditure. This section outlines some of the key aspects of creating summary measures from self-reports of physical activity.

Often the behavioral characteristics of physical activity are the most descriptive and effective ways to translate self-reported physical activity data into meaningful exercise prescriptions or public health recommendations. As noted previously, the behavioral characteristics of physical activity are the type, intensity, frequency, and duration of activity reported. The type of physical activity that is selected for special consideration depends on the research question being investigated. For example, types of physical activity of interest could be aerobic, anaerobic (strengthening), and flexibility activities as well as weight-bearing physical activity that loads the skeletal system and strengthens bone. Types of activity can also be obtained by stratification of physical activity by intensity (e.g., inactivity, light, moderate, or vigorous). Typically, the type of physical activity serves as a logical grouping variable that subsequent characteristics of physical activity behavior seek to quantify. Two of these characteristics quantify the total duration of physical activity in the time frame of the assessment as the frequency (e.g., per day or week) and duration (e.g., per occasion, day, or week) of the activity reported.

Physical activity intensity, or the energy cost of the activity reported, is now most commonly quantified in multiples of resting metabolic rate (RMR), otherwise known as metabolic equivalents (METS). Prior to the publication of the first compendium of physical activities by Ainsworth, Haskell et al. (1993), researchers around the world were using a variety of types (i.e., kcal/min, Passmore & Durnin, 1955; or METS, Taylor et al., 1978) and sources of energy cost estimates for common activities. This lack of uniformity in research methods greatly hindered the comparability of results across studies. Publication of the original compendium in 1993 (Ainsworth, Haskell et al., 1993) and its update in 2000 (Ainsworth, Haskell et al., 2000) has resulted in more uniformity in the way physical activity intensity has been used to estimate the energy cost of physical activity and in the standardized classification of physical activity by purpose, type, and intensity. The compendium is the recommended resource for obtaining the energy cost of physical activities when measurement of the actual energy costs of individual activities for each person in a research study is not feasible.

One MET is equivalent to RMR, and 4 METS represents an energy cost of physical activity that is four times RMR (Ainsworth, Haskell et al., 1993). This approach evolved from the work-to-basal metabolic rate formulation employed by physiologists to normalize physical activity energy cost measures by metabolic size (Buskirk, 1960; Dill, 1936). MET values represent energy cost values that are made relative to metabolic size by using body mass (kg) as a normalizing factor. One MET is equivalent to 3.5 ml/kg/min of oxygen consumption, or roughly 1 kcal/kg/min for an individual with a resting metabolic rate of 60 kcal/hr (e.g., body mass $\cong$ 60 kg). The MET approach was employed in the development of questionnaire-based physical activity assessment methods in early cardiovascular health studies (Buskirk et al., 1971; Montoye, 1971; Taylor et al., 1978) as a way of systematically classifying physical activity energy cost in large populations of individuals. While the MET approach has many strengths, it also has several limitations that have been continually highlighted by researchers that advocate their use (Ainsworth, Haskell et al., 1993; Montoye, 1971; Taylor et al., 1978). MET values represent the average value of the energy cost of the physical activity for a population of subjects and cannot easily account for interindividual differences in energy costs within the population in which they are utilized (e.g., differences in mechanical efficiency or absolute work rates). Nor can they always account for differences in conditions in which individuals participate in certain activities (e.g., terrain or % grade). Clearly there are limitations to the precision of these measures at the individual level, but they remain the best available way to systematically apply energy cost estimates in self-report measures.

While the MET approach advocated by the compendium suggests perfect standardization of these methods, there are several different summary measures of both relative (to metabolic size) and absolute physical activity energy expenditure that can be calculated. Figure 7.4 presents the calculation of relative and absolute physical activity energy expenditure from reported intensity, total duration, and an estimate of RMR. The two indicators of total duration for the time period (i.e., frequency and duration) combine with the MET intensity value to calculate relative energy expenditure values that are expressed in the two elemental units of measure: MET-hours for the time period (e.g., week).

Similarly, MET-minutes have also been employed in this formulation and are identical to MET-hours, except that the values are multiplied by 60 (Ainsworth et al., 1999; Ainsworth, Leon et al., 1993). The MET-minutes metric has the added feature of approximating kcal values of physical activity energy expenditure for an individual who has an RMR of 60 kcal/hr.

Absolute energy expenditure measures can be calculated from the relative physical activity expenditure values by multiplying by an estimate of RMR. Once this calculation has been made, the summary expenditure values reflect both the elemental data obtained on the self-reports (i.e., intensity and duration) as well as metabolic size reflected by an

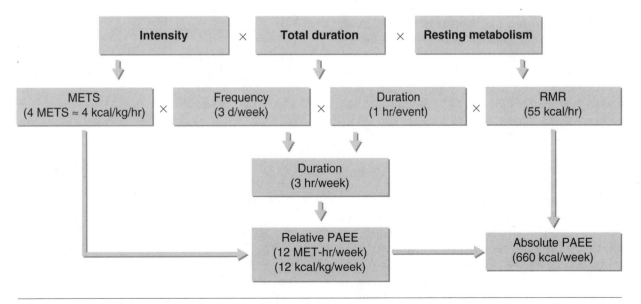

Figure 7.4   Computation of physical activity energy expenditure summary measures.

individual's RMR. Inclusion of metabolic size information in measures of absolute activity energy expenditure must be considered carefully when comparing expenditure values between individuals of different metabolic sizes and in correlation analyses with variables that also reflect body size (Ainsworth, Haskell et al., 1993). Body mass (kg) has often been used as an RMR estimate using the rough equivalence of 1 MET to 1 kcal/kg/hr, but this equivalence may not hold if an individual's RMR is not proportional to body mass (Philippaerts et al., 1999; Racette et al., 1995). Racette et al. (1995) have noted that using the equivalence of 1 kcal/kg/hr to estimate RMR (i.e., 1 MET) can result in significant overestimation of RMR when body mass is not proportional to RMR. Researchers interested in obtaining the most realistic estimates of absolute physical activity energy expenditure (kcal/day), as suggested in the original compendium paper (Ainsworth, Haskell et al., 1993), should employ the best RMR prediction method that is feasible for them.

An electronic version of the revised compendium (Ainsworth, Haskell et al., 2000) is available from the University of South Carolina's Prevention Research Center Web site (**http://prevention.sph.sc.edu**).

Measurement Properties: Reliability and Validity

As previously noted, several reports have provided detailed reviews of published reliability and validity information for a number of the published instruments (Baranowski, 1988; Kohl et al., 2000; Laporte et al., 1985; Sallis & Saelens, 2000; Washburn & Montoye, 1986). In general, reliability measures, typically Pearson or Spearman correlation values, for the most popular instrument are reasonably high and in the range of 0.70 to 0.95 (Pereira et al., 1997; Sallis & Saelens, 2000). In their recent review of reliability and validity studies of self-report instruments published in the 1990s, Sallis and Saelens (2000) evaluated the validity of various instruments against objective measures of activity, primarily accelerometers and doubly labeled water. They noted that validity coefficients for global measures of activity were relatively low (i.e., r = 0.14 to 0.36), but they were higher for certain instruments, such as the 7-day physical activity recall (i.e., r = 0.50 to 0.56).

Similarly, two recent studies suggest that a number of instruments explained about 50% of the variance in doubly labeled water derived estimates of physical activity energy expenditure (Philippaerts et al., 1999; Schuit et al., 1997). These studies suggest that current assessment methods, as Durante and Ainsworth (1996) have noted, explain only about half the variance in objective measures of activity. The modest strength of the correlation coefficients demonstrates a significant linearity of the relationship between the self-report and objective measures, but the precision of measure at the individual level remains elusive. The next section reviews the major factors that may be contributing to variation between measured activity values and the hypothetically "true" values obtained from validation criterion.

The major components of measurement error in self-reports can be characterized as either systematic or random errors (see chapter 1). Systematic variation represents the difference between the true value of the physical activity behavior for the measurement period and that of the value measured using a particular instrument. In contrast, random errors represent variation of the measured values around the underlying true value. Recall that random variation can be further decomposed into terms describing behavioral and analytic variation. Behavioral variation represents naturally occurring changes in physical activity (i.e., day-to-day or intra-individual variation), while analytic variation represents differences in the way data are recorded from one administration of an instrument to the next, but that are not due to a true change in physical activity.

Systematic Variation in Physical Activity Behavior

Many factors that result in systematic variation in self-reported physical activity measures are well known, such as reporting biases (over- or underreporting) (Lichtman et al., 1992; Racette et al., 1995; Taylor et al., 1984), seasonal variation (Dannenberg et al., 1989; Levin et al., 1999; Matthews, Freedson et al., 2001), and day-of-the-week effects (Levin et al., 1999; Matthews, Hebert et al., 2001). Table 7.2 describes the level of seasonal and day-of-the-week effects in the Seasonal Variation of Blood Cholesterol Study, a purely observational study conducted in a large population of healthy, middle-aged adults (age 20 to 70 years) from the northeastern region of the United States. These data, obtained using a series of 15 24-hour activity recalls over the course of one year, demonstrate strong weekday, and even weekend day, ef-

TABLE 7.2   Mean Values for Physical Activity Energy Expenditure (MET-hours/day) by Season and Day of the Week in Men and Women

	Total activity Mean	(SE)	Occupational Mean	(SE)	Nonoccupational Mean	(SE)
Men (n = 300)						
Season						
Winter	11.8	(0.5)[1]	6.0	(0.5)	5.8	(0.3)[1]
Spring	13.1	(0.5)	5.9	(0.5)	7.3	(0.3)
Summer	12.9	(0.5)	6.1	(0.5)	6.8	(0.3)
Fall	12.1	(0.5)	6.0	(0.5)	6.1	(0.3)
Day of the week						
Sunday	10.3	(0.5)[2,4]	1.9	(0.5)[2,4]	8.4	(0.3)[2]
Monday	13.7	(0.5)	7.8	(0.5)	5.9	(0.4)
Tuesday	12.7	(0.5)	7.6	(0.5)	5.1	(0.4)
Wednesday	13.6	(0.5)	7.9	(0.5)	5.7	(0.4)
Thursday	12.8	(0.6)	7.8	(0.5)	5.0	(0.4)
Friday	13.2	(0.6)	8.0	(0.6)	5.3	(0.4)
Saturday	12.1	(0.6)	2.8	(0.5)	9.3	(0.4)
Women (n = 280)						
Season						
Winter	9.2	(0.3)[1]	3.2	(0.2)[1]	6.0	(0.3)[1]
Spring	10.1	(0.3)	3.2	(0.2)	7.0	(0.3)
Summer	9.9	(0.3)	2.7	(0.3)	7.2	(0.3)
Fall	9.4	(0.3)	3.2	(0.2)	6.3	(0.3)
Day of the week						
Sunday	8.5	(0.3)[2,4]	1.2	(0.3)[2]	7.3	(0.3)[2,3,4]
Monday	10.1	(0.4)	3.9	(0.3)	6.2	(0.3)
Tuesday	9.9	(0.3)	4.0	(0.3)	5.9	(0.3)
Wednesday	10.4	(0.4)	4.2	(0.3)	6.3	(0.3)
Thursday	9.7	(0.4)	4.3	(0.3)	5.4	(0.3)
Friday	9.8	(0.4)	3.5	(0.3)	6.4	(0.4)
Saturday	9.7	(0.4)	1.1	(0.3)	8.7	(0.3)

[1]Main effect for season (p < 0.05).

[2]Main effect for all days (Sunday-Saturday, p < 0.05).

[3]Main effect for weekdays (Monday-Friday, p < 0.05).

[4]Main effect for weekend days (Sunday versus Saturday, p < 0.05).

Adapted from Matthews, Hebert et al., 2001. The Seasonal Variation of Blood Cholesterol Study, Worcester, MA 1994-1998.

fects as well as seasonal variation in total, occupational, and nonoccupational (i.e., household and leisure) activities (Matthews, Hebert et al., 2001). The values in this table are MET-hours/day of activity energy expenditure, excluding expenditure from sleep and inactivity. Additional longitudinal analyses from this study have revealed that seasonal variations in physical activity behaviors are complex, varying by the type (i.e., household, occupational, and leisure) and intensity of the physical activity behavior (i.e., light, moderate, or vigorous) (Matthews, Freedson et al., 2001).

Systematic variation in self-report measures can also come from the failure to ascertain, in the structure of the instrument, all of the salient activities done in the course of a day. For example, many of the early self-reported physical activity instruments that were developed to capture the physical activity patterns of middle-aged men enrolled in studies of cardiovascular disease (Jacobs, 2000) did not routinely assess physical activity behaviors that contribute significantly to the physical activity energy expenditure of women (Ainsworth, Richardson et al., 1993; Ainsworth, 2000). Similarly, many early physical activity assessments focused specifically on leisure activity rather than on physical activity done in other social domains, such as occupation and household pursuits. In the 1990s, the importance of moderate intensity activity on health and the rise of the obesity epidemic have resulted in a refocusing of physical activity questionnaires on the full range of moderate-to-vigorous physical activity that we encounter in our daily lives. Indeed, many instruments developed in this decade include the assessment of household, occupational, transportation, and leisure activities (Ainsworth, Sternfeld et al., 2000; Dipietro et al., 1993; Washburn et al., 1993).

Systematic reporting errors, or biases, in physical activity measures have been related to characteristics of both the activities being reported and characteristics of the respondents (Durante & Ainsworth, 1996). Activities of high-intensity physical activity (e.g., running, exercise training) have generally been recalled with reasonable accuracy over periods ranging from 7 days (Jacobs et al., 1993; Taylor et al., 1984) up to many years (Blair et al., 1991; Slattery & Jacobs, 1995). In contrast, light- to moderate-intensity activities have been less accurately obtained. In one of the few detailed studies of recall accuracy for moderate-intensity activities, Taylor et al. (1984) reported that 40% of moderate-intensity activities recorded in a 7-day activity diary were not recalled at the end of the 7-day period by 30 middle-aged Caucasian men. This systematic underreporting resulted in an underestimation of physical activity energy expenditure by about 160 kcal/day (or about 2 kcal/kg/day). Higher levels of adiposity have been associated with overreporting (Buchowski et al., 1999; Jakicic et al., 1998; Lichtman et al., 1992), underreporting (Klesges et al., 1990), and accurate reporting of physical activity (Racette et al., 1995). In addition, gender, habitual activity levels, and memory skills have been associated with reporting biases (Klesges et al., 1990). To date, most work has focused on body fat as a potential source of bias in self-reported physical activity information, and relatively little work has been done to systematically examine the full range of factors associated with reporting errors, such as socioeconomic status and ethnic/cultural differences, literacy (comprehension), or on response set biases such as social desirability (Marlowe & Crowne, 1961). In recent years, it has been observed that comprehension of the terminology in some physical activity questions differ by ethnicity (Tortolero et al., 1999; Warnecke et al., 1997). More systematic inquiry into the phenomenon of reporting bias and specific factors or contexts (e.g., interview vs. self-administered questionnaire, with and without an intervention) associated with either the over- or underreporting of physical activity behaviors may suggest ways to improve the accuracy of physical activity self-reports.

Random Variation

Random variation in measures of activity behavior has been observed to be an important source of variance in several short-term physical activity instruments (Baranowski et al., 1999; Levin et al., 1999; Matthews, Hebert et al., 2001). These studies have suggested that natural variations in activity (i.e., intra-individual variation) are often larger than the true differences between individuals in the population (i.e., inter-individual variation). In other words, the fidelity of the signal we are trying to detect (true differences between subjects) can be obscured to a significant degree by static or noise in the system (i.e., random variation). Consider the 2000 presidential election in the state of Florida. The level of error in the voting system within the state was greater than the percentage of difference between the votes for the two candidates; therefore, identifying true differences between the candidates was extraordinarily difficult (if not impossible).

An understanding of the nature and magnitude of the random variability in physical activity levels is necessary for the proper design, analysis, and interpretation of studies of physical activity and health. Perhaps most important, estimates of biologic variation in physical activity behaviors are required to reliably identify the number of days of assessment required to characterize activity patterns of individuals in a population (Beaton et al., 1979; Gretebeck & Montoye, 1992; Levin et al., 1999). In turn, the number of days of assessment required directly influences the study design (e.g., assessment methods and sample sizes) and statistical power available to detect differences between groups or to quantify the relationships between measured values for physical activity and an outcome of interest (Beaton, 1994).

Results from the Seasonal Variation of Blood Cholesterol Study have suggested that 7 to 10 and 14 to 21 days of assessment are required for the reliable measure (i.e., $R = 0.80$) of total activity in men and women, respectively. In contrast, 21 to 28 days appear to be required for the reliable measure of nonoccupational activities in men and women (Matthews, Hebert et al., 2001). These data are entirely consistent with several smaller studies conducted in more homogenous study populations using an array of activity assessment instruments (Baranowski et al., 1999; Coleman & Epstein, 1998; Lakka & Salonen, 1992). Levin et al. (1999), using similar methods and data from the Survey of Activity, Fitness, and Exercise study (SAFE) (Jacobs et al., 1993), have estimated that nine administrations of 48-hour physical activity records and two to three administrations of a 4-week physical activity history were required to achieve a reliability (R) value of 0.80. It should be noted that nonoccupational activity, in particular, appears to require a longer period of assessment than previously has been suggested from estimates derived from proxy measures of physical activity energy expenditure of approximately 3 and 7 days, respectively, for total energy expenditure (Bouchard et al., 1983) and energy intake (Beaton et al., 1979).

The most significant effect of random variation in measures of physical activity is the attenuation of statistical measures of effect (e.g., correlation and regression coefficients, odds ratios) (Beaton et al., 1979; MacMahon et al., 1990). That is, because of the "noise" in the system, we may have a reduced ability to identify the true relationship between measures of physical activity and outcomes of interest (e.g., other physical activity measures, health outcomes, or measures of the determinants of physical activity). Attenuation of statistical measures of effect may result in several analytic problems, including

1. an erroneous conclusion that there is no effect,
2. that observed effects are "weak," and
3. a significant limitation in our ability to carefully characterize the dose-response relationships between physical activity and the outcome of interest (Beaton, 1994).

Test-retest reliability estimates are available for many self-reported physical activity instruments in many different populations (Pereira et al., 1997; Sallis & Saelens, 2000). Among instruments seeking to capture long-term physical activity behaviors (e.g., the last 12 months), short-term test-retest coefficients (i.e., 2 weeks to a month) reflect analytic variation in the instrument rather than behavioral variation in physical activity. It should be noted that even a modest and generally acceptable level of random variation (i.e., intra-class correlation = 0.80) results in some attenuation in statistical measures. Separation of the source of random variation (i.e., behavioral vs. analytic) is more difficult when examining self-reported physical activity instruments with shorter time frames such as 24-hour or 7-day recalls. For example, a 2-week test-retest of a 7-day recall will reflect both sources of random variation; therefore, understanding the level of actual analytic variation is often difficult. Random variation in activity measures can have a profound effect on the ability of the instruments to detect and carefully quantify important relationships between physical activity and health or the determinants of physical activity behavior. The importance of the effects of random errors varies with the type of analyses being conducted (e.g., evaluation of linear effects, group means, dose responses, or individual prediction). Therefore careful consideration of the effects of random variation should be given to the particular application at hand (Beaton, 1994).

Summary

This chapter has sought to provide practical information about the most effective ways to collect and interpret self-reported physical activity in research settings. By reviewing cognitive models of memory storage and retrieval and by considering the measurement properties of our assessment instruments (e.g., systematic and random variation), this chapter has sought to provide a conceptual framework for implementing and interpreting information derived from these instruments. Self-report methods

gather information about the characteristics of physical activity behaviors for specific periods of time and in specific activity domains. These behavioral characteristics (i.e., type, frequency, intensity, and duration of activity) can be reported as outcome variables alone or translated into an estimate of physical activity energy expenditure using standardized methods. Because of their simplicity, flexibility, and validity (albeit modest relative to objective measures), self-report instruments will remain an invaluable tool for physical activity research in the coming decades.

References

Ainsworth, B.E. (1997). A collection of physical activity questionnaires for health-related research: Lipid Research Clinics Questionnaire. In A.M. Kriska & C.J. Casperson (Eds.), *A collection of physical activity questionnaires. Medicine and Science in Sports and Exercise, 29,* S59-S61.

Ainsworth, B.E. (2000). Issues in the assessment of physical activity in women. *Research Quarterly for Exercise and Sport, 71,* S37-S42.

Ainsworth, B., Haskell, W., Leon, A., Jacobs, D., Montoye, H., Sallis, J., & Paffenbarger, R. (1993). Compendium of physical activities: classification of energy costs of human physical activities. *Medicine and Science in Sports and Exercise, 25,* 71-80.

Ainsworth, B., Haskell, W., Whitt, M., Irwin, M., Swartz, A., Strath, S., O'Brien, W., Bassett, D., Schmitz, K., Emplaincourt, P., Jacobs, D., & Leon, A. (2000). Compendium of physical activities: An update of activity codes and MET intensities. *Medicine and Science in Sports and Exercise, 32,* S498-S516.

Ainsworth, B.E., Irwin, M.L., Addy, C.L., Whitt, M.C., & Stolarczyk, L.M. (1999). Moderate physical activity patterns of minority women: The Cross-Cultural Activity Participation Study. *Journal of Women's Health & Gender-Based Medicine, 8,* 805-813.

Ainsworth, B.E., Jacobs, D.R., Jr., Leon, A.S., Richardson, M.T., & Montoye, H.J. (1993). Assessment of the accuracy of physical activity questionnaire occupational data. *Journal of Occupational Medicine, 35,* 1017-1027.

Ainsworth, B.E., Leon, A.S., Richardson, M.T., Jacobs, D.R., & Paffenbarger, R.S., Jr. (1993). Accuracy of the College Alumnus Physical Activity Questionnaire. *Journal of Clinical Epidemiology, 46,* 1403-1411.

Ainsworth, B., Richardson, M., Jacobs, D., & Leon, A. (1993). Gender differences in physical activity. *Women in Sport and Physical Activity Journal, 2,* 1-16.

Ainsworth, B.E., Sternfeld, B., Richardson, M.T., & Jackson, K. (2000). Evaluation of the Kaiser Physical Activity Survey in women. *Medicine and Science in Sports and Exercise, 32,* 1327-1338.

Baecke, J., Burema, J., & Frijters, J. (1982). A short questionnaire for the measurement of habitual physical activity in epidemiological studies. *American Journal of Clinical Nutrition, 36,* 936-942.

Baranowski, T. (1988). Validity and reliability of self-report measures of physical activity: An information-processing perspective. *Research Quarterly for Exercise and Sport, 59*, 314-327.

Baranowski, T., & Domel, S.B. (1994). A cognitive model of children's reporting of food intake. *American Journal of Clinical Nutrition, 59*, S212-S217.

Baranowski, T., Smith, M., Thompson, W.O., Baranowski, J., Hebert, D., & de Moor, C. (1999). Intra-individual variability and reliability in a 7-day exercise record. *Medicine and Science in Sports and Exercise, 31*, 1619-1622.

Beaton, G. (1994). Approaches to analysis of dietary data: Relationship between planned analyses and choices of methodology. *American Journal of Clinical Nutrition, 59*, S253-S261.

Beaton, G., Milner, J., Corey, P., McQuire, V., Cousins, M., Stewart, E., Ramos, M.D., Hewitt, D., Grambsch, V., Kassim, N., & Little, J. (1979). Sources of variance in 24-hour dietary recall data: Implications for nutrition study design and interpretation. *American Journal of Clinical Nutrition, 32*, 2546-2559.

Blair, S.N., Dowda, M., Pate, R.R., Kronenfeld, J., Howe, H.G., Jr., Parker, G., Blair, A., & Fridinger, F. (1991). Reliability of long-term recall of participation in physical activity by middle-aged men and women. *American Journal of Epidemiology, 133*, 266-275.

Bouchard, C., Tremblay, A., Leblanc, C., Lortie, G., Savard, R., & Theriault, G. (1983). A method to assess energy expenditure in children and adults. *American Journal of Clinical Nutrition, 37*, 461-467.

Brewer, W.F. (1994). Autobiographical memory and survey research. In N. Schwarz & S. Sudman (Eds.), *Autobiographical memory and the validity of retrospective reports* (pp. 11-20). New York: Springer-Verlag.

Buchowski, M.S., Townsend, K.M., Chen, K.Y., Acra, S.A., & Sun, M. (1999). Energy expenditure determined by self-reported physical activity is related to body fatness. *Obesity Research, 7*, 23-33.

Buskirk, E.R. (1960). Problems related to the caloric cost of living. *Bulletin of the New York Academy of Medicine, 36*, 365-388.

Buskirk, E.R., Harris, D., Mendez, J., & Skinner, J. (1971). Comparison of two assessments of physical activity and a survey method for calorie intake. *American Journal of Clinical Nutrition, 24*, 1119-1125.

Casperson, C., Powell, K., & Christenson, G. (1985). Physical activity, exercise, and physical fitness: Definitions and distinctions for health-related research. *Public Health Reports, 100*, 125-131.

Casperson, C.J. (1997). A collection of physical activity for health-related research: Behavioral risk factor surveillance system. *Medicine and Science in Sports and Exercise, 29*, S1-S203.

Coleman, K.J., & Epstein, L.H. (1998). Application of generalizability theory to measurement of activity in males who are not regularly active: A preliminary report. *Research Quarterly for Exercise and Sport, 69*, 58-63.

Dannenberg, A., Keller, J., Wilson, P., & Castelli, W. (1989). Leisure-time physical activity in the Framingham Offspring Study: Description, seasonal variation, and risk factor correlates. *American Journal of Epidemiology, 129*, 76-88.

Dill, D.B. (1936). The economy of muscular exercise. *Physiologic Reviews, 16*, 263-291.

Dipietro, L., Caspersen, C.J., Ostfeld, A.M., & Nadel, E.R. (1993). A survey for assessing physical activity among older adults. *Medicine and Science in Sports and Exercise, 25*, 628-642.

Durante, R., & Ainsworth, B. (1996). The recall of physical activity: Using a cognitive model of the question-answering process. *Medicine and Science in Sports and Exercise, 28*, 1282-1291.

Friedenreich, C.M., Courneya, K.S., & Bryant, H.E. (1998). The lifetime total physical activity questionnaire: Development and reliability. *Medicine and Science in Sports and Exercise, 30*, 266-274.

Gretebeck, R.J., & Montoye, H.J. (1992). Variability of some objective measures of physical activity. *Medicine and Science in Sports and Exercise, 24*, 1167-1172.

Jacobs, D., Ainsworth, B., Hartman, T., & Leon, A. (1993). A simultaneous evaluation of 10 commonly used physical activity questionnaires. *Medicine and Science in Sports and Exercise, 25*, 81-91.

Jacobs, D.J. (2000). Comment on issues in the assessment of physical activity in women. *Research Quarterly for Exercise and Sport, 71*, 43-46.

Jakicic, J.M., Pilley, B.A., & Wing, R.R. (1998). Accuracy of self-reported exercise and the relationship with weight loss in overweight women. *Medicine and Science in Sports and Exercise, 30*, 634-638.

Jobe, J., Tourangeau, R., & Smith, A. (1993). Contributions of survey research to the understanding of memory. *Applied Cognitive Psychology, 7*, 567-584.

Jobe, J.B., & Hermann, D.J. (1996). Implications of models of survey cognition for memory theory. In D.J. Hermann, C. McEvoy, C. Hertzog, P. Hertel, & M.K. Johnson (Eds.), *Basic and applied memory research: Practical applications* (vol. 2, pp. 193-205). Mahwah, NJ: Lawrence Erlbaum.

Jobe, J.B., & Mingay, D.J. (1991). Cognition and survey measurement: History and overview. *Applied Cognitive Psychology, 5*, 175-192.

Klesges, R.C., Eck, L.H., Mellon, M.W., Fulliton, W., Somes, G.W., & Hanson, C.L. (1990). The accuracy of self-reports of physical activity. *Medicine and Science in Sports and Exercise, 22*, 690-697.

Kohl, H.W., Fulton, J.E., & Casperson, C.J. (2000). Assessment of physical activity among children and adolescents: A review and synthesis. *Preventive Medicine, 31*, S54-S76.

Kriska, A., Sandler, R., Cauley, J., Laporte, R., Hom, D., & Pamianco, G. (1988). The assessment of historical activity and its relation to adult bone parameters. *American Journal of Epidemiology, 127*, 1053-1063.

Lakka, T.A., & Salonen, J.T. (1992). Intra-person variability of various physical activity assessments in the Kuopio Ischaemic Heart Disease Risk Factor Study. *International Journal of Epidemiology, 21*, 467-472.

Laporte, R., Montoye, H., & Casperson, C. (1985). Assessment of physical activity in epidemiologic research: Problems and prospects. *Public Health Reports, 100*, 131-146.

Levin, S., Jacobs, D.R., Jr., Ainsworth, B.E., Richardson, M.T., & Leon, A.S. (1999). Intra-individual variation and estimates of usual physical activity. *Annals of Epidemiology, 9*, 481-488.

Lichtman, S.W., Pisarska, K., Berman, E.R., Pestone, M., Dowling, H., Offenbacher, E., Weisel, H., Heshka, S., Matthews, D.E., & Heymsfield, S.B. (1992). Discrepancy between self-reported and actual caloric intake and exercise in obese subjects. *New England Journal of Medicine, 327*, 1893-1898.

Macera, C.A., Jones, D.A., Kimsey, C.D., Ham, S., & Pratt, M. (2000). New directions in surveillance of physical activity among U.S. adults: A pilot study. *Medicine and Science in Sports and Exercise, 32*, S260 [Abstract].

MacMahon, S., Peto, R., Cutler, J., Collins, R., Sorlie, P., Neaton, J., Abbott, R., Godwin, J., Dyer, A., & Stamler, J. (1990). Blood pressure, stroke, and coronary heart disease. Part 1, Prolonged differences in blood pressure: Prospective observational studies corrected for the regression dilution bias. *Lancet, 335*, 765-774.

Marlowe, D., & Crowne, D.P. (1961). Social desirability and response to perceived situational demands. *Journal of Consulting Psychology, 25*, 109-115.

Matthews, C.E., Freedson, P.S., Stanek, E.J., III, Hebert, J.R., Merriam, P.A., Rosal, M.C., Ebbeling, C.B., & Ockene, I.S. (2001). Seasonal variation of household, occupational, and leisure-time physical activity: Longitudinal analyses from the Seasonal Variation of Blood Cholesterol Study. *American Journal of Epidemiology, 153*, 172-183.

Matthews, C.E., Hebert, J.R., Freedson, P.S., Stanek, E.J., III, Merriam, P.A., Ebbeling, C.B., & Ockene, I.S. (2001). Sources of variance in daily physical activity levels in the Seasonal Variation of Blood Cholesterol Study. *American Journal of Epidemiology, 153*, 987-995.

Matthews, C.E., Hebert, J.R., Freedson, P.S., Stanek, E.J., III, Ockene, I.S., & Merriam, P.A. (2000). Comparing physical activity assessment methods in the Seasonal Variation of Blood Cholesterol Levels Study. *Medicine and Science in Sports and Exercise, 32*, 976-984.

Montoye, H., Kemper, H., Saris, W., & Washburn, R. (1996). *Measuring physical activity and energy expenditure.* Champaign, IL: Human Kinetics.

Montoye, H.J. (1971). Estimation of habitual physical activity by questionnaire and interview. *American Journal of Clinical Nutrition, 24*, 1113-1118.

Paffenbarger, R.S., Wing, A.L., Hyde, R.T., & Jung, D.L. (1981). Physical activity and index of hypertension in college alumni. *American Journal of Epidemiology, 117*, 245-257.

Passmore, R., & Durnin, J.V.G. (1955). Human energy expenditure. *Physiological Reviews, 35*, 801-840.

Pereira, M.A., Fitzgerald, S.J., Gregg, E.W., Joswiak, M.L., Ryan, W.J., Suminski, R.R., Utter, A.C., & Zmuda, J.M. (1997). A collection of physical activity questionnaires for health-related research. *Medicine and Science in Sports and Exercise, 29*, S1-S203.

Philippaerts, R.M., Westerterp, K.R., & Lefevre, J. (1999). Doubly labelled water validation of three physical activity questionnaires. *International Journal of Sports Medicine, 20* (5), 284-289.

Racette, S.B., Schoeller, D.A., & Kushner, R.F. (1995). Comparison of heart rate and physical activity recall with doubly labeled water in obese women. *Medicine and Science in Sports and Exercise, 27* (1), 126-133.

Sallis, J., Haskell, W., Wood, P., Fortmann, S., Rodgers, T., Blair, S., & Paffenbarger, R. (1985). Physical activity assessment methodology in the Five-City Project. *American Journal of Epidemiology, 121* (1), 91-106.

Sallis, J.E. (1997). A collection of physical activity questionnaires for health-related research: Seven-day physical activity recall. In A.M. Kriska & C.J. Casperson (Eds.), *A collection of physical activity questionnaires. Medicine and Science in Sports and Exercise, 29*, S89-S103.

Sallis, J.F., & Saelens, B.E. (2000). Assessment of physical activity by self-report: Limitations and future directions. *Research Quarterly for Exercise and Sport, 71*, S1-S14.

Schuit, A.J., Schouten, E.G., Westerterp, K.R., & Saris, W.H. (1997). Validity of the Physical Activity Scale for the Elderly (PASE): According to energy expenditure assessed by the doubly labeled water method. *Journal of Clinical Epidemiology, 50*, 541-546.

Slattery, M.L., & Jacobs, D.R., Jr. (1995). Assessment of ability to recall physical activity of several years ago. *Annals of Epidemiology, 5*, 292-296.

Sudman, S., Bradburn, N.M., & Schwarz, N. (1996). *Thinking about answers: The application of cognitive processes to survey methodology.* San Francisco: Jossey-Bass.

Taylor, C., Coffey, T., Berra, K., Iaffaldo, R., Casey, K., & Haskell, W. (1984). Seven-day activity and self-report compared to a direct measure of physical activity. *American Journal of Epidemiology, 120*, 818-824.

Taylor, H., Jacobs, D., Schucker, B., Knudsen, J., Leon, A., & Debacker, G. (1978). A questionnaire for the assessment of leisure-time physical activities. *Journal of Chronic Disease, 31*, 741-755.

Tortolero, S.R., Masse, L.C., Fulton, J.E., Torres, I., & Kohl, H.W., III. (1999). Assessing physical activity among minority women: Focus group results. *Women's Health Issues, 9*, 135-142.

Tourangeau, R. (1984). Cognitive sciences and survey methods. In T.B. Jabine, M.L. Straf, J.M. Tanur, & R. Tourangeau (Eds.), *Cognitive aspects of survey methodology: Building a bridge between disciplines* (pp. 73-101). Washington, DC: National Academy Press.

U.S. Department of Health and Human Services. (1996). *Physical activity and health: A report of the Surgeon General.* Atlanta, GA: U.S. Department of Health and Human Services, Centers for Disease Control and Prevention.

U.S. Department of Health and Human Services. (2001). Physical activity trends—United States, 1990-1998. *Morbidity and Mortality Weekly Report, 50*, 166-169.

Voorips, L., Ravelli, A., Dongelmans, P., Durenberg, P., & Stavren, W.V. (1991). A physical activity questionnaire for the elderly. *Medicine and Science in Sports and Exercise, 23*, 974-979.

Warnecke, R.B., Johnson, T.P., Chavez, N., Sudman, S., O'Rourke, D.P., Lacey, L., & Horm, J. (1997). Improving question wording in surveys of culturally diverse populations. *Annals of Epidemiology, 7*, 334-342.

Washburn, R., & Montoye, H. (1986). The assessment of physical activity by questionnaire. *American Journal of Epidemiology, 123*, 563-576.

Washburn, R.A., Smith, K.W., Jette, A.M., & Janney, C.A. (1993). The Physical Activity Scale for the Elderly (PASE): Development and evaluation. *Journal of Clinical Epidemiology, 46*, 153-162.

Willis, G.B. (1994). *Cognitive interviewing and questionnaire design: A training manual (working paper series, no. 7).* Hyattsville, MD: National Center for Health Statistics.

Willis, G.B., Royston, P., & Bercini, D. (1991). The use of verbal report methods in the development and testing of survey questionnaires. *Applied Cognitive Psychology, 5*, 251-267.

Wolf, A.M., Hunter, D.J., Colditz, G.A., Manson, J.E., Stampfer, M.J., Corsano, K.A., Rosner, B., Kriska, A., & Willett, W.C. (1994). Reproducibility and validity of a self-administered physical activity questionnaire. *International Journal of Epidemiology, 23*, 991-999.

8

Use of Accelerometry-Based Activity Monitors to Assess Physical Activity

Gregory J. Welk, PhD
Iowa State University

Accelerometry-based physical activity monitors have become one of the most commonly used methods for assessing physical activity under field conditions. They are small, are noninvasive, and provide an objective record of overall movement. Because most units can store considerable amounts of data, it is possible to obtain detailed information about a person's activity patterns over an extended period of time. Research on these devices has been accumulating rapidly and considerable progress has been made to advance their utility for field-based research. There is general consensus that accelerometry-based monitors provide a valid indicator of overall physical activity but a less accurate prediction of energy expenditure (EE), particularly under free-living conditions. A number of issues associated with the reliability and validity need to be resolved to further enhance their utility. The purpose of this chapter is to provide background on important measurement issues with activity monitors and to offer practical information about the application of accelerometry-based moni-

tors for various physical activity research applications.

Principles of Method

Accelerometry-based devices have been used to study human movement for more than 30 years, but the sophistication of the monitors and direct applications to physical activity has increased dramatically in the past 10 years. Acceleration is the change in velocity over time and is usually expressed in terms of multiples of gravitational force (g = 9.8 m/s^2, or 32 ft/s^2). The degree of acceleration provides an index of movement since it incorporates the rate at which distance is covered.

According to a historical review of these devices (Montoye et al., 1996), one of the first monitors, the Large-Scale Integrated (LSI) Motor Activity Monitor, was developed in the early 1970s to study movements of psychiatric patients. This device used a mercury switch mechanism to detect movement and a telemetry-based recorder to receive the data. The application of the LSI monitor for physical activity research was first described by LaPorte et al. in the late 1970s and early 1980s (1979, 1982). These early studies demonstrated the potential of these devices for physical activity research and initiated other developments and research in the field. Subsequent research demonstrated some limitations associated with the use of mercury switches, and emphasis was placed on the use of piezoelectric electric bender elements as transducers for accelerometers (Montoye et al., 1983; Redmond & Hegge, 1985). These elements consist of a ceramic material with a brass core. When acceleration occurs, a charge is produced that is proportional to the force exerted. Piezoresistive elements are also used to sense acceleration. With this sensor, the element changes resistance when under load from acceleration, and the electrical signal is altered. The properties of these two types of sensing elements are different, but it is not clear if they present any particular advantages or disadvantages for physical activity research. Nearly all contemporary monitors now utilize this type of sensor as the basis for the accelerometry readings.

Processing and filtering of the raw accelerometry signals is needed to exclude accelerations and signals that are outside the range of human movement. High-pass filters remove low frequency signals (e.g., normal g force). This allows the monitors to record accelerations of zero while a person is stationary. Low-pass filters remove high frequency accelerations and electrical interference. Accelerations of the human body are typically less than 10 Hz and between −6 and +6 g, but higher forces are found during impacts and landings. Most contemporary monitors have low-pass filters set at 20 Hz to cover the full range of movement.

An important general concept for accelerometers is that they measure segment or limb acceleration rather than overall body acceleration. During normal locomotor movements, the hips (and other limbs) accelerate and de-accelerate with each stride. This repeated pattern allows accelerometers to record a constant acceleration signal or count during steady-state activity. Accelerations that occur when a person moves through space in a car are typically lower in magnitude and of short duration and not contribute much to the overall signal. Most undimensional accelerometers also are oriented vertically and would not measure linear acceleration. It is important to consider the nature of the acceleration when interpreting data from these monitors.

There are a variety of factors that can influence accelerometer output, including acceleration due to body movement, gravitational acceleration, external vibration not produced by the body, or the jolting of the sensor due to a less-than-secure attachment to the body (Bouten, Sauren et al., 1997). Because only the first source reflects acceleration from human movement, the other sources essentially contribute error. Guidelines for reducing sources of error are discussed in the section on data collection, which emphasizes the various factors that specifically influence the output recorded from human movement.

Effect of Positioning on Accelerometer Output

One major factor influencing accelerometer output is the relative position of the monitor on the body. Researchers have tested monitors worn at the hip, waist, and ankle (as well as combinations of these positions) to identify the optimal site for wearing the monitors. In a comprehensive mechanical analysis of placement and orientation, Bouten, Sauren et al. (1997) concluded that monitor placement position does not influence the prediction of energy expenditure, despite differences in the extent of movement detected at each site. Essentially, movement (and energy expenditure) can be estimated accurately (for group comparisons) with the monitors positioned at a variety of positions. The study mentioned some advantages of attachment on the back (at the waist) for general monitoring, but the preferred site for most investigators is the hip. This position is well suited for picking up the accelerations that occur during normal locomotor movements and participants usually find it less obtrusive for sitting and moving around. Because the output varies between monitors placed at different sites, researchers should position the devices in a consistent way across the study. If an existing calibration equation is used to process the data, care should be used to position the device in the same way as in the published study.

Effect of Activity Type on Accelerometer Output

A second factor influencing output is the type of activity performed. A well-known limitation of activity monitors is that they are not able to account for the increased energy cost of walking uphill or the extra energy required to carry a weight or move a load. This is because the acceleration patterns remain essentially unchanged under these conditions, despite the increase in effort required. This is true regardless of where the monitor is positioned, but positioning can differentially influence the ability to assess other activities. For example, a waist-worn monitor is not able to accurately assess activity or energy during activities such as cycling (Jakicic et al., 1999) since little hip motion occurs in this movement. A knee-mounted unit would likely pick up this cyclical movement, but this is a less than desirable position for assessing other activities. The same is true for upper-body activities that involve throwing. A wrist-worn monitor may assess movement in the upper body better than a waist-worn monitor, but it would not be as useful for picking up movement during normal locomotion. While the concept of wearing multiple accelerometers simultaneously may seem advantageous, the degree of improvement in predicting energy cost may not be worth the increased cost and the added burden for participants (Bouten, Koekkoek et al., 1997; Swartz et al., 2000). The inability to measure all activities equally well is thus a limitation of accelerometry but perhaps not a large one because activity monitors are reasonably accurate for measuring locomotor movement, which constitutes the bulk of daily activity (at least for adults).

Effect of Orientation on Accelerometer Output

A final factor influencing accelerometer output is the orientation of the accelerometer on the body. Because the accelerometer picks up movement along a particular axis, it will record movement proportional to the direction of orientation. If the accelerometer is designed to measure vertical accelerations, then the device must be positioned in this plane to accurately assess movement in this dimension. Standardization of position on a participant is important to reduce unwanted variability in the data. The use of three-dimensional accelerometers may minimize the extent of this error because the composite magnitude value from the three directions would always reflect the movement that occurs. (Studies comparing uniaxial and triaxial monitors are discussed in the section on measurement properties.)

Despite some limitations, accelerometry-based activity monitors have been found to provide valuable information about physical activity patterns under free-living conditions. The various commercially available monitors are reviewed in the next

section to provide a better understanding of the different features of these devices.

Instrumentation

There are a variety of commercially available activity monitors. Most contemporary models are based on the same accelerometry principles but differ in the amount of memory, the software capabilities, and the type, size, and sensitivity of the sensor. They also vary considerably by price (roughly $200 to $500). Contemporary units range in size and weight, but most are about the size of electronic pagers. Most units allow for data to be stored and downloaded for subsequent processing on a computer. The accelerometers function using high frequency signals, but the information recorded by the monitors is typically integrated over a user-specified epoch to facilitate data processing. The amount of data that can be stored is inversely related to this sampling interval. With a 60-second epoch, the maximum capacity for most monitors ranges from 10 to 30 days.

While some differences may exist between the monitors, studies have generally indicated that they provide highly similar information (Welk, Blair et al., 2000; Welk & Corbin, 1995). They do differ in the way they process and report the accelerometry signals, so decisions regarding the selection of an appropriate monitor may depend more on the ease with which the devices can be used for various field applications. The characteristics and features of the most commonly used monitors are reviewed in this section (in essentially the order they were developed). Issues with the reliability and validity of the different instruments are discussed in a later section. Images of some of the most commonly used devices are shown in figure 8.1.

Caltrac

The Caltrac was one of the first commercially available accelerometers. It was originally designed as a "personal exercise monitor" that could provide an estimate of activity-related EE and total EE. A person's height, weight, age, and gender are entered, and an internal algorithm computes a measure of resting EE. Activity-related EE is then calculated based on the person's weight, although specific entries on the unit allow it also to function in "counts mode." A number of studies have examined the utility of the Caltrac for measuring different types

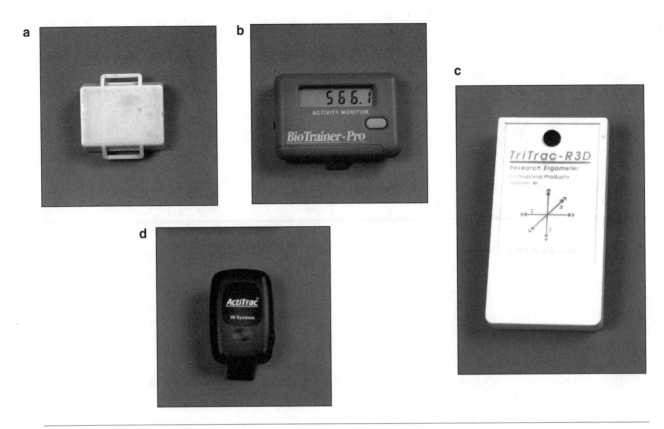

Figure 8.1 Some of the more commonly used accelerometry-based activity monitors: *(a)* Computer Science and Applications (CSA) monitor, *(b)* Biotrainer, *(c)* Tritrac, and *(d)* Actitrac. Both the CSA and the Tritrac are now manufactured by other companies.

and intensities of physical activity. Because it does not allow data to be stored on a minute-by-minute level, studies with the Caltrac can only examine total activity accumulated over a specific period of time. The Caltrac is used less today but is still available through a company called Muscle Dynamics Fitness Network (Torrance, CA).

Tritrac

The Tritrac-R3D was developed using the same principles as the Caltrac monitor but with features that made it more useful for research. It was designed with solid-state technology to improve reliability and includes an interface to allow data to be sent to a computer. An additional distinction is that the Tritrac records data in three directions through the use of three separate accelerometers, positioned internally at 90° to one another. Output from each accelerometer is reported along with a composite three-dimensional signal called the *vector magnitude*. The software also provides an estimate of physical activity and total EE based on the same internal algorithms as the Caltrac. Because the Tritrac has three

accelerometers and requires a separate power source (because of the solid-state components), the unit is somewhat larger than other monitors (4.72″ × 2.58″ × 0.87″). The latest version of the Tritrac, called the RT3, is now smaller, but no published literature is currently available on this monitor. Tritrac products are available from the Web site **www.stayhealthy.com**.

CSA

The Computer Science and Applications (CSA) monitor is currently the most widely used monitor in physical activity research. It is smaller than most other monitors (2″ × 1.5″ × 0.6″), which makes it easier for participants to wear over several days. It also has a hard shell case and an infrared computer interface that has been shown to be very reliable for data transfer purposes. A prototype version of the CSA was first studied by Janz et al. (1995), but the more commonly used version has been the Model 7164 Actigraph. This monitor has 64k of memory, which allows continuous recording for 22 days at 1-minute intervals. An expanded memory model (71256) has 256k of memory to allow for even longer recording durations. According to the manufacturer, the sensitivity threshold is 0.05g and the full scale can accommodate 2.13g. A limitation of the CSA monitor is that the software requires a DOS platform to work. Various macros are available for processing the files in Microsoft Excel, but the two-step process is a bit more cumbersome than the Windows®-based software used with most other monitors. The CSA is currently available from Manufacturing Technology Incorporated (MTI), Fort Walton Beach, FL.

Biotrainer

The Biotrainer is a uniaxial monitor but is considered to be bidirectional, since the accelerometer is positioned 45° to vertical in the sagittal plane. It uses Windows®-based software and connects to the computer with a standard serial cable. The original Biotrainer monitor was developed to track time retrospectively from the time it was downloaded (rather than prospectively from initialization). This offered some considerable advantages for field research because the monitors could be passed out to groups of participants and then processed when returned. Subtle timing errors occurred during download, however, because the internal clock in the monitor was inactivated during the download procedure (personal communication with IM Systems). The loss of time (roughly 2 to 3 minutes) would not hamper results from a large-scale project, but it seriously challenges efforts to assess activity at specific time intervals.

The new release of the Biotrainer, called the Biotrainer Pro, is smaller than the original unit and has been designed as more of a true research instrument. It allows for user-defined recording epochs ranging from 15 seconds to 2 minutes as well as various user-defined sensitivity settings that allow the user to maximize the range of scores for monitoring. Like other monitors, it also relies on a computer initialization step to activate the unit and initiate the recording of data. It includes a built-in belt clip that allows the unit to be attached to a participant's waistband or belt. The Biotrainer Pro is available from IM Systems in Baltimore, MD (**www.imsystems.net**).

Actitrac

The Actitrac monitor is produced by the same company as the Biotrainer (IM Systems, Baltimore, MD) but was originally developed as a wrist-worn de-

vice for sleep research. It was designed to be very small (smaller than the CSA) and highly sensitive to movement. Because these are also desirable characteristics for physical activity research, it has also been tested as a waist-worn device like the other physical activity monitors. The Actitrac uses the same serial cable as the Biotrainer to connect to the computer and operates with a similar software interface. The Actitrac has similar memory capabilities to the CSA and is also priced about the same. A recent study (Welk et al., 2002) reported high correlations between the Biotrainer and the Actitrac and strong predictive validity for estimating energy expenditure, but this is the only study that has evaluated the Actitrac thus far.

Tracmor

The Tracmor is a three-dimensional monitor composed of three uniaxial accelerometers mounted orthogonally in a small resin block. Several studies (Bouten et al., 1994; Bouten, Koekkoek et al., 1997) reported on the design and technical specifications for a three-dimensional monitor and did not specify this name, but subsequent studies have begun calling this device the *Tracmor* (*tr*iaxial *ac*celerometer for *mo*vement *r*egistration). It has been used predominantly in Europe and less is known about how it relates to the other devices more commonly used in the United States.

Assessment Procedures

Obtaining accurate information from accelerometry-based activity monitors requires careful attention to the way the data are collected, processed, and analyzed. Monitors from most companies come factory calibrated, and many researchers assume that the monitors will continue to provide reliable measures of activity for an extended period of time. Evidence is accumulating that this is not the case, however. While guidelines are not currently available for how often calibration of monitors should be done, researchers intending to use the monitors over an extended period (i.e., several months) are encouraged to incorporate some type of quality control into their measurement protocol to ensure that the monitors function consistently over time. Output from monitors may also vary as a function of battery life (personal communication with IM Systems). To minimize this potential error, batteries should be checked and replaced in a systematic manner.

Data Collection

Most contemporary activity monitors require computer initialization prior to being used in the field. This process is unique for each of the devices. The internal clock settings on the computer are used as the basis for the start time, so this time should be accurate and checked periodically to avoid "drift" in the time sampling.

After initialization, the monitors are typically secured to the participant's waist with a belt clip or inserted into a waist-worn pouch. As described earlier, a variety of positions have been tested, but the most common site has been the waist because it is easy to affix monitors at that location and is more comfortable for participants. Standardization of the waist position is still important to reduce unwarranted variance in the monitor output. A study comparing three different monitors tested the positional effects by comparing output of the monitors when worn at the anterior, mid-, and posterior axillary lines (Jones et al., 1999). Significant differences were found in output from the CSA ($p < 0.05$) but not the Biotrainer or the Tritrac. The implications of these results are not clear because they haven't been confirmed by other studies. As described earlier, slight differences in accelerometer orientation at the three positions may lead to systematic differences in accelerometer output. Thus, care should be taken to minimize unwanted variability. The manufacturer guidelines for the CSA recommend positioning the device at the anterior axillary line, whereas the most common position for the Biotrainer is the mid-axillary line.

Securing the monitor tightly against the body is another important consideration for obtaining reproducible results. Because the monitors are highly sensitive, extraneous movement produced from bouncing or jostling can lead to spurious results in the output. By tightly securing the monitor against the waist, this type of movement can be minimized. An issue that becomes more relevant when monitoring young children is to minimize subject tampering. This can be accomplished either by blanking the display to decrease curiosity or by securing the monitors inside a sealed pouch.

After passing out the monitors, it is important to provide participants with clear guidelines about how the device should be used. If the goal is to obtain data over a series of days, participants should be instructed on how to attach the monitor and given clear instructions on what to do when showering, swimming, and changing clothes. Most researchers include logs that allow the participant to record the times they put on and took off the monitor. This type of information can facilitate subsequent data processing.

Data Processing

Most monitors provide a way to transfer data from the monitor to the computer for data processing. Because of the volume of data produced from the short, time-based recordings (1,440 lines of data for 1 day of recording at 1-minute intervals), considerable work is needed to process and reduce the data into meaningful summary units. The software from the various companies have incorporated different ways to access the data, but most allow for raw data to be exported or converted into a comma-delimited format to allow for subsequent processing. The graphical interface used with the Biotrainer monitor is shown in figure 8.2.

A challenge for research with activity monitors is screening data for noncompliance with monitoring protocols. The presence of a few outliers can severely influence the results from correlational analyses, so it is essential to check for outliers in these types of studies. Failure to wear the monitor as directed can also lead to measures of activity that are artificially low, an error that would significantly bias results of intervention or surveillance studies. Researchers are encouraged to develop screening criteria that are appropriate for their particular study and the activity monitor being used. The CSA monitor is highly sensitive and will usually record a few counts even if a person is sitting down or stationary. By searching for extended blocks of data with zero counts, it is relatively easy to identify cases in which a person may have removed the monitor. The output from the Biotrainer is smaller in magnitude for a given amount of activity. Thus, it is harder to discern in a given minute if a person was wearing the monitor. In this case, the sum of counts accrued over a specified interval can be examined for the presence of zero counts. In recent research with children (Welk et al., 2001), data were processed in 30-minute time blocks and the presence of four consecutive blocks with zero counts was used to screen for noncompliant participants. (See figure 8.3 for an example of how the data were screened.)

Outcome Variables

A unique advantage of accelerometers is that they can segment activity into discrete time periods,

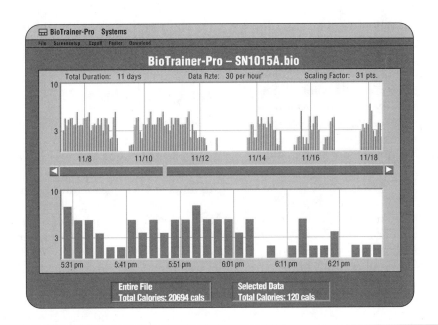

Figure 8.2 Software interface used to view and process data from the Biotrainer Pro. The top panel shows all detail for whole monitoring period, while the bottom panel can be changed to view different segments of time in more detail.

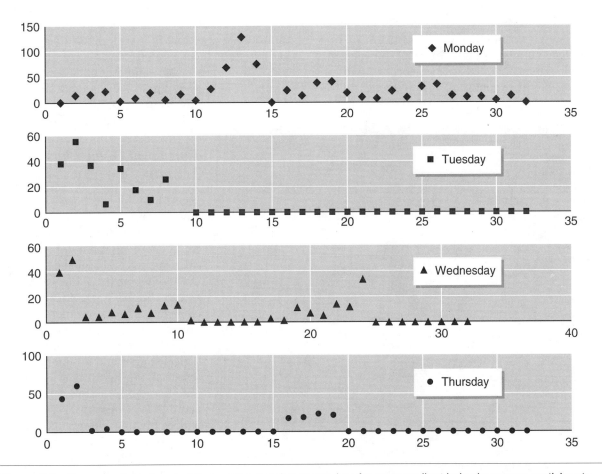

Figure 8.3 Example of graphs used to examine accelerometry data for noncompliant behavior among participants. Each block represents the summed activity counts for a 30-minute interval. Note the "flat line" responses on days 2 and 3, suggesting poor compliance on these days.

usually at a user-specified time interval. This allows the different characteristics of activity (e.g., frequency, intensity, duration) to be examined independently. A researcher may be interested in examining activity patterns at a specific point in time or the total amount of activity accumulated throughout the day. Some degree of processing is required to reduce the interval-based data into smaller segments that can be examined more systematically. Most users have developed statistical programs or spreadsheet-based macros that can be used to facilitate data processing. A sample program developed for the CSA processes multiple days of data from the CSA and summarizes the data in various output measures. Sample output from this program is included in figure 8.4 as an example of the types of data processing that are typically needed to track and analyze data from several days of monitoring. Descriptions of different outcome measures are provided in the following sections.

Total Movement Counts

A number of studies have expressed accelerometry-based activity monitor data in the form of raw movement counts. The advantage of this measure is that it avoids potential errors associated with estimating energy expenditure or time spent in activity. The disadvantage is that the units are not easily interpretable against established public health guidelines.

Energy Expenditure (EE) Prediction

One of the most common outcome measures of physical activity is energy expenditure. Many investigators have developed prediction equations that allow counts from the monitor to be converted to units of energy expenditure. The most common approach is to develop a regression equation that defines the linear relationship between counts and energy expenditure. The equations either include body weight and report the data as kcal/min or use a weight-relative measure of METS (3.5 ml/kg/min or about 1 kcal/kg/hr). Some investigators have developed individual calibration curves to reduce individual error and improve the accuracy of EE estimates. Others have integrated accelerometry data with heart rate data to further enhance the accuracy of these estimations (see chapter 13). Aside from the intuitive appeal of an easily understood unit, the advantage of using energy expenditure as an outcome measure is that it allows all types of activity to be incorporated into one overall indicator. A limitation is that estimates of energy expenditure from free-living situations have not been found to be highly accurate, particularly for individual comparisons.

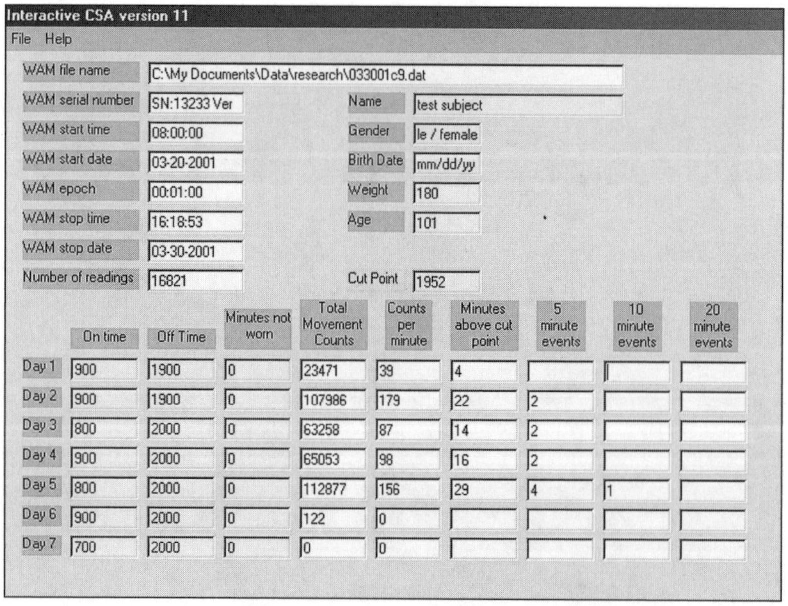

Time Spent Doing Physical Activity

There has recently been considerable interest in assessing physical activity behavior as a primary outcome. The time spent doing physical activity is probably the most relevant outcome for public health research because it provides links to current activity recommendations. The total exercise duration is also a common indicator for epidemiological studies and for intervention research. An advantage of using time-based approaches is that they avoid the various errors associated with estimating energy expenditure and instead provide a measure that is more behaviorally based. Determining time spent in activity requires that the monitors be calibrated to determine the relationship

Interactive CSA version 11

File   Help

WAM file name: C:\My Documents\Data\research\033001c9.dat
WAM serial number: SN:13233 Ver
WAM start time: 08:00:00
WAM start date: 03-20-2001
WAM epoch: 00:01:00
WAM stop time: 16:18:53
WAM stop date: 03-30-2001
Number of readings: 16821

Name: test subject
Gender: le / female
Birth Date: mm/dd/yy
Weight: 180
Age: 101

Cut Point: 1952

	On time	Off Time	Minutes not worn	Total Movement Counts	Counts per minute	Minutes above cut point	5 minute events	10 minute events	20 minute events
Day 1	900	1900	0	23471	39	4			
Day 2	900	1900	0	107986	179	22	2		
Day 3	800	2000	0	63258	87	14	2		
Day 4	900	2000	0	65053	98	16	2		
Day 5	800	2000	0	112877	156	29	4	1	
Day 6	900	2000	0	122	0				
Day 7	700	2000	0	0	0				

Figure 8.4 Screen image of an executable program for processing 7 days of minute-by-minute data from the CSA.

Software developed by R. Paulos and K.F. Janz as part of the Iowa Bone Development Study (University of Iowa). For information contact Kathleen Janz at **kathleen-janz@uiowa.edu**. Software is available from MTI, the current provider of the CSA monitor.

between counts and different intensities of activity. Empirical relationships between activity counts and some metabolic indicator (e.g., METS) are typically used for these calculations, but there have been inconsistencies in the ways in which activity levels have been operationalized. Nichols et al. (1999) categorized light, moderate, and vigorous activity as 2, 4, and 7 METS, whereas Freedson et al. (1998) used categories of light (<3 METS), moderate (3 to 6 METS), hard (6 to 9 METS), and very hard (>9 METS). The differences in these levels make it difficult to compare the results of studies that may have used different monitors. A schematic illustrating the use of cutpoints is provided in figure 8.5. Assuming that the regression equation yields a straight line, the level of counts for the different intensity levels can be extrapolated from the figure or solved mathematically.

While the cutpoint method provides a systematic way to reduce large amounts of data into meaningful units for public health research, there are some major limitations associated with this approach. One limitation is that the use of an experimentally derived cutpoint essentially imposes somewhat artificial categories on data that are continuous in nature. This can lead to some errors, particularly if data are examined for individual participants. Accelerometer output has been shown to be related to various anthropometric measurements, so considerable inter-individual variability in the measured counts is likely for a standard bout of activity. For example, individuals who are taller or who have longer legs will generally have fewer

movement counts for a standardized bout of activity. Thus, the use of cutpoints may systematically bias the results in some studies. If based on a large, representative sample, the cutpoints would likely be adequate for group comparisons, but individual comparisons should be avoided.

Another issue with the use of cutpoints is that the threshold level of counts for the different intensity levels assumes steady-state exercise. This is particularly problematic in studies with children because they typically have very sporadic activity patterns and rarely exercise in a continuous manner. Since the counts for a given minute may be lower than the threshold, failure to adjust or account for this will lead to a systematic underestimation of activity in children. (For additional clarification on this topic, see a recent article by Welk, Corbin et al. [2000] that specifically describes measurement issues for the assessment of physical activity in children.)

Measurement Properties: Validity

A number of studies have investigated the validity of accelerometry-based activity monitors under both laboratory and field conditions. The most common criterion measure for these studies is from metabolic carts or portable gas analysis systems that allow for minute-by-minute values of VO_2, or some indicator of energy expenditure (i.e., kcal/min or METS). A number of studies have also compared the EE estimates from activity monitors with the doubly labeled water technique (Bouten et al., 1996; Ekelund et al., 2001; Fogelholm et al., 1998). While this technique provides a precise estimate of energy expenditure, data cannot be parsed by time to allow for an examination of the temporal relationships in the data. Another criterion measure used in behavioral research is direct observation. This technique allows behavior to be coded at more discrete time intervals than even metabolic data can be processed. Current computer-based observation systems allow for continuous, real-time coding of behavior. Because children have highly intermittent activity patterns, this may be the preferred criterion measure for use with children (Welk, Corbin et al., 2000). While the use of some criterion measure is best for criterion-related or predictive validity (see chapter 3), a number of other studies have examined concurrent validity by making comparisons against other measures of physical activity.

This section provides an overview of some of the validity studies that have been conducted with

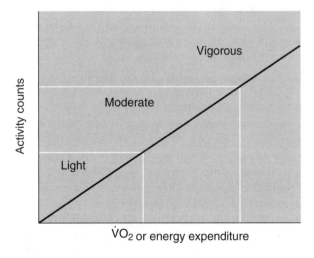

Figure 8.5 Concept behind the use of cutpoints to convert movement counts into time spent in different levels of physical activity.

accelerometry-based devices. Since it is impossible to systematically review every study, those studies that may have particular relevance for field-based research applications are emphasized. Studies that examined the overall utility of accelerometry are first described because these studies helped to establish some of the limitations associated with the use of accelerometers for field studies. Particular attention is given to comparisons of uniaxial and triaxial monitors because that is still an unresolved issue. Calibration studies that used multiple regression techniques to predict energy expenditure are then compared because these equations have been used in a variety of studies to process accelerometry data. The last part of this section reviews studies that have specifically sought to assess various lifestyle activities in the field. These findings are important for evaluating the potential of monitors to assess moderate-intensity activities that are frequently cited in public health guidelines for physical activity.

Criterion and Concurrent Validity Studies

A number of early studies examined the validity of the Caltrac monitor to estimate energy expenditure under laboratory conditions against measured EE from metabolic carts (Balogun et al., 1989). These studies provided important evidence about the utility of accelerometers for assessing physical activity. The general consensus was that the Caltrac was sensitive to changes in speed but overestimated energy expenditure across treadmill speeds (Fehling et al., 1999; Haymes & Byrnes, 1993; Pambianco et al., 1990). Research also demonstrated that the Caltrac and other monitors were insensitive to increases in grade (Fehling et al., 1999; Swan et al., 1997). Because of the general limitations associated with the use of these predictions, most users opted to program the Caltrac to record raw movement counts.

The development of triaxial monitors led some investigators to examine the possibility that the use of three-dimensional measures may improve the ability to assess certain types of activities. A study by Fehling et al. (1999) reported that the Caltrac and Tritrac both underestimated energy expenditure for treadmill and bench stepping exercises. Both monitors were equally insensitive to changes in grade. Jakicic et al. (1999) tested whether the Tritrac could accurately assess cycling and sliding activity and found that the Tritrac underpredicted the energy expenditure of both activities. The consistency of these findings suggests that activity monitors may

have inherent limitations in their ability to assess some forms of movement.

A number of studies examined the relative validity of the Caltrac and Tritrac under different conditions and with different criterion measures. One study reported high correlations (i.e., r = 0.88) between a Caltrac and Tritrac when assessing physical activity in children (Welk & Corbin, 1995). A more recent study found correlations of r = 0.90 between the CSA and Tritrac under free-living conditions (Leenders et al., 2000). The high correlations observed in these studies could suggest that uniaxial and triaxial monitors provide essentially the same information about physical activity. Other studies, however, have reported some potential advantages of triaxial devices. A study with children (Eston et al., 1998) found that the three-dimensional measure of vector magnitude provided a better prediction of energy expenditure than any of the unidimensional outputs from the Tritrac or from the CSA. Results from the prototype of the current Tracmor three-dimensional monitor also reported a higher correlation with energy expenditure for the three-dimensional value than for any of the single plane measurements (Bouten, Koekkoek et al., 1997). Similarly, Coleman et al. (1997) observed higher correlations between vector magnitude and both heart rate and self-report measures than similar comparisons with the vertical component of the Tritrac signal. The potential benefits of three-dimensional monitors require further clarification. While it is true that most movements in the sagittal and frontal planes will also involve movements in the vertical plane, the three-dimensional output may provide a more stable indicator of overall body movement. (See the section on reliability starting on page 137 for further discussion on this issue.)

Laboratory-Based Calibration Studies

The early studies on the Caltrac and Tritrac evaluated the accuracy of the manufacturer-based equations for predicting energy expenditure. Because monitors also provide output in raw movement counts, some investigators have developed and tested new prediction equations for estimating energy expenditure. These studies have used multiple regression techniques to essentially calibrate movement counts in terms of EE equivalents. Detailed calibration studies have been conducted with the CSA, Tritrac, and Biotrainer/Actitrac using similar treadmill-based protocols and the same criterion measure (energy expenditure assessed by metabolic carts). Comparisons of these different studies provide useful information about the relative validity of these instruments.

In a study on the CSA monitor (Freedson et al., 1998), the equation predicting energy expenditure (kcal/min) from activity counts and body mass yielded a regression variance of $R^2 = 0.82$ and a standard error of estimate of SEE = 1.40 kcal/min (n = 35). Cross-validation with a separate sample (n = 15) found no difference between estimated and predicted energy expenditure for all three speeds (SEE = 0.50 to 1.40). Nichols et al. (1999) developed an equation for predicting energy expenditure (kcal/kg/min) from the Tritrac using the same treadmill speeds. The validation results for the equation were quite strong (i.e., $R^2 = 0.90$, SEE = 0.014 kcal/kg/min). Similar results were obtained from a separate sample of 20 participants who repeated the same protocol.

Cross-validation of the Nichols et al. (1999) and Freedson et al. (1998) equations were also tested by Welk, Blair et al. (2000) as part of a study that directly compared the CSA, Tritrac, and Biotrainer under both treadmill and field conditions. This study found no differences in EE estimates for the CSA at all three treadmill speeds. There was a significant difference in estimates at one speed for the Tritrac, but examination of the Bland-Altman plots revealed that the errors for individual estimates were less than those for the CSA. The Biotrainer results were less favorable, but this was attributed to the use of the manufacturer-based formula for these estimations.

A subsequent study conducted with the same treadmill protocol reported high predictive validity for both the Biotrainer and Actitrac monitors (Welk et al., 2002). Correlations between the raw counts from each monitor and the measured metabolic variables (i.e., METS, VO_2, and EE) ranged from r = 0.62 to r = 0.85 for the Biotrainer and from r = 0.79 to r = 0.91 for the Actitrac. The equations predicting energy expenditure from counts yielded strong validation results for both the Biotrainer (i.e., $R^2 = 0.93$, SEE = 1.47) and the Actitrac (i.e., $R^2 = 0.89$, SEE = 1.38). When the equations were applied to the data from a separate cross-validation sample, the correlation between measured and predicted EE was r = 0.93 and r = 0.94 for the Biotrainer and the Actitrac, respectively. Collectively, the results of these studies suggest that energy expenditure can be predicted with reasonable accuracy with all of the monitors under controlled conditions. (See table 8.1 for a compilation of the various prediction equations.)

Field-Based Validation Studies

Because the most common application for physical activity monitors is to assess free-living field conditions, a number of studies have also evaluated activity monitors in the field. There has been considerable interest in the ability of these devices to assess moderate forms of "lifestyle" physical activity, since these are commonly recommended in most public health guidelines. The International Life Sciences Institute (ILSI) recently commissioned three different research teams to specifically investigate the accuracy of activity monitors to assess moderate-intensity, lifestyle-based activities. All three studies utilized portable metabolic units for collecting data on a variety of different activities. Welk, Blair et al. (2000) evaluated the ability of various monitors to estimate the energy costs of three indoor chores (i.e., sweeping, stacking, and vacuuming) and three outdoor chores (i.e., shoveling, mowing grass, and raking leaves). The monitors consistently underestimated the measured EE of the six tasks by 38 to 48% (see figures 8.6 and 8.7). The high correlations among the three different monitors (i.e., r = 0.78) revealed that the three

TABLE 8.1   Comparison of Validation Criteria From Various Treadmill Calibration Studies

Reference	Monitor	Prediction equation
Freedson, Melanson, and Sirard (1998)	CSA	EE (kcal/min) = 0.00094 × counts/min + (0.1346 × kg weight) − 7.37418
Nichols, Morgan, Sarkin, Sallis, and Calfas (1999)	Tritrac	EE (kcal/kg/min) = 2.90512E-05 × vector magnitude + 0.018673
Welk, Morss, Church, Differding (2002)	Biotrainer	EE (kcal/min) = 0.7863 × counts/min + (0.0659 × kg weight) − 3.3377
Welk, Morss, Church, Differding (2002)	Actitrac	EE (kcal/min) = 0.04266 × counts/min + (0.0636 × kg weight) − 3.182

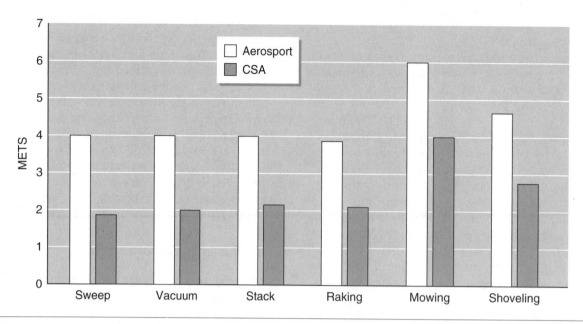

Figure 8.6 Example of the underprediction of measured EE for various lifestyle tasks. The data use the CSA as an example, but all monitors yielded similar results.

Based on data from Welk, Blair et al. (2000).

monitors had similar limitations in their ability to assess these types of tasks. Hendelman et al. (2000) reported a similar degree of underprediction (i.e., –31 to –67%) for six different recreational (e.g., golf) and lifestyle tasks. Bassett et al. (2000) compared three different prediction equations for the CSA (including one derived specifically from lifestyle tasks in the Hendelman et al. [2000] study) on 28 different lifestyle tasks. The correlations between measured and predicted EE were low to moderate for the three different equations (i.e., r = 0.32 to r = 0.62). Similar to the other studies, the equations were generally found to underpredict the measured EE values. The Hendelman et al. (2000) equation (based on lifestyle activities) yielded non-significant differences when averaged across the different activities, but the Bland-Altman plots revealed that this equation significantly overestimated low-intensity activities and underestimated high-intensity activities.

In interpreting these results, it is important to note that these three studies sought to examine the ability to assess specific lifestyle tasks. Point estimates of the energy costs for specific activities are likely to be much less accurate than more general assessments of overall activity levels because there is less variability in the data. It is likely that errors associated with assessing total daily activities would be lower.

These studies collectively demonstrate the inherent challenges associated with measuring physical activity in the field. The relationship between counts and measured EE are clearly dependent on the type and intensity of the activities performed. While this is a potential limitation of accelerometers, it is important to remember that activity monitors are reasonably accurate for general locomotor tasks that constitute the bulk of most people's activity. The overall accuracy of the monitors for field-based research, then, is largely determined by the extent to which daily activities involve locomotor activity, as opposed to upper-body movements. While the use of field-based equations may improve the accuracy with which some activities can be measured, they could lead to other sources of bias and limit the ability of the equations to accurately assess locomotor activity. It is doubtful that any equation can accurately account for all the different movements that take place in daily life. Field-based equations may demonstrate reasonable estimates for group data, but the accuracy of the equations for individual data may be limited. Research should aim to identify the limits associated with the use of different laboratory- and field-based equations to determine the best approaches for field-based research.

Measurement Properties: Reliability

While considerable effort has been spent on evaluating the validity of monitors, less attention has been placed on determining the reliability. One issue with reliability is the consistency with which data can be collected from participants and downloaded. Early studies with the Caltrac reported failure rates of as high as 10%, but rates have been found to be much lower with most of the newer devices (i.e., 2 to 3% of the trials). The use of solid-state technology in the manufacturing of these devices would likely improve their overall reliability and utility, but to date, only the Tritrac uses these types of components.

Comparisons of inter-instrument reliability are also important because small differences in the sensitivity or positioning of the sensors may influence the output. Several studies have tested the inter-unit reliability of the Tritrac with a mechanical shaker table. Kochersberger et al. (1996) compared nine different Tritrac units and reported intra-class correlation coefficients of R = 0.97. In a test of four different Tritrac units, Nichols et al. (1999) reported no significant differences in output and a small coefficient of variation (SD/mean) of 1.79% across the monitors. While these studies reveal reasonable reliability, fewer studies have examined the reliability of monitors under less-constrained conditions. Results from shaker table comparisons have also not been reported for most of the other monitors.

Some researchers have reported intra-class correlation coefficients between two units worn on opposite hips. Jakicic et al. (1999) reported reliability coefficients ranging from R = 0.44 to R = 0.92 for two Tritrac monitors across a variety of activities. The reliability was higher for walking and running (i.e., R = 0.76 to 0.92) than for the stepping, sliding, or cycling tasks (i.e., R = 0.54 to 0.88). In t-tests, significant differences were revealed between the two units for walking, stepping, and cycling activities. The Nichols et al. (1999) study on the Tritrac reported intra-class correlations ranging from R= 0.73 to R = 0.87 for matched left and right monitors. While most results reveal reasonable correspondence, basing estimates of reliability on the findings from two selected monitors may not be appropriate, since it is possible that the results could vary if other units were used. Similar to the shaker table results, it was not possible to identify studies evaluating the reliability of other commonly used monitors.

In a recent pilot study (Welk, 2002), the reliability of the CSA monitor was systematically examined by having participants wear 10 new units (obtained directly from the manufacturer) on three different occasions while walking on a motorized treadmill (3 mph). The coefficient of variation across monitors was approximately 10 to 15% for the vari-

ous instruments tested. Interestingly, we observed similar amounts of variability when the output from the same monitor was compared across three different trials. This suggests that the variability we observed may be due more to subtle differences in positioning between the devices than to variability among the monitors themselves.

Technical studies of the effects of monitor positioning have indicated that gravitational acceleration may vary depending on the relative position of the monitors with respect to the gravitational field (Bouten, Sauren et al., 1997). Thus, if the monitors are positioned or angled in slightly different ways between participants and across trials, the error in measurement may be greater than the inter-unit differences in output. The high reliability observed with the Tritrac in the two mechanical shaker studies may be due to the three-dimensional nature of the assessment. By using a composite three-dimensional measure of movement (vector magnitude), the Tritrac may be less influenced by slight differences in the orientation of the monitor. Output from uniaxial devices such as the CSA and Biotrainer, in contrast, may be more susceptible to gravitational influences because the relative positioning more directly alters the magnitude of the acceleration in the one dimension. This possible difference in reliability of the monitors deserves further study.

Another measurement issue associated with the reliability of activity monitors is the stability of calibration. All monitors are typically tested at the factory to ensure some measure of quality control. However, little is known about how frequently calibration must be checked to ensure consistent responses over time. A specific calibration tool is available for use with the CSA monitor, and calibration procedures are also available from IM Systems for the Biotrainer and the Actitrac. Unfortunately, there is no mention about the need for calibration in the manuals from the manufacturers and few, if any, references about calibration in the scientific literature. Many anecdotal observations from various researchers around the country indicate that monitors do go out of calibration and that failure to correct for this can lead to spurious data. This type of error is troubling in cross-sectional studies taking place over an extended period of time, and especially problematic for longitudinal interventions or tracking studies. Because reliability is an essential prerequisite for validity, researchers are encouraged to use appropriate quality control measures to ensure that the devices are calibrated prior to use and positioned in a consistent way across trials.

Acceptability and Utility of Accelerometers

One of the major advantages of accelerometers is that they are easy to use for field research. Like pedometers, accelerometry-based activity monitors are easy for participants to wear and do not interfere with normal movement patterns. An issue with activity monitoring research is whether the act of wearing a monitor leads to changes in normal activity patterns. While this is still not resolved, the absence of direct feedback to the participant for both the CSA and Tritrac may help to minimize this effect. Like the original Caltrac, the Biotrainer has an LCD display, but the investigator can blank this out during the initialization phase.

The raw accelerometry data from the various monitors is provided as a dimensionless unit typically referred to as *counts*. A challenge in this line of research is that the value of counts is specific for each monitor. Unlike heart rate monitors and pedometers, different monitors produce output that is measured in different units. This has made it difficult to compare the output from different monitors or the results from different studies. An advance for this area of research would be a systematic comparison of output from each monitor to some standard unit such as g forces. This would allow for more direct comparisons across studies employing different monitors. The literature from several manufacturers (CSA and Biotrainer) makes reference to g forces in their information on calibration, although it is not clear how the internal processing influences the integrated outputs reported as counts.

Applications for Activity Monitors

Activity monitors can provide an objective record of physical activity patterns over an extended period of time. Because of the cost and time involved in processing the data, they are probably most effective for studies requiring detailed information on physical activity. The most commonly used applications for using activity monitors are described in the following sections.

Validation of Other Self-Report Instruments

Because activity monitors provide an objective measure of movement, they have often been used as a criterion measure in studies validating other physical activity instruments, particularly self-report instruments (Going et al., 1999; Matthews et al., 2000; Matthews & Freedson, 1995; McMurray et al., 1998; Trost et al., 1999; Welk et al., 2001). Because of the inherent limitations, some investigators have not specified the activity monitor as a criterion measure and instead have just examined convergent validity of the two assessments. Although activity monitors are certainly not a perfect criterion, the ability to provide detailed, temporal information about activity over extended periods makes them perhaps the most suitable instrument available for this purpose.

A major advantage of activity monitor data is the flexibility with which the data can be processed. If the instrument produces an index of total activity, raw counts can be used to reflect overall amounts of activity. If the goal of the self-report instrument is to estimate energy expenditure, this can be estimated by one of the available calibration equations. Alternately, if the self-report instrument reports minutes of moderate or vigorous activity, the accelerometer data can be processed with cutpoints to estimate time spent in activity. By processing the data from activity monitors in the same manner as the self-report instrument, it is possible to examine the validity of the instrument in considerable detail.

An example of how activity monitors can be used for validation purposes can be found in a study by Welk et al. (2001). In this study, participants wore a Tritrac activity monitor for the week prior to completing a 7-day physical activity recall (7DPAR). The data from the monitor were temporally matched to the self-reported exercise levels across all 7 days to examine the convergent validity of these assessments. To facilitate these comparisons, the activity monitor data were processed in the same way as the MET-based algorithms used in scoring the 7DPAR. This allowed the minutes of activity reported on the two devices to be directly compared. Correlations between the total activity scores from both instruments were compared across all 7 days to examine the accuracy of recall over the entire time span. The correlations were found to be consistently high across all 7 days of monitoring. This study is described mainly to highlight how temporal processing of activity monitoring data can be used to systematically evaluate the accuracy of self-report instruments. If only total counts or EE over the week were used in the comparisons, the comparisons wouldn't be as effective.

Correlates and Behavioral Research

A number of studies have begun using activity monitors as the dependent variable in studies designed to predict levels of physical activity. Current measures have been unable to predict much more than 35% of the variance in activity behavior, but this may be because of limitations in the accuracy of the physical activity assessment. A challenge with using accelerometry data is that data collected over several days may not be representative of what a person usually does (Baranowski & de Moor, 2000; Gretebeck & Montoye, 1992). As previously described, care must also be used in the collection, processing, and interpretation of the data to get accurate results. Failure to follow consistent protocols for wearing the device or failure to screen for noncompliance can lead to systematic bias in the results. Because correlational studies are highly influenced by the presence of outliers, these sources of bias can significantly impact the results.

Investigations have reported different relationships when the correlates of activity are measured with either self-report or objectively recorded activity from a monitor (Sallis et al., 1998). Because this line of research is still relatively new, it is not clear what to make of these distinctions. Either the two activity assessments assess different types of activity patterns or the psychosocial mediators have differential effects on the two types of indicators, depending on one's perspective. In either case, the differences complicate our view of the correlates of physical activity. The continued use of triangulation techniques for improving our understanding of correlates of activity is promising.

Intervention Research

Because of their objectivity and ease of use, accelerometers are increasingly being considered for use in intervention studies designed to promote physical activity. There are a number of unique factors that must be considered when using a monitor for this type of study. The measure must obviously be reliable and valid, but it must also be sensitive enough to detect differences in activity levels between participants and/or changes over time.

In the Trial for Activity in Adolescent Girls (TAAG), a multisite, randomized, controlled physical activity intervention funded by the National Institutes of Health, the CSA monitor is being used to collect data for the primary outcome measure. Ancillary studies were conducted to determine how many days of monitoring are needed to collect representative data on physical activity in children. While previous reports have provided guidelines for generalizability (Baranowski & de Moor, 2000) it was deemed important to examine the stability of activity patterns within the specific target population for the study (i.e., 10- to 12-year-old girls). The variances in scores across schools were also critical for determining the sample sizes required in the study so that enough power was available to detect a meaningful change in activity levels.

Another ancillary study compared the CSA output for various physical activities against measured oxygen consumption and EE (from a portable metabolic system). Here the goal was to determine appropriate cutpoints that would represent levels of moderate and vigorous activity in the study. Special consideration was given to selecting activities that middle school girls would likely participate in so that the relationships were appropriate for the specific application. The eventual publication of these studies will add to the literature on applications of activity monitors for intervention research.

Summary

Accelerometry-based activity monitors offer considerable promise for improving current physical activity assessment techniques. They provide an objective indicator of overall movement that allows the various parameters associated with activity (e.g., frequency, intensity, duration) to be examined in detail. Contemporary units are small enough to wear for several days and durable enough to tolerate the bumping and jostling that may occur during physical activity. These characteristics make accelerometry-based devices very promising for field-based research on physical activity. Most of the research on the application of these devices for research purposes has been compiled in the past 10 years. The progression of studies has clearly increased our understanding of the validity, reliability, and overall utility of these monitors. The increasing sophistication of the designs used in these studies has also helped to improve the techniques associated with collecting, processing, and interpreting this type of data.

While considerable progress has been made, the existing studies also point to new issues that need to be resolved to continue to advance this research. Work is needed to better understand the influence of body size on the output from accelerometers. Work is also needed to address differences in the approaches used to analyze and interpret data from children and adults. In addition to the size differences between children and adults, differences in the types and intensities of activities performed by children necessitate the development of separate calibration equations and validation studies for children. Laboratory research is also needed to examine the mechanical properties of these devices in more detail.

Future studies should seek to equate the output from the different monitors to a common metric of g forces so that all monitors can be more readily compared. Studies are also needed to test the reliability of the different monitors under controlled and field conditions. This type of research will increase our understanding of the impact of various sources of error in monitoring studies and help to further refine measurement protocols. As the results of future studies become integrated into the collective body of literature, the application of activity monitors for physical activity research should continue to advance.

References

Balogun, J.A., Martin, D.A., & Clendenin, M.A. (1989). Calorimetric validation of the Caltrac accelerometer during level walking. *Physical Therapy, 69,* 501-509.

Baranowski, T., & de Moor, C. (2000). How many days was that? Intra-individual variability and physical activity assessment. *Research Quarterly for Exercise and Sport, 71,* S74-S78.

Bassett, D.R., Jr., Ainsworth, B.E., Swartz, A.M., Strath, S.J., O'Brein, K.O., & King, G.A. (2000). Validity of four motion sensors in measuring moderate intensity physical activity. *Medicine and Science in Sports and Exercise, 32,* S471-S480.

Bouten, C.V.C., Koekkoek, K.T.M., Verduin, M., Kodde, R., & Janssen, J.D. (1997). A triaxial accelerometer and portable data processing unit for the assessment of daily physical activity. *IEEE Transactions on Biomedical Engineering, 44,* 136-147.

Bouten, C.V.C., Sauren, A.A.H.J., Verduin, M., & Janssen, J.D. (1997). Effects of placement and orientation of body-fixed accelerometers on the assessment of energy expenditure during walking. *Medical and Biological Engineering and Computing, 35,* 50-56.

Bouten, C.V.C., Verboeket-Van de Veene, W.P.H.G., Westerterp, K.R., Verduin, M., & Janssen, J.D. (1996). Daily physical activity assessment: Comparison between movement registration and doubly labeled water. *Journal of Applied Physiology, 81,* 1019-1026.

Bouten, C.V.C., Westerterp, K.R., Verduin, M., & Janssen, J.D. (1994). Assessment of energy expenditure for physical activity using a triaxial accelerometer. *Medicine and Science in Sports and Exercise, 26,* 1516-1523.

Coleman, K.J., Saelens, B.E., Wiedrich-Smith, M.D., Finn, J.D., & Epstein, L.H. (1997). Relationships between Tritrac-R3D vectors, heart rate, and self-report in obese children. *Medicine and Science in Sports and Exercise, 29,* 1535-1542.

Ekelund, U., Sjostrom, M., Yngve, A., Poortvliet, E., Nilsson, A., Froberg, K., Wedderkopp, N., & Westerterp, K. (2001). Physical activity assessed by activity monitors and doubly labeled water in children. *Medicine and Science in Sports and Exercise, 33,* 275-281.

Eston, R.G., Rowlands, A.V., & Ingledew, D.K. (1998). Validity of heart rate, pedometry, and accelerometry for predicting the energy cost of children's activities. *Journal of Applied Physiology, 84,* 362-371.

Fehling, P.C., Smith, D.L., Warner, S.E., & Dalsky, G.P. (1999). Comparison of accelerometers with oxygen consumption in older adults during exercise. *Medicine and Science in Sports and Exercise, 31,* 171-175.

Fogelholm, M., Hilloskorpi, H., Laukkanen, P.O., Lichtenbelt, W.V.M., & Westerterp, K. (1998). Assessment of energy expenditure in overweight women. *Medicine and Science in Sports and Exercise, 30,* 1191-1197.

Freedson, P.S., Melanson, E., & Sirard, J. (1998). Calibration of the Computer Science and Applications (CSA), Inc. accelerometer. *Medicine and Science in Sports and Exercise, 30,* 777-781.

Going, S.B., Levin, S., Harrell, J., Stewart, D., Kushi, L., Cornell, C.E., Hunsberger, S., Corbin, C., & Sallis, J. (1999). Physical activity assessment in American Indian schoolchildren in the Pathways study. *American Journal of Clinical Nutrition, 69,* S788-S795.

Gretebeck, R.J., & Montoye, H.J., (1992). Variability of some objective measures of physical activity. *Medicine and Science in Sports and Exercise, 24,* 1167-1172.

Haymes, E.M., & Byrnes, W.C., (1993). Walking and running energy expenditure estimated by Caltrac and indirect calorimetry. *Medicine and Science in Sports and Exercise, 25,* 1365-1369.

Hendelman, D., Miller, K., Bagget, C., Debold, E., & Freedson P.S. (2000). Validity of accelerometry for the assessment of moderate intensity physical activity in the field. *Medicine and Science in Sports and Exercise, 32,* S442-S449.

Jakicic, J.M., Winters, C., Lagally, K., Ho, J., Robertson, R.J., & Wing, R.R. (1999). The accuracy of the Tritrac-R3D accelerometer to estimate energy expenditure. *Medicine and Science in Sports and Exercise, 31,* 747-754.

Janz, K.F., Witt, J., & Mahoney, L.T. (1995). The stability of children's physical activity as measured by accelerometry and self-report. *Medicine and Science in Sports and Exercise, 27,* 1326-1332.

Jones, S.L., Wood, K., Thompson, R., & Welk, G.J. (1999). Effect of monitor placement on output from three different accelerometers. *Medicine and Science in Sports and Exercise, 31,* S142 [Abstract].

Kochersberger, G., McConnell, E., Kuchibhatla, M.N., & Pieper, C. (1996). The reliability, validity, and stability of a measure of physical activity in the elderly. *Archives of Physical Medicine and Rehabilitation, 77,* 793-795.

LaPorte, R.E., Cauley, J.A., Kinsey, C.M., Corbett, W., Robertson, R., Black-Sandler, R., Kuller, L.H., & Falkel, J. (1982). The epidemiology of physical activity in children, college students, middle-aged men, menopausal females and monkeys. *Journal of Chronic Disease, 35,* 787-795.

LaPorte, R.E., Kuller, L.H., Kupfer, D.J., McPartland, R., Matthews, G., & Caspersen, C. (1979). An objective measure of physical activity for epidemiologic research. *American Journal of Epidemiology, 109,* 158-168.

Leenders, N.Y.J.M., Sherman, W.M., & Nagaraja, H.N. (2000). Comparison of four methods of estimating physical activity in adult women. *Medicine and Science in Sports and Exercise, 32,* 1320-1326.

Matthews, C.E. & Freedson, P.S. (1995). Field trial of a three-dimensional activity monitor: Comparison with self report. *Medicine and Science in Sports and Exercise, 27,* 1071-1078.

Matthews, C.E., Freedson P.S., Hebert, J.R., Stanek, E.J., III, Merriam, P.A., & Ockene, I.S. (2000). Comparing physical activity assessment methods in the Seasonal Variation of Blood Cholesterol Study. *Medicine and Science in Sports and Exercise, 32*, 976-984.

McMurray, R.G., Harrell, J.S., Bradley, C.B., Webb, J.P., & Goodman, E.M. (1998). Comparison of a computerized physical activity recall with a triaxial motion sensor in middle-school youth. *Medicine and Science in Sports and Exercise, 30*, 1238-1245.

Montoye, H.J., Kemper, H.C., Saris, W.H.M., & Washburn, R.A. (1996). *Measuring physical activity and energy expenditure.* Champaign, IL: Human Kinetics.

Montoye, H.J., Washburn, R.A., Servais, S., Ertl, A., Webster, J.G., & Nagle, F.J. (1983). Estimation of energy expenditure by a portable accelerometer. *Medicine and Science in Sports and Exercise, 15*, 403-407.

Nichols, J.F., Morgan, C.G., Sarkin, J.A., Sallis, J.F., & Calfas, K.J. (1999). Validity, reliability, and calibration of the Tritrac accelerometer as a measure of physical activity. *Medicine and Science in Sports and Exercise, 31*, 908-912.

Pambianco, G., Wing, R.R., & Robertson, R. (1990). Accuracy and reliability of the Caltrac accelerometer for estimating energy expenditure. *Medicine and Science in Sports and Exercise, 22*, 858-862.

Redmond, D.P., & Hegge, F.W. (1985). Observations on the design and specification of a wrist-worn human activity monitoring system. *Behavior Research Methods, Instruments, & Computers, 17*, 659-669.

Sallis, J.F., McKenzie, T.L., Elder, J.P., Hoy, P.L., Galati, T., Berry, C.C., Zive, M.M., & Nader, P.R. (1998). Sex and ethnic differences in children's physical activity: Discrepancies between self-report and objective measures. *Pediatric Exercise Science, 10*, 277-284.

Swan, P.D., Byrnes, W.C., & Haymes, E.M. (1997). Energy expenditure estimates of the Caltrac accelerometer for running, race walking, and stepping. *British Journal of Sports Medicine, 31*, 235-239.

Swartz, A.M., Strath, S.J., Bassett, D.R., Jr., O'Brien, W.L., King, G.A., & Ainsworth, B.E. (2000). Estimation of energy expenditure using CSA accelerometers at hip and wrist sites. *Medicine and Science in Sports and Exercise, 32*, S450-S456.

Trost, S.G., Ward, D.S., McGraw, B., & Pate, R.R. (1999). Validity of the Previous Day Physical Activity Recall for fifth-grade children. *Pediatric Exercise Science, 11*, 341-348.

Welk, G.J. (2002). Reliability of the CSA activity monitor for assessing physical activity. *Research Quarterly for Exercise and Sport, 73*, A14.

Welk, G.J., Blair, S.N., Wood, K., Jones, S., & Thompson, K.W. (2000). A comparative evaluation of three accelerometry-based physical activity monitors. *Medicine and Science in Sports and Exercise, 32*, S489-S497.

Welk, G.J., & Corbin, C.B. (1995). The validity of the Tritrac-R3D activity monitor for the assessment of physical activity in children. *Research Quarterly for Exercise and Sport, 66*, 202-209.

Welk, G.J., Corbin, C.B., & Dale, D. (2000). Measurement issues for the assessment of physical activity in children. *Research Quarterly for Exercise and Sport, 71*, 59-73.

Welk, G.J., Morss, G., Church, T., & Differding, J. (2002). Laboratory calibration of the Biotrainer and Actitrac activity monitors. *Medicine and Science in Sports and Exercise, 34*, S140.

Welk, G.J., Thompson, R.W., Galper, D.I. (2001). A temporal validation of scoring algorithms for the 7-Day Physical Activity Recall. *Measurement in Exercise Science and Physical Education, 5*, 123-128.

9

Use of Heart Rate Monitors to Assess Physical Activity

Kathleen F. Janz, EdD

University of Iowa

Investigators have been interested in assessing physical activity using a physiological variable that is easy to measure since at least the early 1900s, when Benedict proposed the use of heart rate as an indirect indicator of energy expenditure (Torun, 1984). This concept was refined in the 1950s when Berggren and Christensen (1950) recognized that the acute response of heart rate to increasing levels of physical activity is directly proportional to the intensity of movement and driven by an increasing need to deliver oxygen to skeletal muscles. However, the practical use of the heart rate method to assess physical activity and energy expenditure in epidemiology only became possible in the 1970s with the development of portable heart rate monitors.

Early portable monitors either (1) accumulated heart rate over time providing an average heart rate or (2) accumulated heart rate over time and stored values into distribution ranges (Andrews, 1971; Saris et al., 1977). These instruments allowed for the assessment of day-to-day variation among and within subjects and for the rank ordering of physical activity levels within groups. However, earlier monitors precluded precise estimates of physical activity and energy expenditure (Andrews, 1971; Montoye et al., 1996; Saris et al., 1977). Advances in microtechnologies during the 1980s and 1990s have provided what is now the standard in ambulatory heart rate monitoring: lightweight, continuous (minute-by-minute) heart rate recorders with extended data storage capacities. Continuous recorders have improved the precision of activity and energy expenditure estimates, and have expanded the use of the heart rate method in epidemiology to include the study of the temporal patterns of physical activity: intensity, duration, and frequency (Armstrong et al., 1990; Armstrong et al., 2000; Kelly, 2000; Vuori, 1998). Recent advances in electrocardiogram (ECG) signal transmission, recording functions, programming, and software features make heart rate monitors versatile and unobtrusive instruments for use in epidemiology as well as in athletics, physical education, and psychophysiology (Berntson et al., 1997; O'Toole et al., 1998; Steptoe, 2000; Stratton, 1996).

Principles of Method: Physiology of Heart Rate

Heart rate measures an individual's response to physical activity by providing an indication of the relative stress placed on the cardiorespiratory system during movement (Armstrong, 1998). Because aerobic physical activity is dependent on oxygen, heart rate monitoring can also be used to estimate the metabolic cost (energy expenditure) of physical activity (Montoye et al., 1996). Physical activity and energy expenditure are interrelated; they are not synonymous. Energy expenditure is an outcome of physical activity since movement produced by skeletal muscle contractions results in energy expenditure (Caspersen et al., 1985).

Heart rate increases during physical activity are controlled by the sinoatrial node (the heart's pacemaker) in response to decreased parasympathetic neural stimulation, increased sympathetic neural stimulation, and increased circulating catecholamine stimulation—a phenomenon termed *cardiac chronotropic regulation* (Robergs, 1996). An increase in heart rate contributes to increasing cardiac output, thus facilitating delivery of oxygen to contracting skeletal muscles (Robergs, 1996). Throughout a large range of physical activity, moderate through vigorous, heart rate (HR) increases linearly and proportionately with the intensity of movement and the volume of oxygen consumed (VO_2) by contracting skeletal muscles (Freedson & Miller, 2000; Robergs, 1996). During sedentary and light activity, heart rate is poorly correlated with VO_2 and strongly influenced by such factors as emotional stress, digestion, nicotine, altitude, and temperature (Freedson & Miller, 2000; Spurr et al., 1988). These factors elevate resting heart rates independently of changes in VO_2 (Van den Berg-Emons et al., 1996). This nonlinearity of $HR-VO_2$ during sedentary and light activity creates some measurement challenges.

There are a variety of other factors that influence the $HR-VO_2$ relationship during physical activity. One major influence is the skeletal muscle mass involved in the movement. The slope of the $HR-VO_2$ association line primarily depends on the skeletal muscle mass being used and on the type of activity. Both lower-body dynamic activity (e.g., running) and upper-body dynamic activity (e.g., rowing) result in nearly linear associations between heart rate and VO_2. However, at any given submaximal VO_2, heart rate is approximately 10% higher for upper-body exercise (compared to lower-body exercise) because of reductions in stroke volume and arterial-venous oxygen difference during upper-body movement (Nieman, 1999). This means that one subject running and another rowing at the same activity level (the same submaximal VO_2) would have different heart rates. If not considered in the construction of heart rate monitoring outcome variables, this difference would result in a

prediction error for physical activity for one of the subjects.

A second factor influencing the $HR-VO_2$ relationship is the type of activity performed. At the same submaximal VO_2, heart rates are higher during static exercise such as weightlifting than during dynamic exercise. This is because during static exercise, heart rate is proportional to the active muscle mass and the percentage of maximum voluntary contraction rather than oxygen consumption (Nieman, 1999). During static exercise, increased vascular resistance reduces stroke volume, so elevated heart rates also reflect the body's efforts to maintain cardiac output (Hurley et al., 1984). This difference is a potential source of error, particularly during field monitoring when it is not possible to determine what activities a person was performing.

Physical fitness is another confounding factor that causes a shift of the $HR-VO_2$ slope. At any given submaximal VO_2, heart rate will be lower for a physically fit individual when compared to an unfit person because stroke volume is higher for the fit individual (Nieman, 1999; Rowlands et al., 1997). Heart rate also lags behind physical activity changes and remains elevated after activity has ceased (Nieman, 1999). In some situations, this lack of synchronicity can create data interpretation problems and systematic errors. Heart rate methods that do not consider inter-subject fitness levels tend to underpredict physical activity for more physically fit individuals. For example, Saris et al. (1980) have observed that physically fit children who were also physically active had lower mean daily heart rates than less-fit children with low activity levels. The authors attributed this apparent discrepancy to their heart rate methodology that did not account for lower heart rates for the more-fit children compared to the less-fit children during similar activity levels.

In summary, specific muscle mass utilization, type of activity, physical fitness levels, and other exercise-related factors can confound the $HR-VO_2$ relationship and create sources of error when using heart rate as an index of physical activity and energy expenditure. Since most people, including children, spend a large proportion of their time in sedentary activity, the nonlinearity of heart rate and oxygen consumption during resting and light activity represents the greatest challenge in using heart rate to measure physical activity and energy expenditure.

A variety of approaches have been developed to overcome some of these potential sources of bias. Detailed descriptions of these approaches are included in subsequent sections of this chapter.

Instrumentation

Almost all of the continuous heart rate monitors use the electrocardiogram (ECG) signal to detect the beat-to-beat heart rate via an ECG transmitter attached to the chest (Leger & Thivierge, 1988). This transmitter consists of circuitry packaged in a small, lightweight unit (approximately 40 g) usually attached using an elastic strap that encircles the chest, a specially designed sport bra, or disposable ECG electrodes (see figure 9.1). Wireless (radio) signals are sent to a microprocessor receiver, also small and lightweight (approximately 45 g), which is designed to resemble a wristwatch. Most receivers must be 1 m or less from the transmitter to pick up this signal (Karvonen et al., 1984; Leger & Thivierge, 1988). Lithium batteries power both units, and battery life is typically three to five years for the transmitter and one to two years for the receiver. The newest Polar heart rate monitors use a sealed transmitter and belt assembly. This means that the entire assembly must be replaced rather than changing a lithium battery. The estimated life for the new Polar transmitters is three to five years or approximately 2,500 hours.

When the ECG signals are transmitted to the receiver, a timing circuit measures the interval between heartbeats. The receiver then processes the values by calculating a moving heart rate average for short periods of time (usually 5 to 15 s), which is updated and displayed as beats per minute (b/min) (Karvonen et al., 1984). Because the number of heart rate values determines the averaging time, the calculation period is shortened during vigorous physical activity and the averaged heart rate value nearly approximates real-time heart rate (Karvonen et al., 1984; Montoye et al., 1996).

Average heart rate is stored in the receiver based on a preprogrammed period or epoch determined by the investigator. The amount of data that can be stored depends on the memory capability of the receiver as well as the length of interval that is selected. At a 60-second epoch, most units can store up to 3 or 4 days. At a 5-second epoch, the maximum capacity might be around 10 to 12 hours. Advanced monitors can even store data on beat-to-beat variability in heart rate, although this option is not relevant for most applications. Because of challenges associated with processing very small units of data for extended time periods, the most common storage epoch for field monitoring of physical activity is 1 minute. Most contemporary monitors use a telemetry or infrared downloading unit to allow the heart rate data to be transferred to a personal computer (see figure 9.2).

Figure 9.1 Examples of chest-worn transmitters for heart rate monitors.
Photos courtesy of Polar.

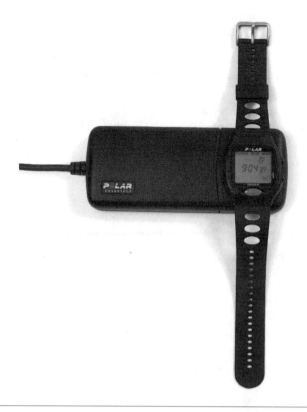

Figure 9.2   Example of equipment required for downloading heart rate monitors.

There are several commercially available heart rate monitors providing options with respect to cost, durability, resistance to water, memory, and additional functions such as alarms and stopwatches. With certain exceptions, most of the additional functions are designed for athletes. Additional functions increase the complexity of programming the monitors for investigators; they also increase the opportunities for subjects to deprogram the monitors. Monitor costs vary, although most range from $200 to $500. The interface, which can be used with multiple monitors, is an additional $200 to $300. Software from manufacturers is available for most monitors, including graphic programs for plotting heart rate against time and sorting heart rate into distribution ranges.

Regardless of the model, heart rate monitors share the common problem of electrical or magnetic interference from common household devices such as televisions, computers, microwaves, and motorized exercise equipment (Montoye et al., 1996). This interference causes erratic readings and lost data. Gretebeck et al. (1991) reported that 1.9 to 2.4% of heart rates are lost because of electrical interference. Keeping heart rate units approximately 1 m away from the electrical circuitry of household devices can minimize interference, although the practicality of this advice with free-ranging subjects is ques-

tionable. Another approach to reducing interference is to position the heart rate receiver close to the transmitter, therefore strengthening the transmitter signal relative to the strength of the interference signal. Since all heart rate monitors tend to share the same radio frequency, collecting simultaneous data on multiple subjects in a small area is also a problem and can result in the "trading" of heart rate data when subjects are in close proximity to each other (e.g., during physical education class). To address this issue, Polar has constructed heart rate monitor models with coded transmission systems (Laukkanen & Virtanen, 1998). This technological advance creates a directed signal from the transmitter to its receiver, therefore preventing the exchange of heart rate data among subjects and reducing interference from electrical devices.

There has been a considerable amount of laboratory research examining the accuracy or precision of wireless heart rate monitors compared to the criterion of ECG recordings (Bar-Or et al., 1996; Bassett, 2000; Karvonen et al., 1984; Kinnunen & Heikkila, 1998; Laukkanen & Virtanen, 1998; Leger & Thivierge, 1988; Treiber et al., 1989). This research has included a variety of exercise modes (e.g., treadmill, bicycle, stair stepping), intensities (e.g., resting, moderate, vigorous), and patterns (e.g., steady state, intermittent). The research has been nearly uniform in supporting good to excellent agreement between the more popular heart rate monitor brands and ECG recordings. This agreement indicates that heart rate monitors are accurate instruments for ambulatory heart rate monitoring. In adults, for example, Karvonen et al. (1984) found Polar monitors to differ from ECG recordings at most by 5 b/min. Leger and Thivierge (1988) reported correlation coefficients (r) between heart rate monitors and ECG recordings ranging from r = 0.93 to 0.98 and Syx = 3.7 to 6.8% for Polar, Exersentry, Pacer, and Monark monitors. These investigators also examined the reliability of monitors by calculating the percentage of doubtful cases and unrealistic heart rate values as compared to 10 s ECG recordings. The more popular monitors had zero to one doubtful cases and 0 to 6% unrealistic values. In the latter study, associations were considerably lower, standard errors were higher, and the percentage of doubtful cases was greater for other heart rate monitor brands, thus suggesting variability in accuracy across manufacturers.

In elementary school children, Trieber et al. (1989) reported strong associations (i.e., r = 0.94 to 0.99) and low standard errors (i.e., SEM = 1.1 to 3.7 b/min) between the Polar monitor and ECG recordings. Again using the Polar monitor, Bar-Or et al. (1996) reported similar associations (i.e., r = 0.86 to 0.99) in preschool children and 0% data loss because of malfunctioning. In this study, the authors noted some exercise and recovery differences in monitor accuracy. Specifically, they found lower heart rate values for exercise and higher heart rate values during recovery for the monitor as compared to ECG recordings. The investigators suggested the discrepancy was a result of lag time between the averaging of heart rate values by the monitor and its display of these values. Bassett (2000) reported a validity correlation coefficient of 1.0 and SEE = 0.6 b/min for the new Polar Vantage NV, which records beat-to-beat intervals. The "perfect" correlation demonstrated by Bassett (2000) provides some support for the Bar-Or et al. (1996) observations that lag time is a source of error.

Assessment Procedures

There is a relatively small number of heart rate monitor brands and models available for research. Most monitor models are similar to each other and require similar preparation and placement procedures. For these reasons, most investigators using heart rate monitoring to measure physical activity use similar approaches for data collection. However, differences in data processing and the construction of outcome variables vary widely among researchers. To improve data comparison across studies, researchers should standardize their approaches for data processing and construction of outcome variables.

Data Collection

The typical sequence of field procedures for heart rate monitoring involves

1. measuring the resting heart rate,
2. preparing the skin for the transmitter,
3. positioning and securing the transmitter and the receiver,
4. programming the receiver, and,
5. after data collection, downloading the receiver.

To ensure effective transmission, the ECG transmitter should be placed directly below the chest muscles or breast tissue. Older heart rate monitor models (e.g., CIC Heartwatch, AMF Quantum XL, Polar Vantage XL) have transmitters that can be attached using disposable, pre-jelled electrode patches or with an elastic strap that includes embedded electrodes. Generally, disposable electrodes improve subject comfort and are preferred over the elastic chest strap. However, the electrodes should be specifically designed for exercise situations and must be able to stick to the chest despite accumulation of sweat and dirt. Cleaning the chest with a gauze pad soaked with bathing alcohol or acetone and then gently abrading the skin with fine sandpaper or an emery board improves the ECG signal by lowering skin resistance. During this process, care should be taken, particularly with children, to minimize irritation and potential infection. In some individuals, removal of chest hair may be required to secure a strong ECG signal and reduce interference. Placing the electrodes directly over bone and avoiding muscle also minimizes interference (Montoye et al., 1996). Surgical tape, transparent tape, and plasters have been used to further secure the electrodes and transmitter to the chest (Eston et al., 1998; Van den Berg-Emons et al., 1996). Nonallergenic adhesives and tapes that allow for aeration are preferred.

Newer heart rate monitor models integrate the electrodes and the transmitter in a thin (i.e., 0.3 to 1 cm), flexible, and relatively long ($\cong$ 32 cm) plastic unit, which must be attached to the chest with an elastic strap (disposable electrodes cannot be used). The thinness of the transmitter unit improves subject comfort and minimizes the bending and cracking of wires connecting the transmitter to the electrodes (Laukkanen & Virtanen, 1998). Damage to wires is a common source of malfunctioning in older model heart rate monitors. However, the newer transmitter design does not fit all subjects, particularly individuals with small ribcages, including children. The increased transmitter-to-skin surface area can also increase problems with chafing during long periods of use. General procedures for skin preparation, including the use of an electrolyte gel mixture, still apply with this newer design because all heart rate monitors require moisture to conduct the signal between the skin and electrodes. Signal transmission problems common to both heart rate monitor styles include lost contact due to dry surface area beneath the transmitter and weakened signal due to aging batteries.

Unlike second-generation accelerometers, most heart rate monitors do not have a "black box" design to make them uninteresting to subjects. Data can easily be lost if investigators do not take precautions to minimize tampering and accidental reprogramming of the receiver (Livingstone et al., 2000). In addition, the continuous display of the heart rate increases the likelihood of subject awareness of activity, which could change activity patterns (Rennie et al., 2000). To minimize problems, receivers are often placed in pouches attached to safety belts that are worn around the waist and beneath the clothing. To minimize tampering and subject feedback, investigators have covered the programming buttons with opaque Plexiglas shells, removed the programming function buttons, or used pouches that lock shut.

Although most heart rate monitors can collect data for extended periods, investigators generally attach, detach, and download data from the units on a daily basis. This routine reduces subject discomfort, particularly because of the transmitter, by providing a break during periods of rest and sleep. It also reduces the amount of potential data that can be lost because of a malfunctioning unit. Heart rate data is downloaded to a personal computer using an interface.

Data Processing

After downloading, erratic heart rate values associated with electrical or magnetic interference can be identified. These values tend to be unrealistically high (e.g., >200 b/min) and repeat for unrealistically long periods of time. Erratic values from poor electrode contact or from too great a distance between transmitter and receiver tend to be unrealistically low (e.g., zero). Heart rates considered physiologically impossible or at least improbable are deleted using varied criteria. In children, Pate et al. (1996) used the criterion of <55 b/min or >215 b/min to delete erratic values. In adolescents, Ekelund et al. (1999) deleted heart rate values above predicted maximal heart rate and below 35 b/min. In adults, Wareham et al. (1997) identified unrealistic values (i.e., >220 and 0 b/min) and replaced them with interpolated heart rate values by averaging the previous and subsequent realistic values. More than five unrealistic readings occurring in a row resulted in the removal of the specific heart rate data segment. Figure 9.3 illustrates unrealistically high and low heart rate values.

Rennie et al. (2000) developed a single-piece heart rate monitor with an embedded accelerometer specifically designed for physical activity research rather than for fitness and sport training. The unit is lightweight, small (i.e., 100 g, 14.5 cm in length, 3 cm in width, and 0.8 cm in depth) and attaches to the chest using an elastic strap. By placing the transmitter and receiver in one unit, electrical interference is greatly reduced and fewer data are lost. This design also reduces tampering by the subject as well as the possibility that physical activity is altered in response to viewing the heart rate. In this prototype, the important features of currently available monitors are maintained, including the ability for continuous minute-by-minute monitoring for relatively long time periods (five days) and data downloading to a PC. However, unnecessary and complicated stopwatch and split-recording functions have been eliminated. This technological advance should reduce the complexity of heart rate monitoring field procedures and improve the feasibility of this method. Combining heart rate monitoring with accelerometry should also minimize overall error because accelerometry readings allow investigators to more accurately identify increases in heart rate due to causes other than physical activity (Meijer et al., 1989; Rennie et al., 2000).

Corrections and Adjustments for Resting Heart Rate

Resting heart rate provides a baseline value with which to compare physical activity heart rates and is often used to adjust for differences in fitness and age among subjects (Freedson, 1989). Compared to less-fit individuals, physically fit individuals typically have lower resting heart rates because of an increase in parasympathetic control. Children have

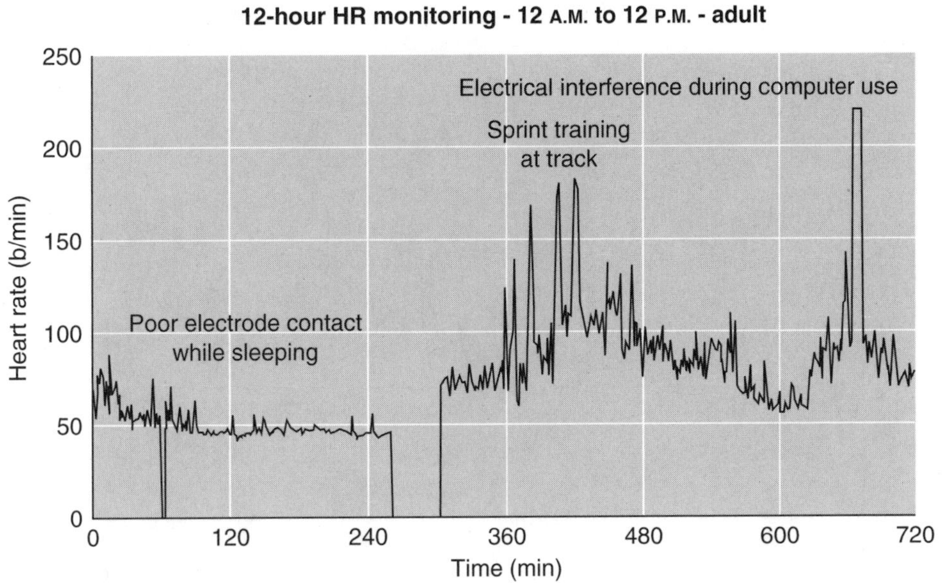

Figure 9.3 Example of heart rate data with unrealistically low and high values from interference.

higher resting heart rates than adults because of a higher resting metabolic rate (Nieman, 1999).

Adjusting for variability in resting heart rate generally improves the quality of heart rate data. Ideally, resting heart rate is measured in the morning prior to significant physical activity, emotional arousal, or the consumption of stimulants such as coffee or tobacco. After a 5- to 10-minute seated or supine rest period, multiple 30-second or 1-minute measures of heart rate are taken, generally in a sitting position, and the mean is used to define resting heart rate (Nieman, 1999). Supine measures are also acceptable, although they result in lower resting heart rates than sitting measures because stroke volume increases and heart rate subsequently decreases while lying down. Radial pulse palpation, auscultation with a stethoscope, ECG recording, or an ambulatory heart rate monitor are all acceptable techniques for measuring the resting heart rate. When possible, the mean from multiple readings over two to three days should be used (Nieman, 1999). When care is taken to standardize conditions, directly measured resting heart rate is reproducible. For example, using automated ECG recording systems, Durant et al. (1992) and Melanson (2000) have reported high (i.e., 0.93) intra-class correlation coefficients for repeated measures of resting heart rate in children and adults. Problems with assessing resting heart rate prior to data collection include measurement error between investigators (i.e., inter-tester reliability), elevated resting values due to subjects'

pre-monitoring apprehension, and the lack of a quiet and distraction-free environment.

An alternative method is to determine resting heart rate during data collection by averaging the lowest observed heart rate with all heart rates within some predefined range of this lowest heart rate (Freedson, 1989, 1991) (see figure 9.4). Investigators have proposed several predefined ranges for calculating this value. These definitions range from the mean of the 3 lowest heart rates to the mean of the lowest 50 heart rates. This method eliminates inter-tester reliability errors and the need for a quiet space. It also reduces setup time. If worn during sleep, however, data can be confounded by sleeping heart rates that are substantially lower than resting heart rates.

Some evidence for acceptable reproducibility of this method exists. Using the mean of the three lowest heart rates, Janz et al. (1992) have shown a between-day agreement of 84% in children and adolescents. Using the mean of the five lowest heart rates, Welk and Corbin (1995) have also shown a relatively small amount of between-day variability (i.e., 6.6 b/min) in children.

Several pediatric studies have compared directly measured resting heart rate with resting heart rate determined during data collection. Coleman et al. (1997) reported directly measured resting heart rates that were significantly lower than values determined during data collection (i.e., 83 b/min compared to 88 b/min). Logan et al. (2000) reported the opposite, higher directly

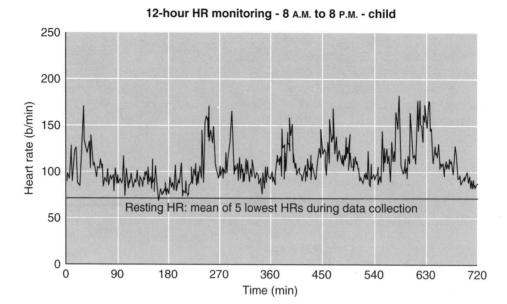

Figure 9.4   Example of an approach used to estimate resting heart rate from data collected in the field.

measured resting heart rates compared to the field determination (i.e., 101 b/min compared to 88 b/min). Logan et al. also examined how differences in resting heart rate definitions influence physical activity outcome variables calculated in relation to resting heart rate. Their results indicate that different resting heart rate definitions create differences in apparent levels of physical activity. They reported high rank-order correlations among the various methods used to establish resting heart rate during data collection (i.e., r = 0.84 to 0.99). This indicates that the measures provide reasonably similar values. In contrast, the rank-order correlation between directly measured resting heart rate and a value defined as the mean of the lowest five heart rates was moderate (i.e., r = 0.66) (Logan et al., 2000). The correlations between physical activity outcome variables calculated using these two resting heart rate methods were also moderate (i.e., r = 0.43 to 0.66). This work indicates that different procedures for the determination of resting heart rate can yield different results. To make absolute comparisons of physical activity levels among studies, it is necessary to standardize resting heart rate definitions.

Outcome Variables

Like other methods of continuous electronic monitoring, minute-by-minute heart rate monitoring provides hundreds of data points per subject per day. To reflect physical activity and energy expenditure in meaningful and comparable ways, these data must be collapsed and summarized. Regardless of the method used to construct outcome variables, heart rate data are best used to measure physical activity and energy expenditure at moderate through vigorous intensity levels, where heart rate predominantly represents oxygen consumption by contracting skeletal muscles. Because estimation errors are magnified at lower and higher activity intensities, most methods are designed to distinguish resting and light activity from moderate activity (Freedson & Miller, 2000). This convention improves the validity of the measure but results in excluding a large portion of the collected heart rate data, therefore limiting our understanding of the complexity of physical activity patterns and their potential health risks and health benefits. This limitation is a particular problem in research examining sedentary behavior such as television viewing. Sedentary behavior can be identified by low energy expenditure requirements; however, other methods are better at measuring low energy expenditure than the heart rate method. The methodological shortcomings of current heart rate monitoring approaches certainly support research efforts directed toward combining methods, such as combining accelerometry and heart rate monitoring (Meijer et al., 1989; Rennie et al., 2000). See chapter 13 for additional details about this approach.

Prediction of Energy Expenditure

Heart rate data are used to predict energy expenditure by establishing subject-specific regression equations between heart rate (HR) and volume of oxygen consumed (VO_2) (Ceesay et al., 1989; Schulz et al., 1989; Spurr et al., 1988; Spurr & Reina, 1990). The outcome variable is kilocalories (kcal) or kJoules (kJ) expended, typically per unit of time (kcal/day) or adjusted for body size per unit of time (kcal/kg/day). This method requires establishing a HR-VO_2 calibration curve using laboratory measures (such as indirect calorimetry) for each individual study subject. Constructing individualized curves reduces between-subject variability in the slope of the HR-VO_2 calibration curve attributable to differences in age, gender, and physical fitness, factors that contribute to measurement error (Bassett, 2000). Data points for the HR-VO_2 calibration curve are constructed by measuring a range of resting and exercise intensities meant to mimic usual activities. The laboratory-determined calibration curves are then used to transform heart rates measured during field conditions to estimates of VO_2 and ultimately predictions of energy expenditure. The latter step uses standardized equations that generally assume 1 L of O_2 is equivalent to 4.8 kcal (20.2 kJ) during aerobic activity (Consolazio et al., 1963; Weir, 1949). Though investigators differ in the mathematical models and laboratory procedures used to predict energy expenditure from heart rate, almost all approaches use some type of segmented linear regression or curvilinear modeling that takes into account the nonlinearity of HR and VO_2 during resting and light activity (Wareham & Rennie, 1998).

One of the more thoroughly examined approaches for estimating energy expenditure from heart rate is the flex heart rate (Flex HR) method (Ceesay et al., 1989; Davidson et al., 1997; Fogelholm et al., 1998; Kalkwarf et al., 1989; Livingstone et al., 2000; Racette et al., 1995; Spurr et al., 1988; Wareham et al., 1997). This approach requires four parameters to be determined during the HR-VO_2 calibration curve lab session:

1. The slope of the linear part of the HR-VO_2 association
2. The intercept of this prediction line
3. The Flex HR
4. The mean resting energy expenditure (EE)

Empirically, Flex HR is defined as the mean of the highest heart rate at rest and the lowest during exercise (see figure 9.5) (Ceesay et al., 1989). Commonly, Flex HR is determined by measuring HR and VO_2 at rest (i.e., lying, sitting, and standing) and during progressive increases in exercise (e.g., bicycle ergometry), and then calculating the mean of the highest resting value and lowest exercise value (Wareham et al., 1997). Exact definitions of highest resting activities and lowest exercise workloads often vary among researchers, thus creating problems in reproducing Flex HR among studies. Reproducibility errors are also created by the assumption that

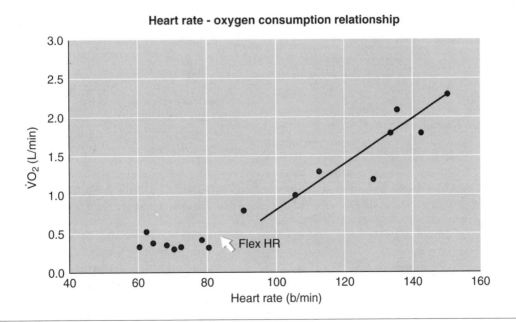

Figure 9.5 Illustration of the use of the Flex HR approach with heart rate monitors.

one HR point differentiates resting $\dot{V}O_2$ from exercise $\dot{V}O_2$ (Livingstone, 1997).

Livingstone et al. (2000) tested different HR-$\dot{V}O_2$ calibration methods designed to account for differences in upper- and lower-body activity in children as well as differences in posture. The results revealed no significant differences in energy expenditure estimates among seven different regression equations that used four to eight different activities to construct HR-$\dot{V}O_2$ calibration curves. This suggests that more complicated calibration procedures do not necessarily result in more precise estimates of energy expenditure. Because children are likely to engage in a greater variation of movement than adults, it seems likely that this observation would also apply to adults.

During field monitoring, energy expenditure above the Flex HR is estimated from the individualized HR-$\dot{V}O_2$ calibration curve. Energy expenditure below the Flex HR is estimated from the mean of the lying and sitting values or sometimes predicted using energy expenditure tables (Racette et al., 1995). Once estimated, total energy expenditure can be divided by resting energy expenditure to create an index of physical activity level (PAL) (Wareham et al., 1997). When compared to other heart rate measures, the need for laboratory calibration makes the Flex HR approach a costly method with high subject burden. In addition, because the HR-$\dot{V}O_2$ curve shifts within an individual in response to changing physiological and environmental conditions, the calibration must occur close to the time of the field heart rate monitoring (Rowlands et al., 1997). This can create significant research planning problems. Yet even with these limitations, Flex HR measures provide the best approach for concurrent measures of energy expenditure and physical activity patterns (Livingstone, 1997; Wareham & Rennie, 1998).

Percent Heart Rate Reserve

Data can also be analyzed using % heart rate reserve (% HRR) methods. These methods are based on the assumption of a linear and equivalent (one-to-one) relationship between % HRR and % $\dot{V}O_2$ reserve (% $\dot{V}O_2$R) throughout a large range of dynamic activity (Swain & Leutholtz, 1997). HRR is the maximal heart rate (HRmax) minus the resting heart rate; $\dot{V}O_2$R is maximal $\dot{V}O_2$ minus resting $\dot{V}O_2$. Ideally, HRmax, maximal $\dot{V}O_2$, resting heart rate, and resting $\dot{V}O_2$ are determined using laboratory procedures including graded exercise testing. One approach to constructing outcome variables using % HRR re-

quires calibrating the HRR to $\dot{V}O_2$R by expressing heart rate values collected during field monitoring using the following equation:

$$\% \text{ HRR} = [(\text{monitored HR} - \text{resting HR})/(\text{HRmax} - \text{resting HR})] \times 100$$

Next, % $\dot{V}O_2$R is calculated using a similar equation:

$$\% \dot{V}O_2R = [(\text{predicted } \dot{V}O_2 - \text{resting } \dot{V}O_2/(\dot{V}O_2\text{max} - \text{resting } \dot{V}O_2)] \times 100$$

The value from % HRR is imputed as the value for the % $\dot{V}O_2$R (based on the assumption that % HRR = % $\dot{V}O_2$R). This imputation permits $\dot{V}O_2$ to be calculated and therefore allows investigators to construct an outcome variable estimating oxygen consumption and variables estimating the relative intensity of physical activity (e.g., the number of minutes $\geq 60\%$ $\dot{V}O_2$R) (Strath et al., 2000). This method has the potential to reduce inter-subject variability associated with age and physical fitness and to provide a physiologically relevant outcome variable: oxygen consumption (Bassett, 2000; Strath et al., 2000; Swain & Leutholtz, 1997). Other investigators (Falgairette et al., 1996; Gavarry et al., 1998; Janz et al., 1992; Pardo et al., 2000) have used a simpler % HRR method by constructing outcome variables defined as the number of monitored minutes greater than or equal to a specific % HRR (e.g., the number of minutes $\geq 60\%$ HRR). This approach provides an estimate of the intensity of physical activity without a prediction of energy expenditure. If maximal heart rates are calculated using valid prediction equations, this method can be used without laboratory-based graded exercise testing, although directly measuring maximal heart rates would reduce measurement error. The American College of Sports Medicine (2000) recommends % HRR for monitoring exercise intensity, making % HRR approaches desirable when examining exercise adherence issues or comparing subjects to recognized exercise guidelines.

Average Net Heart Rate and Physical Activity Heart Rate

Heart rate data can also be analyzed using a net heart rate method in which the monitored heart rate is subtracted from the resting heart rate (Andrews, 1971). Data are described in b/min. Using this approach, physical activity during specific time periods can be estimated by summing the net heart rates and dividing by the duration of monitored time

(Average Net HR). This procedure quantifies the area under the monitored heart rate versus time curve by setting the baseline for the area as the resting heart rate (Freedson, 1989). In a sense, it creates a measure of volume or total physical activity. It also minimizes some of the inter-subject variability in physical fitness and age. However, this method has drawbacks because of the relatively large amount of time that adults and even children spend in sedentary and light activity where heart rate is not predictive of energy expenditure.

To circumvent the faulty interpretation of low heart rates, investigators often construct an additional variable: physical activity heart rates (PAHR), defined as the time spent at or above a specific % of resting heart rate (e.g., number of minutes ≥ 1.25 of resting heart rate) (Durant et al., 1992; Pate et al., 1996). The PAHR approach allows investigators to express data relative to resting heart rate without using the nonlinear component of the HR-$\dot{V}O_2$ relationship, therefore making the PAHR method easier to interpret than the Avg Net HR method and probably more valid. Unlike the Avg Net HR method, PAHR can also be used to reflect activity patterns and varying exercise intensity, such as moderate (i.e., ≥1.25 of resting heart rate) or vigorous activity (i.e., ≥1.50 of resting heart rate). For example, investigators can count the number of minutes spent ≥125 or 150% of resting heart rate and contrast these data for selected periods of the day (e.g., morning versus afternoon heart rates).

Heart Rate Threshold

A fourth method for constructing outcome variables is to determine the number of minutes or % of a time period greater than or equal to a selected target heart rate without adjusting for resting heart rate (Armstrong, 1998; Falgairette et al., 1996; Gilbey & Gilbey, 1995; Gilliam et al., 1981). The cut point or threshold heart rate is selected to correspond to known heart rate values for moderate and vigorous intensities during common forms of locomotion (e.g., brisk walking or running). When subjects are relatively similar in age and physical fitness and when the threshold is empirically selected for the cohort, this method provides an unambiguous approach to data analysis and, similarly to other approaches, circumvents the use of heart rate data in the nonlinear proportion of the HR-$\dot{V}O_2$ curve.

These various approaches for constructing outcome measures of physical activity and energy expenditure intensity can also be used to create frequency and duration outcome measures (Armstrong & Bray, 1991; Welk et al., 2000). Outcome variables reflecting the frequency of physical activity variables can be constructed by counting the number of bouts greater than or equal to a specific measure (Riddoch et al., 1991). For example, using the HR Threshold method, vigorous activity may be defined as heart rates ≥140 b/min. The *frequency* of vigorous activity is determined by counting the number of daily minutes of heart rates ≥ 140 b/min. Similarly, using the Flex HR method, moderate through vigorous activity may be defined as heart rates ≥ Flex HR. The *frequency* of moderate through vigorous activity is determined by counting the number of daily minutes ≥ Flex HR (Grund et al., 2000). Outcome variables to reflect duration of physical activity are constructed by counting the number or length of sustained periods greater than or equal to a specific measure (Durant et al., 1992). For example, using the % HRR method, vigorous activity may be defined as heart rates ≥60% HRR. The *duration* of vigorous activity is measured by counting the number of minutes that heart rates are sustained ≥ 60% HRR.

Measurement Properties: Validity

Validating heart rate monitoring for assessing physical activity requires comparing heart rate methods to established methods. Ideally, these established methods objectively measure the true exposure of physical activity, energy expenditure, or some subset of these variables without correlating measurement error with the comparison method (Rennie & Wareham, 1998). Most studies examining validity issues have used concurrent designs in which heart rate monitoring and a criterion method are simultaneously measured in field settings or in laboratory and field settings. Generally, investigators use the validation criterion of doubly labeled water, although they have also used accelerometry, activity diaries, direct observation, and indirect calorimetry. Certainly, when it comes to measuring the true exposure of interest and for validating heart rate methods, some of these criteria are more appropriate than others. Therefore, evidence of validity is limited by the quality of the selected criterion. In addition to providing a measure of association between the heart rate method and the criterion method (correlation coefficients) and providing measures of the exactness of the estimate (standard errors), the more valuable validation studies examine sources of error and methodological assumptions. Usually, these latter studies identify errors that confound the

HR-$\dot{V}O_2$ relationship or that test threshold criterion meant to sort resting and light activity from moderate activity (Fogelholm et al., 1998; Kashiwazaki, 1999; Livingstone et al., 2000). However, we need more empirical work examining criterion-referenced thresholds in children and adults. Without these additional validation studies, interpreting heart rate data in relationship to health outcomes and comparing results across studies remain difficult and imprecise tasks.

In adults, the Flex HR procedure has been shown to provide highly accurate estimates of energy expenditure. Reported correlation coefficients between Flex HR methods and energy expenditure measured using doubly labeled water or whole-body calorimetry are moderate to high, generally ranging from r = 0.54 to r = 0.98 (Ceesay et al., 1989; Schulz et al., 1989). When compared to criterion measures, adult validation studies demonstrate acceptable group estimates of energy expenditure using the Flex HR method. Group mean differences generally range from –5.8 to 17.2%. (Ceesay et al., 1989; Kashiwazaki, 1999). Most studies show group mean differences within ±10%. Livingstone et al. (1990) showed that very low group mean errors (i.e., 2%) are possible. However, at the individual level, prediction errors vary widely, with differences from the criterion measure ranging from –36.2 to 73.5% (Ceesay et al., 1989; Kashiwazaki, 1999; Livingstone, et al., 1990; Schulz et al., 1989; Wareham et al., 1997).

Flex HR validation studies in children have reported results similar to those found in the adult literature (Ekelund et al., 1999; Ekelund et al., 2000; Eston et al., 1998; Kohl et al., 2000). For example, correlation coefficients between energy expenditure estimated using Flex HR methods and criterion measures of energy expenditure (i.e., doubly labeled water or whole body calorimetry) are moderate to high. The group estimates of energy expenditure are acceptable with mean differences from the criterion ranging from –16.9 to 20%; however, the precision of individual estimates of energy expenditure varies widely. Most of the reported group mean differences for children are within ±10% of the criterion (Ekelund et al., 1999; Ekelund et al., 2000; Livingstone et al., 1992; Van den Berg-Emons et al., 1996). This is similar to adult validation studies.

There is limited research examining the validity of other heart rate measures. The work that does exist primarily uses a convergent design in which heart rate methods are compared to field methods such as accelerometry, activity diaries, or observation. This study design allows for a comparison of methods by testing associations. However, because there is no true criterion, the exactness of the estimate (i.e., standard errors) cannot be assessed. In addition, the field methods used to assess the efficacy of heart rate monitoring contribute their own errors to the validity correlation coefficient. One approach that minimizes some of these problems is the work by Strath et al. (2000) examining the validity of the % HRR method for predicting energy expenditure during moderate-intensity activity patterns in adults. These authors reported a strong association (i.e., r = 0.87) for estimates of oxygen consumption from heart rate data to oxygen consumption measured using indirect calorimetry. The % HRR method also resulted in a low standard error of the estimate (i.e., 0.76 METS) (Strath et al., 2000).

Reported associations between various heart rate methods (i.e., Avg Net HR, PAHR, % HRR, and HR Threshold) and other field methods are generally moderate and similar in magnitude between adults and children (Janz, 1994; Washburn & Montoye, 1986; Welk & Corbin, 1995). Washburn and Montoye (1986) examined the validity of the Avg Net HR method in adults by comparing predicted METS calculated from Avg Net HR to predicted METS calculated from an activity diary. The authors reported a moderate association (i.e., r = 0.49) between the two methods and a mean difference of 0.22 METS. In children, correlation coefficients between Avg Net HR and accelerometry-derived measures of movement counts are generally moderate, ranging from r = 0.42 to r = 0.71 (Coleman et al., 1997; Janz, 1994; Sallis et al., 1990; Welk & Corbin, 1995). Moderate correlation coefficients have also been reported between Avg Net HR and activity diaries (i.e., r = 0.50) (Janz et al., 1992), and between Avg Net HR and observation (i.e., r = 0.64) in children (Welk & Corbin, 1995). Reported associations between other outcome variable methods (% HRR, HR Threshold, and PAHR) and accelerometry-derived measures of movement counts are similar (i.e., 0.39 to 0.74) (Janz, 1994; Welk & Corbin, 1995). Table 9.1 summarizes advantages and disadvantages of constructing outcome variables using various approaches.

Measurement Properties: Stability

Determining the stability of habitual or usual physical activity using heart rate monitoring requires examining the variability of individual physical activity patterns under normal conditions. The degree to which intra-individual variability increases

TABLE 9.1 Summary of Heart Rate Monitoring Methods

Method	Outcome units (intensity)	Advantages	Disadvantages
Flex HR	kcal/day, PAL (total energy expenditure/resting energy expenditure), number of min Flex HR	Predicts energy expenditure; reduces between-subject variability in slope of HR-$\dot{V}O_2$ relationship (including fitness, age); accounts for nonlinearity in HR and $\dot{V}O_2$ at rest; most thoroughly validated HR method	Requires laboratory calibration and close timing between calibration and field monitoring; prediction errors for individual estimates of energy expenditure are high; expensive
% HRR calibrated to $\dot{V}O_2R$	METS, number of min selected % $\dot{V}O_2R$	Predicts oxygen consumption; reduces between-subject variability due to fitness and age; variables have relative and absolute physiological meaning; consistent with guidelines for exercise prescription	Requires measures or estimates of HRmax and $\dot{V}O_2$max (expensive when laboratory measures are used); does not address nonlinearity of HR and $\dot{V}O_2$ at lower HR; limited validation work in adults; no reported validation work in children
% HRR	Number of min selected % HRR	Reduces between-subject variability due to fitness and age; avoids nonlinearity of HR and $\dot{V}O_2$ at rest; % HR cut-offs are well established; consistent with guidelines for exercise prescription; simple to use	Requires measures or estimates of HRmax
Avg Net HR	b/min	Reduces between-subject variability due to fitness and age; simple to use	Difficult to interpret because of inclusion of low HR; difficult to construct duration and frequency variables
Physical activity HR	Number of min a selected % of resting HR	Reduces between-subject variability due to fitness and age; avoids nonlinearity of HR and $\dot{V}O_2$ at rest; simple to use	Little empirical data to guide the selection of what % resting HR to use
HR Threshold	Number of min a selected HR	Avoids nonlinearity in HR and $\dot{V}O_2$ at rest; simple to use	Little empirical data to guide the selection of what HR cut point to use

across time influences the time required to estimate usual activity and produce stable measures (Baranowski & de Moor, 2000). Failure to assess the stability of heart rate measures can result in selecting a monitored period that is too short, therefore adding measurement error. On the other hand, periods of heart rate monitoring longer than necessary to produce a stable measure increase both the expense of data collection and the participant burden.

The typical approach for determining the stability of heart rate monitoring is to measure consistency across time (generally across days) using a test-retest design and the intra-class correlation coefficient (ICC). Measurement error and true inter-individual variability in physical activity behavior are both reflected in this correlation coefficient, with ICC of at least 0.75 and preferably 0.80 considered acceptable (Baranowski & de Moor, 2000). In addition to determining the amount of monitored time needed to capture usual activity, the test-retest design can be used to assess whether specific time periods are representative or, conversely, atypical of overall time periods. For example, knowing that heart rates monitored in the afternoon are highly representative of all-day heart rates would allow investigators to limit assessment time. Knowing that weekends are atypical when compared to the rest of the week would indicate that both weekdays and weekends should be included within the monitored period. Seasonal factors affecting inter-individual variability can also be tested with a test-retest design.

Investigators with The Family Health Project (Baranowski & de Moor, 2000; Durant et al., 1991; Durant et al., 1992) have examined the stability of heart rate monitoring in young children. In their work, heart rates were monitored in young children for at least two nonconsecutive days for approximately 12 hours/day. Outcome variables included PAHR, HR Threshold, and the longest duration during a time period where heart rate was greater than a selected threshold (HR Duration). Single (i.e., between-day) ICC ranged from 0.39 to 0.68 for these heart rate variables. Using the Spearman-Brown Prophecy Formula, the investigators estimated multiple-day reliability, specifically, the number of days needed to achieve a reliability of 0.80. Generally, the heart rate variables reached ICC of 0.80 with four days of monitoring. The ICC was higher for variables constructed to adjust for resting heart rate (i.e., PAHR Index). In addition, between-hour stability of heart rate monitoring was examined and ICC of 0.26 to 0.31 reported. Because of the relatively low reliability for between-hour data, obtaining an ICC of 0.80 required 9 to 11 hours of monitoring per day. These findings indicate greater stability between days than between hours within a day and support the convention of whole-day monitoring to assess usual physical activity patterns.

Using similar heart rate measures, Falgairette et al. (1996) and Gavarry et al. (1998) have also shown greater variability between hours than between days in children and adolescents. In addition, these investigators and others (Armstrong & Bray, 1991; Grund et al., 2000) have demonstrated some differences between weekdays and weekend days. In children and adolescents, Janz (1994) and Sallis et al. (1990) have reported low to moderate between-day associations for various heart rate measures (e.g., Avg Net HR, HR Threshold, and % HRR). This work suggests that, at the very least, one day of heart rate monitoring is not stable in this age group.

In adult men, using a HR Threshold measure, Mueller et al. (1986) compared the average of three days of heart rates collected during two consecutive weeks. The reported correlation coefficient was r = 0.33. Also using an HR Threshold measure, Bassey et al. (1980) reported a moderate correlation (i.e., r = 0.61) between two days of heart rate moni-

toring in adult men and women. Gretebeck and Montoye (1992) monitored heart rates in male adults for seven consecutive days and reported a moderate association (i.e., r = 0.44) for between-day HR Threshold values. In this study, the Spearman-Brown Prophecy Formula predicted that four days of monitoring would be needed to obtain an ICC of 0.78, and five days for 0.82. Between-day reliability coefficients for energy expenditure predicted using Flex HR methods appear similar to reported variation using other heart rate measures (Davidson et al., 1997; Wareham et al., 1997). For example, using Flex HR methods, Wareham et al. (1997) reported a 0.69 reliability correlation coefficient for four days of heart rate monitoring sorted by weekdays and weekend days. Generally, the adult literature supports no mean differences between weekday and weekend heart rate data (Davidson et al., 1997; Gretebeck & Montoye, 1992; Wareham et al., 1997). As a whole, studies examining how many days of heart rate monitoring are needed to produce an acceptable measure of usual activity suggest at least four days. In children and adolescents, there is some evidence that patterns of activity differ between weekend days and weekdays, indicating that both should be included during monitoring.

Acceptability and Utility of Heart Rate Monitors

With appropriate methods and great attention to detail, heart rate monitoring can provide a valid and reliable means of assessing physical activity and energy expenditure, particularly for moderate through vigorous activity. However, feasibility of heart rate monitoring and acceptability by subjects are also important factors for the successful use of this method. Heart rate monitoring methods are more cumbersome than current accelerometry methods because of heart rate monitor designs that require securing the transmitter to the chest and maintaining contact between the chest and transmitter. Receivers that are easily reprogrammed by subjects are also problematic (Livingstone, 1997). All three of these issues are particularly troublesome in children who have sensitive skin, small ribcages, and innate curiosities. Securing the transmitter to the chest and maintaining transmitter to chest contact is also an issue with obese subjects. Many of the commercially available elastic straps are too small for this population, and skin and adipose tissue draping over the transmitter can heighten irritation. During active periods or during hot weather, sweat and dirt accumulate beneath transmitters causing skin irritation, chafing, and contact dermatitis in all populations. Anecdotal evidence suggests that heart rate monitoring disrupts sleeping patterns when subjects roll over the units. Many of these problems can be addressed by reducing the amount of continuous time heart rate is monitored, but this compromise requires greater subject-investigator interaction, therefore increasing the cost of data collection as well as inconveniencing subjects.

Although problems exist, investigators have reported high compliance rates with heart rate monitoring procedures. For example, Wareham et al. (1997) examined the feasibility of heart rate monitoring in a random sample of adults and reported that only 3 of 157 subjects chose not to complete a four-day trial. This is a 98% compliance rate. However, in their pilot work, subjects reported a general awareness of the monitors and specific problems associated with skin irritations after four days. In a random sample of school children, Janz et al. (1992) reported an 89% subject compliance rate for a one-day trial. In a smaller cohort, these same investigators reported that children and adolescents uniformly described decreased comfort during hot days and skin irritation after 8 to 10 hours of continuous wear regardless of the weather. Armstrong et al. (1990) reported a compliance rate of 71% for school children in a one-day trial. In their pilot work, these investigators observed children during physical education classes, some wearing heart rate monitors and some not, and could not discern behavioral differences. Using a similar approach but working with preschool children, Bar-Or et al. (1996) rated responses to wearing heart rate monitors during a 90-minute daycare activity. Approximately 90% of the subjects were rated as being enthusiastic or agreeable to wearing the monitors and the refusal rate for wearing the monitor was low (i.e., 4%). However, pilot work by these investigators indicated that preschoolers might not be willing to wear heart rate monitors while sleeping. In summary, even though problems exist during periods of extended wear and hot weather, heart rate monitoring is feasible in small- and medium-sized epidemiological studies and well tolerated by most subjects.

Applications for Heart Rate Monitoring

In epidemiology, the heart rate method is used to assess physical activity and energy expenditure in health-related observational and intervention studies. This method also has other applications. For example, heart rate monitoring can be used to optimize training programs for athletic performance. Setting training intensity based on heart rate rather than speed or pace is particularly useful in bicycling, cross-country skiing, and other sports where an athlete's speed varies depending on environmental conditions and other athletes (e.g., drafting). Although various heart rate intensity indicators are used in athletic training, % HRmax and % HRR are the most common (Jeukendrup & Van Diemen, 1998). Monitoring submaximal and resting heart rate is also used by athletes to gauge post-training recovery (O'Toole et al., 1998). Many heart rate monitors are designed specifically for these tasks and include functions that count split/lap times, sound alarms when heart rates deviate from training zones, and mark recovery heart rates. There is also some indication that heart rate monitoring can be used to detect the onset of overtraining (Jeukendrup et al., 1992; Jeukendrup & Van Diemen, 1998). For example, Jeukendrup and Van Diemen (1998) observed increases in average sleeping heart rate and less-regular heart rate patterns in athletes during evenings after overtraining.

Beat-to-beat variability (R-R) in heart rate (heart rate variability [HRV]) provides a noninvasive index of cardiac autonomic activity, specifically the cardiac sympathetic-parasympathetic interaction (Berntson et al., 1997). Increasingly, this measure is being used to examine the integrity of the autonomic nervous system, the link between psychological processes and autonomic control, and the pathophysiology of diseases that are influenced by autonomic function, such as sudden cardiac death (Berntson et al., 1997, Gutin et al., 2000). Studies examining HRV in field settings commonly use Holter monitors (Massin et al., 2000; Pardo et al., 1996; Pardo et al., 2000). These instruments are hard wired and create tape recordings that are digitized to produce QRS complexes. They are also bulky and have a limited recording time. Polar has introduced a wireless heart rate monitor that measures and stores beat-to-beat intervals (rather than rolling averages) for up to 36 hours. Kinnunen and Heikkila (1998) reported differences between this monitor and ECG recording within 2 msec in 96.4% and within 5 msec in 99.9% in over 80,000 R-R. These results suggest that researchers may be able to monitor HRV or at least some components of it with a small, relatively inexpensive (approximately $350) monitor.

Monitoring heart rates with or without equipment has been an important component of physical education programs for decades. However, as the cost of wireless monitors decreases, using continuous heart rate monitoring to teach students about the cardiovascular responses to exercise, to provide exercise intensity feedback, and to individualize aerobic training programs has become a real option for many physical education programs. For example, curricula have been successfully developed for elementary and secondary students that integrate heart rate monitors into experiential learning assignments on the effects of exercise on heart function and the regulation of exercise through biofeedback (Hinson, 1994). Using heart rate monitors as teaching tools is also appropriate in corporate health promotion programs and fitness clubs.

Summary

During moderate through vigorous physical activity, heart rate increases are directly proportional to oxygen consumption of contracting skeletal muscles. Because of this relationship, heart rate monitoring has been used for several decades to provide a physiological indicator of physical activity and energy expenditure in free-living populations. Recent technological advances have produced lightweight, small, and relatively inexpensive heart rate monitors capable of extended minute-by-minute data collection, therefore allowing investigators to examine duration and frequency of physical activity as well as the intensity of physical activity. When care is given to constructing heart rate measures that account for the disassociation between HR and VO_2 during sedentary and light physical activity, heart rate monitoring provides a valid measure of physical activity. When appropriate laboratory procedures are used to establish individualized calibration curves, heart rate monitoring also provides a good approximation of group energy expenditure. Since factors other than physical activity also affect heart rate, various sources of error influence the validity of the heart rate method for both outcomes (physical activity and energy expenditure). At least four or more days of whole-day monitoring are needed for stable measures of usual physical activity. Unfortunately, extended periods of monitoring have been associated with subject discomfort and skin irritation. Future epidemiological use of the heart rate method may depend on (1) refining methods for constructing outcome variables, (2) standardizing approaches to allow data to be compared across studies, (3) improving subject comfort during extended periods of wear, and (4) combining this method with techniques better suited for detecting sedentary and light activity.

References

American College of Sports Medicine (ACSM). (2000). *ACSM's guidelines for exercise testing and prescription.* Philadelphia: Lippincott, Williams, & Wilkins.

Andrews, R.B. (1971). The heart rate as a substitute for respiratory calorimetry. *American Journal of Clinical Nutrition, 24,* 1139-1147.

Armstrong, N. (1998). Young people's physical activity patterns as assessed by heart rate monitoring. *Journal of Sports Sciences, 16,* S9-S16.

Armstrong, N., Balding, J., Gentle, P., & Kirby, B. (1990). Estimation of coronary risk factors in British schoolchildren: A preliminary report. *British Journal of Sports Medicine, 24,* 61-66.

Armstrong, N., & Bray, S. (1991). Physical activity patterns defined by continuous heart rate monitoring. *Archives of Disease in Childhood, 66,* 245-247.

Armstrong, N., Welsman, J.R., & Kirby, B.J. (2000). Longitudinal changes in 11-13 year olds' physical activity. *Acta Paediatrica, 89,* 775-780.

Baranowski, T., & de Moor, C. (2000). How many days was that? Intra-individual variability and physical activity assessment. *Research Quarterly for Exercise and Sport, 71,* 74-78.

Bar-Or, T., Bar-Or, O., Waters, H., Hirji, A., & Russell, S. (1996). Validity and social acceptability of the Polar Vantage XL for measuring heart rate in preschoolers. *Pediatric Exercise Science, 8,* 115-121.

Bassett, D.R., Jr. (2000). Validity and reliability issues in objective monitoring of physical activity. *Research Quarterly for Exercise and Sport, 71* (2), 30-36.

Bassey, E.J., Bryant, J.C., Fentem, P.H., MacDonald, I.A., & Patrick, J.M. (1980). Customary physical activity in elderly men and women using long-term ambulatory monitoring of ECG and football. In F.D. Stott, E.B. Raftery, & L. Goulding (Eds.), *Proceedings of the 3rd International Symposium on Ambulatory Monitoring* (pp. 425-432). London: Academic Press.

Berggren, G., & Christensen, E.H. (1950). Heart rate as a means of measuring metabolic rate in man. *Arbeitsphysiologie, 14,* 255-260.

Berntson, G.G., Bigger, J.T., Jr., Eckbert, D.L., Grossman, P., Kaufmann, P.G., Malik, M., Nagaraja, N., Porges, S.W., Saul, J.P., Stone, P.H., & van der Molen, M.W. (1997). Heart rate variability: Origins, methods, and interpretive caveats. *Psychophysiology, 34,* 623-648.

Caspersen, C.J., Powell, K.E., & Christenson, G.M.P. (1985). Physical activity, exercise, and physical fitness: Definitions and distinctions for health-related research. *Public Health Reports 100,* 126-131.

Ceesay, S.M., Prentice, A.M., Day, K.C., Murgatroyd, P.R., Goldberg, G.R., & Scott, W. (1989). The use of heart rate monitoring in the estimation of energy expenditure: A validation study using indirect whole-body calorimetry. *British Journal of Nutrition, 61,* 175-186.

Coleman, K.J., Saelens, B.E., Wiedrich-Smith, M.D., Finn, J.D., & Epstein, L.H. (1997). Relationships between TriTrac-R3D vectors, heart rate, and self-report in obese children. *Medicine and Science in Sports and Exercise, 29,* 1535-1542.

Consolazio, C.F., Johnson, R.E., & Pecora, L.J. (1963). *Physiological measurements of metabolic functions in man.* New York: McGraw-Hill.

Davidson, L., McNeill, G., Haggarty, P., Smith, J.S., & Franklin, M.F. (1997). Free-living energy expenditure of adult men assessed by continuous heart rate monitoring and doubly-labeled water. *British Journal of Nutrition, 78,* 695-708.

Durant, R.H., Baranowski, T., Davis, H., Rhodes, T., Thompson, W.O., Greaves, K.A., & Puhl, J. (1992). Reliability and variability of indicators of heart rate monitoring in children. *Medicine and Science in Sports and Exercise, 25,* 389-395.

Durant, R.H., Baranowski, T., Davis, H., Thompson, W.O., Puhl, J., Greaves, K.A., & Rhodes, T. (1991). Reliability and variability of heart rate monitoring in 3-, 4-, or 5-yr-old children. *Medicine and Science in Sports and Exercise, 24,* 265-271.

Ekelund, U., Sjostrom, M., Yngve, A., & Nilsson, A. (2000). Total daily energy expenditure and pattern of physical activity measured by minute-by-minute heart rate monitoring in 14-15 year old Swedish adolescents. *European Journal of Clinical Nutrition, 54,* 195-202.

Ekelund, U., Yngve, A., & Sjostrom, M. (1999). Total daily energy expenditure and patterns of physical activity in adolescents assessed by two different methods. *Scandinavian Journal of Medicine and Science in Sports, 9,* 257-264.

Eston, R.G., Rowlands, A.V., & Ingledew, D.K. (1998). Validity of heart rate, pedometry, and accelerometry for predicting the energy cost of children's activities. *Journal of Applied Physiology, 84,* 362-371.

Falgairette, G., Gavarry, O., Bernard, T., & Hebbelinck, M. (1996). Evaluation of habitual physical activity from a week's heart rate monitoring in French school children. *European Journal of Applied Physiology, 74,* 153-161.

Fogelholm, M., Hiilloskorpi, H., Laukkanen, R., Oja, P., Van Marken Lichtenbelt, W., & Westerterp, K. (1998). Assessment of energy expenditure in overweight women. *Medicine and Science in Sports and Exercise, 30,* 1191-1197.

Freedson, P.S. (1989). Field monitoring of physical activity in children. *Pediatric Exercise Science, 1,* 8-18.

Freedson, P.S. (1991). Electronic motion sensors and heart rate as measures of physical activity in children. *Journal of School Health, 61* (5), 215-219.

Freedson, P.S., & Miller, K. (2000). Objective monitoring of physical activity using motion sensors and heart rate. *Research Quarterly for Exercise and Sport, 71,* 21-29.

Gavarry, O., Bernard, T., Giacomoni, M., Seymat, M., Euzet, J.P., & Falgairette, G. (1998). Continuous heart rate monitoring over 1 week in teenagers aged 11-16 years. *European Journal of Applied Physiology, 77,* 125-132.

Gilbey, H., & Gilbey, M. (1995). The physical activity of Singapore primary school children as estimated by heart rate monitoring. *Pediatric Exercise Science, 7,* 26-35.

Gilliam, T.B., Freedson, P.S., Greenen, D.L., & Shahraray, B. (1981). Physical activity patterns determined by heart rate monitoring in 6-7 year-old children. *Medicine and Science in Sports and Exercise, 13,* 65-67.

Gretebeck, R.J., & Montoye, H.J. (1992). Variability of some objective measures of physical activity. *Medicine and Science in Sports and Exercise, 24,* 1167-1172.

Gretebeck, R.J., Montoye, H.J., Ballor, D., & Montoye, A.P. (1991). Comment on heart rate recording in field studies. *Journal of Sports Medicine and Physical Fitness, 31,* 629-631.

Grund, A., Dilba, B., Forberger, K., Krause, H., Siewers, M., Riecert, H., Mueller, M.J. (2000). Relationships between physical activity, physical fitness, muscle strength and nutritional state in 5- to 11-year-old children. *European Journal of Applied Physiology, 82,* 425-438.

Gutin, B., Barbeau, P., Litaker, M.S., Ferguson, M., & Owens, S. (2000). Heart rate variability in obese children: Relations to total body and visceral adiposity, and changes with physical training and detraining. *Obesity Research, 8,* 12-19.

Hinson, C. (1994). Pulse power: A heart physiology program for children. *JOPERD, 65,* 62-68.

Hurley, B.F., Seals, D.R., Ehsani, A.A., Cartier, L.J., Dalsky, G.P., Hagberg, J.M., & Holloszy, J.O. (1984). Effects of high-intensity strength training on cardiovascular function. *Medicine and Science in Sports and Exercise, 16,* 483-488.

Janz, K.F. (1994). Validation of the CSA accelerometer for assessing children's physical activity. *Medicine and Science in Sports and Exercise, 26,* 369-375.

Janz, K.F., Golden, J.C., Hansen, J.R., & Mahoney, L.T. (1992). Heart rate monitoring of physical activity in children and adolescents: The Muscatine study. *Pediatrics, 89,* 256-261.

Jeukendrup, A., Hesselink, M.K.C., Snyder, A.C., Kuipers, H., and Keizer, H.A. (1992). Physiological changes in male competitive cyclist after two weeks of intensified training. *International Journal of Sports Medicine, 13,* 534-541.

Jeukendrup, A., & Van Diemen, A. (1998). Heart rate monitoring during training and competition in cyclists. *Journal of Sports Sciences, 16,* S91-S99.

Kalkwarf, H.J., Hass, J.D., Belko, A.Z., Roach, R.C., Roe, D.A. (1989). Accuracy of heart-rate monitoring and activity diaries for estimating energy expenditure. *American Journal of Clinical Nutrition, 49,* 37-43.

Karvonen, J., Chwalbinska-Moneta, J., & Saynajakangas, S. (1984). Comparison of heart rates measured by ECG and microcomputer. *Physician and Sportsmedicine, 12,* 65-69.

Kashiwazaki, H. (1999). Heart rate monitoring as a field method for estimating energy expenditure as evaluated by the doubly labeled water method. *Journal of Nutritional Science Vitaminol, 45,* 79-94.

Kelly, L.E. (2000). Patterns of physical activity in 9-10 year-old American children as measured by heart rate monitoring. *Pediatric Exercise Science, 12,* 101-110.

Kinnunen, H., & Heikkila, I. (1998). The timing accuracy of the Polar Vantage NV heart rate monitor. *Journal of Sports Sciences, 16,* S107-S110.

Kohl, H.W., III, Fulton, J.E., & Caspersen, C.J. (2000). Assessment of physical activity among children and adolescents: A review and synthesis. *Preventive Medicine, 31,* S54-S76.

Laukkanen, R.M.T., & Virtanen, P.K. (1998). Heart rate monitors: State of the art. *Journal of Sports Sciences, 16,* S3-S7.

Leger, L., & Thivierge, M. (1988). Hear rate monitors: Validity, stability and functionality. *Physician and Sportsmedicine, 16,* 143-151.

Livingstone, M.B.E. (1997). Heart-rate monitoring: the answer for assessing energy expenditure and physical activity in population studies? *British Journal of Nutrition, 78,* 869-871.

Livingstone, M.B., Coward, W.A., Prentice, A.M., Davies, P.S.W., Strain, J.J., McKenna, P.G., Mahoney, C.A., White, J.A., Stewart, C.M., & Kerr, M.J. (1992). Daily energy expenditure in free-living children: Comparison of heart-rate monitoring with the doubly labeled water method. *American Journal of Clinical Nutrition, 56,* 343-352.

Livingstone, M.B.E., Prentice, A.M., Coward, W.A., Ceesay, S.M., Strain, J.J., McKenna, P.G., Nevin, G.B., Barker, M.E.,

Hickey, R.J. (1990). Simultaneous measurement of free-living energy expenditure by the doubly labeled water method and heart-rate monitoring. *American Journal of Clinical Nutrition, 52,* 59-65.

Livingstone, M.B.E., Robson, P.J., & Totton, M. (2000). Energy expenditure by heart rate in children: An evaluation of calibration techniques. *Medicine and Science in Sports and Exercise, 32,* 1513-1519.

Logan, N., Reilly, J.J., Grant, S., & Paton, J.Y. (2000). Resting heart rate definition and its effect on apparent levels of physical activity in young children. *Medicine and Science in Sports and Exercise, 32,* 162-166.

Massin, M.M., Maeyns, K., Withofs, N., Ravet, F., and Gerard, P. (2000). Circadian rhythm of heart rate and heart rate variability. *Archives of Disease in Childhood, 83,* 179-182.

Meijer, G.A., Westerterp, K.R., Koper, H., & Ten Hoor, F. (1989). Assessment of energy expenditure by recording heart rate and body acceleration. *Medicine and Science in Sports and Exercise, 21,* 343-347.

Melanson, E.L. (2000). Resting heart rate variability in men varying in habitual physical activity. *Medicine and Science in Sports and Exercise, 32,* 1894-1901.

Montoye, H.J., Kemper, H.C.G., Saris, W.H.M., & Washburn, R.A. (1996). *Measuring physical activity and energy expenditure* (pp. 97-115). Champaign, IL: Human Kinetics.

Mueller, J.K., Gossar, D., Adams, F.R., Taylor, D.G., Haskell, W.L., Kraemer, H.C., Ahn, D.K., Burnett, K., & DeBusk, R.G. (1986). Assessment of prescribed increases in physical activity: Application of a new method for microprocessor analysis of heart rate. *American Journal of Cardiology, 57,* 441-445.

Nieman, D.C. (1999). *Exercise testing and prescription: A health-related approach.* Mountain View, CA: Mayfield Publishing.

O'Toole, M.L., Douglas, P.S., & Hiller, W.D.B. (1998). Use of heart rate monitors by endurance athletes: Lessons from triathletes. *Journal of Sports Medicine and Physical Fitness, 38,* 181-187.

Pardo, Y., Bairey Merz, C.N., Paul-Labrador, M., Velasquez, I., Gottdiener, J.S., Kop, W.J., Krantz, D.S., Rozanski, A., Klein, J., & Peter, T. (1996). Heart rate variability, reproducibility and stability using commercially available equipment in coronary artery disease with daily life myocardial ischemia. *Americal Journal of Cardiology, 78,* 866-870.

Pardo, Y., Merz, C.N.B., Velasquez, I., Paul-Labrador, M., Agarwala, A., & Peter, C.T. (2000). Exercise conditioning and heart rate variability: Evidence of a threshold effect. *Clinical Cardiology, 23,* 615-620.

Pate, R.R., Baranowski, T., Dowda, M., & Trost, S.G. (1996). Tracking of physical activity in young children. *Medicine and Science in Sports and Exercise, 28,* 92-96.

Racette, S.B., Schoeller, D.A., & Kushner, R.F. (1995). Comparison of heart rate and physical activity recall with doubly labeled water in obese women. *Medicine and Science in Sports and Exercise, 27,* 126-133.

Rennie, K., Rowsell, T., Jebb, S.A., Holburn, D., & Wareham, N.J. (2000). A combined heart rate and movement sensor: Proof of concept and preliminary testing study. *European Journal of Clinical Nutrition, 54,* 409-414.

Rennie, K., & Wareham, N. (1998). The validation of physical activity instruments for measuring energy expenditure: Problems and pitfalls. *Public Health Nutrition, 1,* 265-271.

Riddoch, C.J., Mahoney, C., Murphy, N., Boreham, A.G., & Cran, G. (1991). The physical activity patterns of North-ern Irish schoolchildren ages 11-16 years. *Pediatric Exercise Science, 3,* 300-309.

Robergs, R.A. (1996). *Exercise physiology: Exercise, performance, and clinical applications.* St. Louis: Mosby-Year Book.

Rowlands, A.V., Eston, R.G., & Ingledew, D.K. (1997). Measurement of physical activity in children with particular reference to the use of heart rate and pedometry. *Sports Medicine, 24,* 258-272.

Sallis, J.F., Buono, M.J., Roby, J.J., Carlson, D., & Nelson, J.A. (1990). The Caltrac accelerometer as a physical activity monitor for school-age children. *Medicine and Science in Sports and Exercise, 22,* 698-703.

Saris, W.H.M., Binkhorst, R.A., Cramwinckel, A.B., Waesbergeh, F., & Van der Vien-Hezemans, A.M. (1980). The relationship between working performance, daily physical activity, fatness, blood lipids and nutrition in schoolchildren. In K. Berg & B.O. Eriksson (Eds.), *Children and exercise: IX* (pp. 166-174). Baltimore: University Park Press.

Saris, W.H.M., Snel, P., & Binkhorst, R.A. (1977). A portable heart rate distribution recorder for studying daily physical activity. *European Journal of Applied Physiology, 37,* 19-25.

Schulz, S., Westerterp, K.R., & Bruck, K. (1989). Comparison of energy expenditure by the doubly labeled water technique with energy intake, heart rate, and activity recording in man. *American Journal of Clinical Nutrition, 49,* 1146-1154.

Spurr, G.B., Prentice, A.M., Murgatroyd, P.R., Goldberg, G.R., Reina, J.C., & Christman, N.T. (1988). Energy expenditure from minute-by-minute heart-rate recording: Comparison with indirect calorimetry. *American Journal of Clinical Nutrition, 48,* 552-559.

Spurr, G.B., & Reina, J.C. (1990). Daily pattern of % $\dot{V}O_2$max and heart rates in normal and undernourished school children. *Medicine and Science in Sports and Exercise, 22,* 643-652.

Steptoe, A. (2000). Stress, social support and cardiovascular activity over the working day. *International Journal of Psychophysiology, 37,* 299-308.

Strath, S.J., Swartz, A.M., Bassett, D.R., O'Brien, W.L., King, G.A., & Ainsworth, B.E. (2000). Evaluation of heart rate as a method for assessing moderate intensity physical activity. *Medicine and Science in Sports and Exercise, 32,* S465-S470.

Stratton, G. (1996). Children's heart rates during physical education lessons: A review. *Pediatric Exercise Science, 8,* 215-233.

Swain, D.P., & Leutholtz, B.C. (1997). Heart rate reserve is equivalent to % $\dot{V}O_2$Reserve, not to % $\dot{V}O_2$max. *Medicine and Science in Sports and Exercise, 29,* 410-414.

Torun, B. (1984). Physiological measurements of physical activity among children under free-living conditions. In E. Pollitt & P. Amante (Eds.), *Current topics nutrition and disease II* (pp. 159-184). New York: Alan R. Liss.

Treiber, F.A., Musante, L., Hartdagan, S., Davis, H.L., Levy, M., & Strong, W.B. (1989). Validation of a heart rate monitor with children in laboratory and field settings. *Medicine and Science in Sports and Exercise, 21,* 338-342.

Van den Berg-Emons, R.J.G., Saris, W.H.M., Westerterp, K.R., & Van Baak, M.A. (1996). Heart rate monitoring to assess energy expenditure in children with reduced physical activity. *Medicine and Science in Sports and Exercise, 28,* 496-501.

Vuori, I. (1998). Experiences of heart rate monitoring in observational and intervention studies. *Journal of Sports Sciences, 16,* S25-S30.

Wareham, N.J., Hennings, S.J., Prentice, A.M., & Day, N.E. (1997). Feasibility of heart-rate monitoring to estimate the Ely young cohort feasibility study 1994-1995. *British Journal of Nutrition, 78,* 889-900.

Wareham, N.J., & Rennie, K.L. (1998). The assessment of physical activity in individuals and populations: Why try to be more precise about how physical activity is assessed? *International Journal of Obesity, 22,* S30-S38.

Washburn, R.A., & Montoye, H.J. (1986). Validity of heart rate as a measure of mean daily energy expenditure. *Exercise Physiology, 2,* 161-172.

Weir, J.B. de V. (1949). New methods for calculating metabolic rate with special reference to protein metabolism. *Journal of Physiology, 109,* 1-9.

Welk, G.J., & Corbin, C.B. (1995). The validity of the Tritrac-R3D activity monitor for the assessment of physical activity in children. *Research Quarterly for Exercise and Sport, 66,* 202-209.

Welk, G.J., Corbin, C.B., & Dale, D. (2000). Measurement issues in the assessment of physical activity in children. *Research Quarterly for Exercise and Sport, 71,* 59-73.

10

Use of Pedometers to Assess Physical Activity

David R. Bassett Jr., PhD
University of Tennessee

Scott J. Strath, PhD
University of Michigan

Pedometers provide information on ambulatory activity (i.e., walking). Researchers are interested in measuring walking and walking-based activities for a number of reasons. First, walking is one of the most popular physical activities in the United States (ranking ahead of gardening, calisthenics, cycling, jogging, weightlifting, swimming, dancing, basketball, tennis, football, softball, golf, and bowling) (Crespo et al., 1996). Second, walking-based activities account for a major portion of physical activity energy expenditure on physical activity questionnaires and logs (Ainsworth et al., 1993; Bassett, Cureton et al., 2000). Third, walking is known to confer substantial health benefits, such as a reduced risk of cardiovascular disease and cancer (Hakim et al., 1999; Hakim et al., 1998; Manson et al., 1999; Paffenbarger et al., 1978). Finally, walking programs can be undertaken on an individual basis, without the need for exercise equipment and facilities. In fact, brisk walking is advocated by the Centers for Disease Control and Prevention (CDC) and the American College of Sports Medicine

(ACSM) (Pate et al., 1995). The American Heart Association Task Force on Risk Reduction (Fletcher, 1997) concluded that walking is *the* recommended form of exercise for secondary prevention of myocardial infarction, unless exercise can be performed in a supervised setting. One advantage of pedometers is that they provide an accurate, objective, and low-cost measure of walking behaviors. In contrast, physical activity questionnaires typically ask people about the distance they walk each day (Ainsworth et al., 1993; Ainsworth et al., 1994). The validity of questionnaires may be limited by subjects' ability to remember how far they walked as well as their perception of distance. Numerous studies have shown that recall of ubiquitous, moderate-intensity activities (e.g., walking) is less accurate than recall of structured, vigorous exercise (Baranowski, 1988; Blair et al., 1991; Richardson et al., 1996; Richardson et al., 1994; Sallis et al., 1985). Thus, there is a need for accurate and objective devices like pedometers to quantify ambulatory activity.

Figure 10.1 Example of an electronic pedometer (Yamax SW-200, Yamasa Corp., Tokyo).

Principles of Method

The pedometer is a device that measures steps and distances in walking or running (Washburn et al., 1980). It is usually worn on the belt or waistband, but models have also been devised for the wrist, ankle, and shoe. Modern waist-mounted pedometers respond to vertical accelerations of the hip that occur when walking (see figure 10.1). Most of them contain a horizontal, spring-suspended lever arm that moves up and down with each step. This action opens and closes an electrical circuit, and the accumulated step count is then shown on a digital display.

The invention of the pedometer is credited to Leonardo da Vinci, whose 15th-century drawings illustrate the principle behind the mechanical pedometer (Gibbs-Smith, 1978). A lever arm was probably affixed to the thigh, and when the lever arm moved back and forth during walking, the gears were rotated and steps were counted (Montoye et al., 1995).

The *World Almanac of Presidential Facts* (Palletta & Worth, 1993) incorrectly credits Thomas Jefferson with inventing the pedometer. Jefferson was, however, an enthusiastic fan of pedometers. He bought a *conte-pas* (step counter) in Paris and used it to keep careful records of his walking. Jefferson noted that an English mile would require 2,066.5 steps, which the "brisk walk of winter" would reduce to 1,735 steps. He measured out the distance to various Paris landmarks in steps (Dumbauld, 1946). Jefferson sent a pedometer to his friend, James Madison, in 1788 along with a one-page letter of detailed instructions on how to use it (Wilson & Stanton, 1999). To another friend, who inquired, "Are you become a great walker?" Jefferson replied, "You know I preach up

that kind of exercise. Shall I send you a *conte-pas*? It will cost you a dozen louis, but be a great stimulus to walking, as it will record your steps" (Nock, 1926).

In early times, pedometers were used primarily for surveying land. In the 1960s, however, researchers became interested in using pedometers to assess habitual physical activity in free-living populations. Stunkard (1960) noted that there was tremendous variation in pedometer readings from one day to the next. Early mechanical pedometers worked on a ratchet and gear mechanism, but variations in spring tension and friction involved in the moving gears limited the validity of these instruments. For the most part, studies of mechanical pedometers found that the errors involved were too large for precise work (Meijer et al., 1991; Montoye, 1988). Gayle et al. (1977) had to perform individual calibrations on mechanical pedometers to make them reflect the true distance. Even so, the range of distances for a 1-mi treadmill walk ranged from 0.7 to 1.4 mi for 8 subjects. It should be noted that a constant stride length of 0.66 cm was assumed for all subjects. In addition, pedometer validity varied in treadmill walking when the speed was varied between 2 and 4 mph (54 and 107 m/min) (Gayle et al., 1977).

Instrumentation

Electronic pedometers were developed in the last decade. Table 10.1 shows some examples of commercially available devices that could be used by researchers. These are battery-operated devices containing a spring-suspended, horizontal lever arm that moves up and down. This motion opens and closes an electrical circuit in response to vertical accelerations of the waist that occur during walking and running. The electronic circuitry accumulates steps and provides a digital display. Some pedometers allow users to input step length and body mass for calculation of distance traveled and energy expended (kcal). Electronic pedometers are more accurate than the older, mechanical-style pedometers.

Bassett et al. (1996) tested several brands of electronic pedometers and found the Yamax DW-500 (Yamasa Corp., Tokyo) to have a high degree of validity and reliability. Although this model has been discontinued, a suitable replacement (the Yamax SW-200 pedometer) was tested in our laboratory and found to be very similar in terms of validity to the original Digi-Walker model. The SW

TABLE 10.1   Commercially Available, Waist-Mounted Electronic Pedometers

Brand, model	Functions	Retail price	Location of manufacturer or importer
Yamax SW-200	S	$25.00	Yamasa Corp.
Yamax SW-401	S, D	$29.95	Tokyo, Japan
Yamax SW-651	S, D, SW, T	$34.95	
Yamax SW-701	S, D, C	$34.95	
Accusplit Alliance AL 1500	S, D, C, SW, T	$19.95	Accusplit
Accusplit Ambassador A200F	D (auto stride adjust), 1-trip memory	$29.95	San Jose, CA
Accusplit Ambassador A2020W	D (auto stride adjust), 1-trip memory	$34.95	
Accusplit Ambassador AA207FW	D (auto stride adjust), 7-trip memory	$29.95	
Accusplit Ambassador A210F	D (auto stride adjust), 7-trip memory	$39.95	
Acumen Jogmate	S, D, SW, T	$26.95	Acumen
Acumen Jogmate Plus	S, D, SW, Speed, T	$29.00	Sterling, VA
Omron HJ-102	S, D, C, SW	$29.95	Omron Tokyo, Japan
NL-2000	S, C, T, 7-day memory	$49.95	New Lifestyles Kansas City, MO
Ultrak 255	S, D, C, SW	$19.00	Ultrak
Ultrak 275	S, D, C, SW, T	$26.00	Laundale, CA
Sportline Calorie Pedometer 340	D	$15.99	Sportline
Sportline Calorie Pedometer 245	S, D, C	$29.95	Campbell, CA
Sportline Calorie Pedometer 347	S, D, C, Speed	$29.95	
Sportline Fitness Pedometer 360	S, D, C, Speed, SW, T, 7-day memory	$39.95	
Sharper Image TP355	S, D, C, Speed, SW, T, 7-day memory	$24.95	Sharper Image San Francisco, CA
Freestyle 598 Pacer Pro	S, D, C, SW, Speed	$22.00	Freestyle Camarillo, CA

S = steps; D = distance; C = calories; SW = stopwatch; T = time of day.

series pedometer works on a different internal mechanism than the DW-500 and uses a coil spring rather than a hairspring device (see figure 10.2). This model is not widely available in retail stores, but it can be purchased for about $25 from New Lifestyles, Inc. (Kansas City, MO, 816-353-1721; **www.digiwalker.com**) and Accusplit (San Jose, CA, 408-432-8228; **www.accusplit.com**). Other Yamax models that work on the same principle as the SW-200 have additional functions, including the SW-701 (steps, calories, and distance) and the SW-651 (steps, stopwatch, time of day, and distance). Battery life for the Yamax SW series pedometer is approximately 3 years. Changing the battery can be accomplished by opening the case with a coin and replacing the lithium watch-type battery.

The Computer Science and Applications (CSA) actimeter (CSA model 7164) is a unidimensional accelerometer with a step-counting function included. This device has real-time data acquisition

Figure 10.2 Mechanism of the Yamax SW-200 electronic pedometer, showing the coil spring affixed to the plastic lever arm.

and an internal clock, enabling it to store information on the number of steps/min over a 21-day period. Thus, it has the advantage of being able to provide information on the "pattern" of physical activity, which normal pedometers cannot.

However, the cost is higher than for ordinary pedometers ($325 for the CSA actimeter and $450 for the computer interface). The device is now produced and supported by Manufacturing Technology Incorporated (Fort Walton Beach, FL).

Assessment Procedures

Electronic pedometers should be positioned by fastening them to the belt or waistband, in the midline of the thigh. If a pedometer were positioned in the mid-axillary line (under the armpit), errors would result. This occurs because the vertical accelerations on the opposite side of the body are often insufficient to cause a step to be recorded (Hatano, 1997). When the pedometer is correctly positioned, it does not matter what side of the body it is worn on (Bassett et al., 1996), unless the person has a gait abnormality.

A common protocol for field research is to have participants wear the pedometer for one week and average the results. This will reduce measurement variability and account for the observation that Sunday has lower step counts (Bassett et al., 1996; Sequeira et al., 1995). Longer sampling periods may be necessary in a group with more variability in steps/day. In training studies, researchers can ask participants to keep a continuous record of their step count, throughout the intervention period. Baseline pedometer measurements should *always* be obtained in training studies so that the increase in steps/day can be quantified.

The simplest method of using pedometers is to hit the reset button, secure the cover shut, and instruct the participant on the correct placement. The participant then resumes normal activity for several days. At the end of the sampling period, the pedometer is opened and the step total is divided by the number of days to yield steps/day. This method has the advantage of achieving a "blinded" measurement. However, it does not allow one to examine between-day variation, and the measurement period is limited by the capacity of the device (99,999 steps for the Yamax pedometer).

Another method of using pedometers is to rely on the participant to record total steps at the end of each day. Participants are told to put the pedometer on when they first awake and wear it until they turn in for the night. At the end of each day, they write down their steps for that day in a log (or calendar) and hit the reset button. It is also helpful if they write down the time of day when the pedometer is put on and taken off, for the purpose of verifying that it was worn throughout the day.

Some investigators have even experimented with mailing pedometers out. Sequeira et al. (1995) examined nearly 500 Swiss residents using this approach. A minor problem was that some steps were accumulated in transit, although this was found to be negligible (i.e., <1%) compared to the total step count for the week.

Outcome Variables

The pedometer records steps taken during activity and uses an electronic counter to accumulate these over time. The output can be dislplayed in various ways, including steps, distance, and calories.

Steps

In its simplest form, the pedometer is used as a step counter. Tudor-Locke and Myers (2001a, 2001b) have argued convincingly that research data should be reported as steps, since this is the most direct expression of pedometer output.

Distance

Pedometers can also estimate the distance traveled by multiplying the number of steps by stride length (distance = steps × stride length). Determinants of stride length include walking speed, height, age, and gender (Welk, Differding, et al., 2000). Stride length is related to walking speed, as demonstrated in table 10.2. Research in gait laboratories has established that the average walking speed for healthy men is about 3.3 mph (88 m/min) (Temes, 1994). At this walking speed, stride length is approximately

TABLE 10.2   Effects of Treadmill Walking Speed on Stride Length (m) in 10 Adults (5 Males, 5 Females)

	54 m/min	67 m/min	80 m/min	94 m/min	107 m/min
Males	0.60 ± 0.01	0.67 ± 0.01	0.74 ± 0.01	0.81 ± 0.02	0.89 ± 0.02
Females	0.59 ± 0.05	0.64 ± 0.03	0.71 ± 0.03	0.77 ± 0.04	0.83 ± 0.05

Values are mean ±SD.
Data from Bassett et al., 1996.

2.6 ft (0.79 m) and it would take approximately 2,000 steps to walk 1 mi. It is interesting to note that the word *mile* is derived from the Latin phrase *milia passuum* meaning "1,000 paces." The Roman mile represented 1,000 paces (or 2,000 steps) of a foot soldier. Japanese researchers have reported that stride length is roughly 42% of a person's height (Hatano, 1993; Ohtsuka, 1997), which concurs with our observations (Bassett et al., 1996) and those of Welk, Differding, et al. (2000).

Calories

Some pedometers have an algorithm to estimate calorie expenditure, based on steps and body weight. It is unclear if the calorie value is intended to reflect gross or net energy expenditure. In our laboratory, we assume that the calorie values displayed reflect the net cost of performing an activity (in excess of resting).

Measurement Properties: Validity

Bassett et al. (1996) studied five brands of electronic pedometers for measuring steps taken and distance walked. The pedometer brands examined were the Freestyle Pacer, Eddie Bauer, L.L. Bean, Yamax DW-500, and Accusplit. All were positioned on the belt or waistband, in the midline of the thigh. In part 1 of the study, the accuracy of five different pedometers was examined over a 3.03-mi (4.88K) sidewalk course. Individual stride length was determined by having the subjects take 20 steps down a hallway, measuring the total distance, and dividing by 20. Twenty subjects (aged 18 to 65 years) then walked a 4.88K sidewalk course while wearing two devices of the same brand (left and right hip) for each of six different trials. There were significant differences among pedometers (p < 0.05), with the Yamax, Accusplit, and Pacer pedometers approximating the actual distance more closely than the other brands (see figure 10.3).

The same study (Bassett et al., 1996) also compared the accuracy of pedometers on different surfaces (concrete sidewalk versus all-weather track). Ten of the original subjects completed additional trials on a 400-m cushioned outdoor track to examine the effect of walking surface on pedometer accuracy. The pedometers generally showed similar values for sidewalk and track surfaces.

The effects of walking speed on pedometer accuracy have been studied (Bassett et al., 1996). Subjects walked on a motorized treadmill at various speeds (i.e., 54, 67, 80, 94, and 107 m/min) to examine the effect of walking speed on the accuracy of pedometers (Eddie Bauer, Yamax, and Pacer). Stride length in this part of the study was determined at 3 mph (80 m/min). The Yamax DW-500 pedometer was significantly better at tracking distance and number of steps than the other models (see figure 10.4). The superior accuracy of the Yamax pedometer, especially evident at slower speeds, appeared to result from a greater sensitivity to vertical acceleration. At faster speeds, all three pedometer brands tended to underestimate distance. However, this was due to a lengthening of strides rather than a miscounting of steps.

Recently, the Yamax DW-500 pedometer was shown to provide a valid measure of total daily walking distance (Bassett, Cureton et al., 2000). Seventeen participants wore a pedometer for an entire

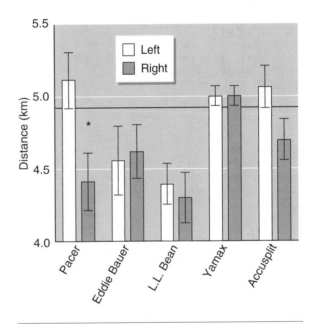

Figure 10.3 Comparison of five brands of electronic pedometers over a 4.88 K sidewalk course (N = 20, mean ± SE). The solid horizontal line in the figure indicates the actual distance walked. Significant differences (p < 0.05) among brands are as follows:

L.L. Bean Eddie Bauer Pacer Accusplit Yamax

Brands that are connected by a solid line are not significantly different, based on an average of data recorded from pedometers worn on the left and right sides of the body. *Significant difference between left and right side (p < 0.05).

Reprinted from Bassett et al., 1996.

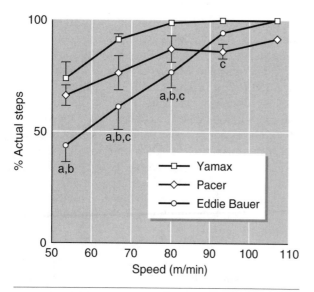

Figure 10.4 Effects of treadmill walking speed on percentage of actual steps taken, as recorded by three electronic pedometers (N = 10, mean ± SE): a = significant difference between Yamax and Eddie Bauer (p < 0.05); b = significant difference between Eddie Bauer and Pacer (p < 0.05); c = significant difference between Yamax and Pacer (p < 0.05).

Reprinted from Bassett et al., 1996.

day. For validation purposes, they rolled a calibrated measuring wheel (affixed to a 1-m aluminum shaft) wherever they walked. Stride length was determined from a 10-step procedure at the subject's usual walking speed. The correlation between the pedometer and measuring wheel values was 0.977. The average distances for the pedometer and measuring wheel were 5.42K ± 2.63 and 5.21K ± 2.58 (mean ± standard deviation [SD]), respectively. This indicates that pedometers can be used to tell how far an individual walks in a day.

Nelson et al. (1998) tested the validity of the Yamax DW-500 pedometer during level walking, comparing it to indirect calorimetry. The pedometer underestimated the gross energy expenditure by 27% at 2 mph and by 7% at speeds ≥ 3.5 mph. Thus, the Yamax pedometer has good validity for measuring energy expenditure for treadmill walking speeds from 3 to 4 mph, but it underestimates energy expenditure at speeds ≤ 2 mph. These findings based on treadmill research also extend to over-ground walking, since the fundamental mechanics of treadmill and overground locomotion are the same (Bassett et al., 1985; van Ingen Schenau, 1980).

Physical activity researchers are interested in using pedometers to measure not only walking but also overall physical activity (including occupational

tasks, activities of daily living, sports, recreation, and so on). Thus, the validity of pedometers for measuring energy expenditure of various activities has been studied. In 1998, the International Life Sciences Institute (ILSI) funded studies on the validity of motion sensors in the field (Bassett, Ainsworth et al., 2000; Hendelman et al., 2000; Welk, Blair et al., 2000). Oxygen consumption (expressed in METS) is one of the best ways to directly measure energy expenditure and, thus, served as a suitable gold standard in these studies. The recent development and validation of small, portable oxygen uptake systems (King et al., 1999; Melanson & Freedson, 1995) allowed these devices to serve as the criterion.

In one of the ILSI studies, participants (aged 19 to 74 years) performed a number of selected tasks from six general categories: yard work, housework, occupation, family care, conditioning, and recreation (Bassett, Ainsworth et al., 2000). During each activity, oxygen consumption was measured with a portable metabolic system. Participants wore the Yamax SW-701 pedometer and three accelerometers. The pedometer values (steps/min) were correlated with energy expenditure (i.e., r = 0.493) across all activities. The pedometer was valid for estimating the net energy cost of slow walking, averaging 78 m/min, and fast walking, averaging 100 m/min. However, the pedometer underestimated the cost of most other activities by an average of 1 MET. This was attributed to the pedometer's inability to detect arm movements and the additional cost incurred in external work (i.e., grade walking, lifting/carrying, and pushing objects).

The concurrent validity of the electronic pedometer in assessing physical activity in free-living subjects has also been examined. Leenders et al. (2000) compared four methods of estimating physical activity in college-aged females. Participants wore a Yamax DW-500 pedometer, Tritrac accelerometer, and CSA accelerometer; they also completed a 7-day physical activity recall (7DPAR). There were significant correlations between pedometer-measured steps/day and activity counts from the Tritrac and CSA (i.e., r = 0.84 to r = 0.93), indicating that a $25 pedometer gives similar information on total accumulated activity as a $325 device. Compared with 7DPAR, the Tritrac, CSA, and Yamax underestimated energy expenditure by 25, 46, and 48% respectively. This underestimation may result from the inability of waist-mounted motion sensors to detect certain types of movement.

Measurement Properties: Reliability

The reliability of pedometer data is also of interest to physical activity researchers. As with any measurement device, lack of reliability can result from analytical variability and biological variability. Analytical variability occurs when a device fails to provide reproducible data under the same set of circumstances. It can also refer to inter-device variability (i.e., differences between devices of the same brand). Biological variability, or behavioral variability, occurs when the actual level of physical activity changes from one measurement period to the next.

To date, only a few studies have examined technical sources of variability. Tryon et al. (1991) tested 19 Digitron pedometers (Gutmann Co., Mt. Vernon, NY) and found an approximate 5% error under laboratory conditions (e.g., shaking them with a mechanical oscillator). Another study (Bassett et al., 1996) had subjects wear two pedometers of the same brand on the left and right sides of the body, during a 4.88K (3.03-mi) walk. In general, inter-device reliability was high among most of the brands. There were significant differences between brands, however, which might pose difficulties for cross-study comparisons of pedometer data.

Tryon et al. (1991) had 39 college students walk a measured half-mile on two occasions and observed that the pedometer readings had good reliability. The average coefficient of variation was 2.75%, with a standard deviation of 3.89%. When pedometers are worn throughout the day, the situation is different. Large day-to-day fluctuations in pedometer data may exist. This lack of "reliability" is actually a strength rather than a weakness, however, because it results from the extreme sensitivity of pedometers to detect even small changes in ambulatory activity (Tudor-Locke & Myers, 2001a).

To improve the reliability of pedometer data, sampling periods longer than 1 day must be used. If pedometer data are collected over several days and then averaged to yield steps/day, a more representative picture of a person's ambulatory activity is obtained. Tudor-Locke and Myers (2001a) note that several studies have addressed the optimal sampling period for pedometer studies. Gretebeck and Montoye (1992) concluded that 5 to 6 days of pedometer data were needed to obtain data representative of "usual" physical activity (i.e., less than 5% error) in active young men. They recommend that both weekdays and weekend days be included.

Schonhofer et al. (1997) studied individuals with pulmonary disease over two 1-week periods, 1 month apart. The intra-class correlation coefficient for steps/day in their study was 0.94. Sieminski et al. (1997) studied individuals with peripheral vascular disease over two 2-day periods, separated by 1 week. They reported an intra-class correlation coefficient of 0.86 for steps/day, suggesting that the shorter sampling period may be sufficient for this patient population. Additional studies are needed to determine the appropriate measurement period for pedometers, although it appears that sedentary groups require shorter measurement periods than active groups (Tudor-Locke & Myers, 2001b).

Another important issue to consider when selecting the sampling period is seasonal variation. The pedometer is sensitive enough to capture seasonal fluctuations in physical activity. For example, Lee et al. (1987) used the pedometer to examine seasonal variation of physical activity in middle-aged and elderly women (aged 51 to 86 years) residing in Kentucky. For a 7-day period in June and July, 147 participants wore the pedometer; the following January and February, 130 of the participants wore the pedometer again for a 7-day period. The results from this study showed that average walking distance decreased significantly from the summer (1.6 ± 1.2 mi/day) to the winter (1.2 ± 0.8 mi/day, mean ± SD).

Acceptability and Utility of Pedometers

Electronic pedometers have gained widespread acceptance among physical activity researchers over the past decade. It has become common for researchers to use them in physical activity interventions, especially those that use walking as the primary mode. In addition, pedometers are often used as an adjunct to physical activity questionnaires since researchers now recognize the advantage to having a simple, accurate, and objective measure of ambulatory activity.

More recently, health promotion experts have begun using pedometers to encourage sedentary children and adults to increase their physical activity. One of the first intervention studies to use electronic pedometers was Project Active at the Cooper Institute for Aerobics Research (Dunn et al., 1999). Participants in the "lifestyle activities" group received pedometers halfway through the 24-week intervention phase. Participants who were inactive reported accumulating 3,000 to 5,000 steps/day. The researchers then had them set goals to gradually work up to 10,000 or more steps/day. Use of the pedometer was voluntary because the researchers wanted the participants to select a self-monitoring technique that worked best for them. However, most of the individuals who used a pedometer managed to accumulate at least 10,000 steps a day (A.L. Dunn, personal communication, December 5, 2000). Behavior modification researchers have hypothesized that pedometers could actually help motivate people to exercise.

Assumptions and Sources of Error With Pedometers

Because of their simplicity, electronic pedometers have some limitations as research tools (Bassett, 2000; Freedson & Miller, 2000). Although the best brands are fairly accurate at counting steps, they are less accurate for estimating distance and energy expenditure. Pedometers cannot discriminate between steps accumulated in walking, running, or stair climbing. Hence, their ability to predict energy expenditure is limited. Pedometers assume that a person expends a constant amount of energy per step. For the Yamax DW-500, this assumed value is about 0.55 cal/kg/step regardless of speed (Bassett, 2000), which is an oversimplification (Hatano, 1993). Like any belt-mounted motion sensor, pedometers cannot detect arm movements and external work done in pushing, lifting, or carrying objects. Another limitation of most pedometers is that they lack internal clocks and data storage ability; thus, they cannot provide information on the pattern of physical activity.

In obese individuals (i.e., body mass index ≥ 30), pedometer accuracy is compromised. For instance, Shepherd et al. (1999) reported that the Sportline pedometer (Campbell, CA) had a greater mean absolute error in obese subjects (i.e., 6.12%) than in non-obese subjects (i.e., 1.56%). Pedometer validity is also compromised in the frail elderly, who walk at slow speeds (Hoodless et al., 1994), and in those with gait abnormalities. To accurately measure walking in these groups, it may be necessary to use a pressure-sensitive footswitch pedometer (Hoodless et al., 1994) or a step-activity monitor that measures acceleration at the ankle (Smith et al., 1999).

Another source of pedometer error is mechanical vibration when operating a motor vehicle. On city streets, the number of steps accumulated in driving is negligible, but they can be a source of error when driving on dirt roads or operating a riding lawn mower. This source of error can be overcome by having participants remove the pedometer or wear it upside down, to prevent extra "steps"

from being recorded. Despite their limitations, pedometers are extremely useful for distinguishing between individuals or groups that vary in their level of ambulatory activity (Hatano, 1993; Sequeira et al., 1995).

Applications for Pedometers

The electronic counter is useful in a variety of settings where an inexpensive, objective measure of physical activity is needed. These range from cross-sectional research studies and investigations relating physical activity to health outcomes to longitudinal interventions designed to increase activity in sedentary groups.

Cross-Sectional and Correlational Studies

Cross-sectional studies are beginning to highlight the concurrent validity of pedometers. Walking activity is positively correlated to cardiovascular fitness and bone density, and negatively correlated with age, strength, and % overweight. Brief summaries of some of these studies are provided in this section.

Ichihara et al. (1996) used a pedometer to quantify weekly physical activity in 513 Japanese adults (40 to 64 years of age). Each participant wore a pedometer (Calorie Counter, Suzuken Co., Nagoya, Japan) for a 7-day period. On average, males took 8,107 ± 2,922 steps/day and females took 7,762 ± 3,301 steps/day (mean ± SD). Peak oxygen uptake values (ml/kg/min) ranged from 15.2 to 49.0 in males and from 12.7 to 38.5 in females. Participants were divided into fitness tertiles based on initial fitness levels. A positive association was found between increasing number of steps/day and fitness levels within both males and females. This association between steps/day and physical fitness has also been shown in children (Rowlands et al., 1999). For 3 to 6 days, 34 children from Bangor, North Wales, aged 8 to 10 years, wore a pedometer (Yamax DW-200). The average number of steps/day for boys and girls was 16,035 ± 5,999 and 12,729 ± 4,026 (mean ± SD), respectively. Steps/day correlated positively (i.e., r = 0.59) to fitness (i.e., endurance time during a Bruce maximal treadmill test) in the children.

Yanagimoto et al. (2000) examined the effect of the number of steps walked on bone quality in the elderly. Fifty-nine female subjects (aged 78 ± 8 years) with no impairments in walking or daily activities participated in the study. Bone quality was measured using an ultrasound bone density measuring device. Physical activity was determined from the total number of steps walked, measured by a pedometer (Kenz Select 2, Suzuken Corp., Japan). Each subject wore the pedometer for approximately 12 hours each day over a 2-week period. The total number of steps walked was 5,003 ± 3,182 steps/day (mean ± SD). The investigators found a significant correlation between the total number of steps walked and bone density (r = 0.606, p < 0.01). These cross-sectional results suggest that the more a person walks in daily life, the greater the benefit to bone quality.

Sequeira et al. (1995) studied 493 men and women, aged 25 to 74 years, from western Switzerland, as part of the World Health Organization Monitoring Trends and Determinants in Cardiovascular Disease (MONICA) project. They showed an age-related decline in activity in this population. The investigators had the participants wear a Pedoboy pedometer (Schwenningen, Germany) during work and leisure time for a 7-day period. The average number of steps/day decreased from 11,900 to 6,700 in men and from 9,300 to 7,300 in women between the age groups of 25 to 34 and 65 to 74 years. The authors also noted differences in the number of steps/hour accumulated by individuals in various occupations (see figure 10.5). An age-related decline in physical activity has also been shown by Hatano (1993) in a cross-sectional study that highlights the inverse relation between steps/ day and age across male and female Japanese subjects (see figure 10.6). Male and female participants in the 30- to 39-year age range accumulated an average of 8,240 and 7,233 steps/day, respectively. These values decreased an average of approximately 700 steps/day, per decade, for participants from 30 to 70 years of age.

Bassey et al. (1988) studied the association of muscle strength and activity in 122 males and females over the age of 65 years in Nottingham, England. All participants were living independently and were free of any disabling diseases. The isometric plantar flexor strength of the triceps surae (calf muscle) was determined. Subjects wore a step counter for all waking hours over a 1-week period. Male subjects averaged approximately 7,100 ± 3,100 steps/day, and the females averaged approximately 6,000 ± 4,000 steps/day (mean ± SD). The investigators found that strength decreased with advancing age. The decrease in strength was associated with step score for males (i.e., r = 0.30, p < 0.05) but not for females.

Tryon et al. (1992) observed an association between activity and % overweight. They studied 127

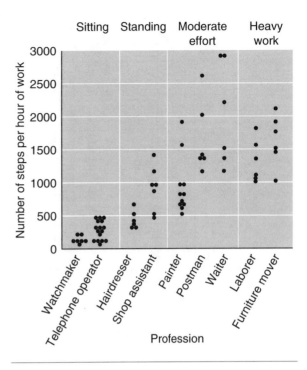

Figure 10.5 Distribution of the number of steps per hour of work in men and women in nine jobs representing different levels of occupational activity (i.e., sitting, standing, moderate effort, and heavy work). Each point represents one individual. Data are from the second MONICA survey in Switzerland.

Reprinted from Sequeira et al., 1995.

women aged 19 to 55 years ranging from 14% underweight to 99% overweight, as determined from the Metropolitan Life Insurance tables of desirable weights. Participants wore a pedometer (Digitron Jog-Walk, Gutmann Co., NY) for 14 consecutive days. Average mileage, along with the amount of time the pedometer was worn each day, was recorded. Because of the cross-sectional design of this study, conclusions about cause-and-effect relationships were not drawn, but regression analysis found activity to be inversely related to % overweight. A similar relationship has been seen in children (Rowlands et al., 1999). Steps/day have been inversely associated with % body fat, as estimated by the seven-site skinfold technique (r = –0.42, p < 0.05).

Use in Patient Populations

Pedometers are also being used to quantify physical activity in diseased and disabled populations. Sieminski et al. (1997) performed a cross-validation study of pedometers and physical activity questionnaires in a group of 43 patients with peripheral arterial occlusive disease. Participants, with a mean

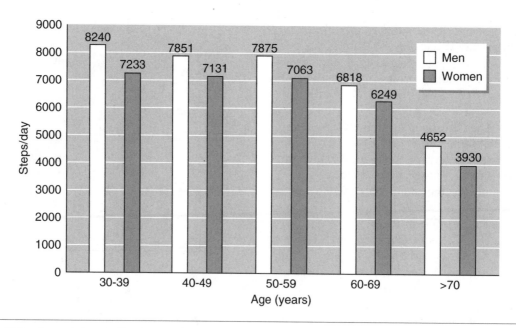

Figure 10.6 Age-related decline in steps/day in Japanese residents.
Reprinted from Hatano, 1997.

age of 65 ± 7 years, wore a pedometer for 2 days. Average steps/day were 4,615 ± 2,839 and 4,498 ± 2,768 (mean ± SD). The relationship between the steps obtained from the pedometer and time spent in leisure activity from the Minnesota Leisure-Time Physical Activity and NASA/Johnson Space Center Physical Activity Scale questionnaire were significantly correlated: $r = 0.46$ and $r = 0.51$, respectively. Shaukat et al. (1995) studied Indo-origin and northern European relatives of patients with coronary artery disease. They found that a 7-day recall of physical activity significantly correlated with weekly pedometer readings (i.e., $r = 0.84$ for Indo-origin, and $r = 0.82$ for northern European).

Gardner et al. (1997) examined the association between cigarette smoking and free-living activity in older patients with claudication. In this study, 34 smokers and 43 nonsmokers (or former smokers) were included. The smokers and nonsmokers had similar levels of peripheral arterial occlusive disease severity. After adjustment for age, weight, body mass index (BMI), % body fat, and calf blood flow, it was found that smokers took an average of 4,039 ± 1,760 steps/day in comparison to the nonsmokers who took an average of 5,684 ± 2,235 steps/day. This 29% difference was statistically significant ($p = 0.003$).

Hoodless et al. (1994) compared normal daily activity in 18 patients with chronic heart failure to 10 age-matched healthy controls (48 to 79 years). Both groups wore a shoe-mounted pedometer during all waking hours for a 7-day period. The aver-

age number of steps/day for the control group was 8,848 ± 956 steps/day, whereas the chronic heart failure patients accumulated 3,540 ± 387 steps/day (mean ± SE). Data from this study highlighted that patients with chronic heart failure had a 60% reduction in customary activity.

To examine the potential effect of physical activity on artificial joint wear, Schmalzried et al. (1998) measured the number of steps/day, for 8 days, in 111 volunteers who had at least one total hip or knee replacement. Patients wore a pedometer for an average of 15 hours a day, and recorded an average of 4,988 steps/day (the range: 395 to 17,718 steps/day). There was a 45-fold range in steps/day among all individuals, which is consistent with the wide variability in rates of polyethylene wear. Therefore, the rate of prostheses wear could be a function of activity and not time. Thus, it would appear that a quantitative approach using steps/day might prove advantageous in the prediction of lower-extremity joint wear. In addition, regression analysis revealed that age was significantly associated with decreasing number of steps/day ($p = 0.048$). This age-related decline was similar to that observed in healthy populations (Hatano, 1997; Sequeira et al., 1995).

McClung et al. (2000) examined the relationship between steps/day and BMI in individuals with a joint replacement (n = 151) and individuals without a joint replacement (n = 58), aged 22 to 82 years. Individuals who did not have a total joint replacement averaged significantly more steps/day than those with a total joint replacement: 7,781 steps/

day versus 5,733 steps/day, respectively (p < 0.001). The average BMI of the subjects without a total joint replacement was 24.9 kg/m², whereas the average BMI within the subjects with a total joint replacement was 29.3 kg/m². This study was not designed to evaluate causal relationships, but it would appear that the main implication from its findings is that an increased BMI among patients with a total joint replacement is associated with decreased ambulatory activity.

In another study, Belcher et al. (1997) examined the functional outcome of operatively treated malleolar fracture patients. This retrospective study compared physical activity levels in 40 patients (16 to 78 years of age) who had undergone operative fixation of a malleolar fracture within the last 8 to 24 months to an age-matched control group of healthy subjects (22 to 68 years of age). The participants wore a pedometer for all waking hours over a 7-day period. The average number of steps/day for the control group was 7,607 ± 2,859, compared to 4,838 ± 3,252 steps/day (mean ± SD) in the malleolar fracture patients. This study did not identify factors contributing to functional outcome, but it did suggest that ambulatory activity is lower in patients who have undergone surgery for a malleolar fracture.

Intervention Studies

Investigators have also used pedometer scores as an outcome measure in physical activity intervention studies. Pedometers can be used to quantify how many steps a person takes, and they are a valid, objective method of tracking ambulatory activity. An early study using mechanical pedometers investigated the effects of an unsupervised walking program in 108 middle-aged factory workers in Great Britain (Bassey et al., 1983). Participants were randomly assigned to three groups and encouraged to begin a 12-week program in the first, second, or third periods of the study. Those individuals who adhered to the program increased from 11,400 ± 600 steps/day (mean ± SE) to 12,700 ± 600 steps/day (p < 0.05). At least some of the increase was maintained 12 weeks after the intervention period was over.

Yamanouchi et al. (1995) performed a physical activity intervention, using a pedometer (H7-7, Omron Industries, Tokyo) in conjunction with diet therapy to improve insulin sensitivity and reduce body weight. The subject population was divided into a diet-only group (N = 10) and a diet-plus-exercise group (N = 14). All subjects were hospitalized and consumed a similar diet (1000 kcal below

their usual self-reported intake) over the 6- to 8-week intervention period. The diet-plus-exercise group was instructed to walk at least 10,000 steps/day. During the intervention, the diet-only group and the diet-plus-exercise group walked a mean of 4,500 ± 290 steps/day and 19,200 ± 2,100 steps/day (mean ± SE), respectively. The diet-plus-exercise group had significantly greater weight loss than the diet-only group: 7.8 ± 0.8 versus 4.2 ± 0.5 kg, respectively. After training, the diet-plus-exercise group showed improvements in insulin sensitivity that were not seen in the diet-only group. This study also found a significant correlation (i.e., r = 0.73) between the change in insulin sensitivity and steps/day. Thus, accumulating 19,200 steps/day in conjunction with diet therapy is an effective strategy for reducing body weight and improving insulin sensitivity.

A study at the University of Tennessee showed the efficacy of using the Yamax SW-200 pedometer in a walking program in postmenopausal women with borderline or mild hypertension, aged 54 ± 2 years (Moreau et al., 2001). Fifteen women in the intervention group increased the number of steps taken from a mean baseline of 5,400 ± 500 steps/day to 9,700 ± 400 steps/day (mean ± SE). The control group (N = 9) was instructed not to change activity habits. Resting systolic blood pressure declined by 11 mmHg over 6 months in the intervention group, with no change noted in diastolic blood pressure. The control group experienced no changes in either systolic or diastolic blood pressure. Body mass was reduced by 1.3 kg in the exercise group, but this was not correlated with the change in blood pressure. It would appear from the results of this study that increasing walking activity was associated with significant reductions in systolic blood pressure in postmenopausal women with borderline or mild hypertension.

Fogelholm et al. (2000) examined the effects of a walking program in 82 obese women. After 12 weeks on a very low energy diet, the women had lost an average of 13.1 kg and they were randomized into three groups. During the ensuing weight-maintenance phase of the study, the control group received diet counseling only, while the walk-1 group and the walk-2 group received diet counseling plus a walking program targeted to expend 4.2 MJ/week and 8.2 MJ/week, respectively. The results showed that walkers regained less weight during the weight-maintenance phase. Prior to weight loss, the women were taking 7,410 ± 3,300 steps/day (mean ± SD), but by the end of the maintenance program, the walk-2 group was taking an average of 3,040 steps/day more than the control group. After 2 years of unsupervised follow-up, all three groups were once again similar in their number of steps/day.

10,000 Steps Per Day: A Public Health Recommendation

Hatano (1993, 1997) has proposed a recommendation of accumulating 10,000 steps/day to confer health benefits. This target was originally established in Japan by evaluating the number of steps/day with electronic pedometers. This investigator found that increased ambulatory activity was associated with lower levels of blood pressure and subcutaneous fat (Hatano, 1997). Hatano also estimated that an average-sized Japanese male would expend 333 kcal/day in walking 10,000 steps/day. This is a level of physical activity that appears to provide some degree of protection against heart attacks (Paffenbarger et al., 1978). In Japan, the term for *pedometer* is *manpo-kei*, which means "10,000 step meter." The pedometer, along with the 10,000 steps/day recommendation, has received considerable media attention over the last few years (Hellmich, 1999; Krucoff, 1999).

As a public health recommendation, a target number of steps/day may have considerable merit. The step recommendation could apply to most adults, regardless of body size. In contrast, recommendations based on kcal do not apply equally to large and small individuals. The step recommendation has a rough adjustment for physical fitness built into it. For example, a young individual could accumulate steps through running or vigorous sports, a middle-aged individual could perform brisk walking, and an elderly person could walk at a more moderate pace. Researchers have observed that overweight individuals who have sedentary occupations and perform little leisure-time physical activity tend to accumulate around 5,000 to 6,000 steps/day (Moreau et al., 2001). Increasing this number to 10,000 steps/day would be approximately equivalent to walking an additional 2 mi a day. This increase in physical activity is in accordance with national recommendations set forth by the ACSM, the CDC, and the U.S. Surgeon General (Pate et al., 1995; U.S. Department of Health and Human Services, 1996). Furthermore, the 10,000 steps/day recommendation has a valuable degree of simplicity, and individuals can receive immediate feedback on activity levels accumulated.

The 10,000 steps/day recommendation could be beneficial to a sedentary population whose barriers to physical activity include a dislike of vigorous exercise and a lack of access to traditional fitness center facilities. The HealthPartners Center for Health Promotion in Minnesota has recently implemented a Getting Started with 10,000 Steps program. Pilot work for this program looks extremely promising for increasing the activity level of previously inactive individuals.

Welk, Differding et al. (2000) compared the 10,000 steps/day recommendation to the traditional public health activity guidelines and found reasonably good correspondence between them. Participants who performed >30 min of total activity in a day reached the 10,000 figure 73% of the time. Conversely, 29% of those who did <30 minutes of self-reported activity (in a 7DPAR) still reached 10,000 steps. In looking at the individual data, they found some people with >10,000 steps/day who reported little or no activity on the 7DPAR and other people who were "active" but recorded <10,000 steps/day. Welk concludes that much of the variability is associated with how many steps a person accrues through activities of daily living. Since these steps are probably of light intensity and scattered throughout the day, they may not count as part of a person's 30 min of moderate-intensity activity.

While the pedometer does not capture all activities (e.g., swimming and cycling), a simple recommendation of accumulating 10,000 steps/day remains attractive for the reasons aforementioned. It should be emphasized, however, that at present this recommendation has not undergone comprehensive scientific investigation to establish its true physiological effect. There is a strong scientific need to establish the health benefits of accumulating a targeted number of steps/day.

Summary

The notion of using a device to measure walking has existed for about five centuries. Early mechanical pedometers were used primarily for surveying, but they were not accurate enough for precise scientific work. The recently developed electronic pedometer is a valid and reliable assessment tool for measuring ambulatory activities. This device is worn on the belt or waistband and responds to vertical accelerations of the body. Data provided by the pedometer include steps, distance, and energy expenditure.

Researchers have used pedometers to measure ambulatory activity patterns in a variety of populations. Studies have already demonstrated the practicality of using pedometers in moderate-sized cohorts of approximately 500 participants (Hatano, 1993; Ichihara et al., 1996; Sequeira et al., 1995). There is a clear age-related decline in walking and some intriguing reports of differences between various groups. Intervention studies have successfully used pedometers to track the increase in activity resulting from walking programs. Preliminary reports suggest that pedometers may help motivate people to exercise by giving them a well-defined goal.

Future studies are needed to establish normative data on steps/day and to define values corresponding to inactive, moderately active, and active groups. Prospective studies are needed to examine the relationship between steps/day and cardiovascular or all-cause mortality. Finally, more research is needed to determine if pedometers are useful in encouraging sedentary individuals to begin and maintain more active lifestyles.

References

Ainsworth, B.E., Leon, A.S., Richardson, M.T., Jacobs, D.R., & Paffenbarger, R.S. (1993). Accuracy of the college alumnus physical activity questionnaire. *Journal of Clinical Epidemiology, 46,* 1403-1411.

Ainsworth, B.E., Montoye, H.J., & Leon, A.S. (1994). Methods of assessing physical activity during leisure and work. In C. Bouchard, R.J. Shepard, & T. Stephens (Eds.), *Physical activity, fitness, and health* (pp. 146-159). Champaign, IL: Human Kinetics.

Baranowski, T. (1988). Validity and reliability of self report measures of physical activity: An information-processing perspective. *Research Quarterly for Exercise and Sport, 59,* 314-327.

Bassett, D.R. (2000). Validity and reliability issues in objective monitoring of physical activity. *Research Quarterly for Exercise and Sport, 71,* 30-36.

Bassett, D.R., Ainsworth, B.E., Leggett, S.R., Mathien, C.A., Main, J.A., Hunter, D.C., & Duncan, G.E. (1996). Accuracy of five electronic pedometers for measuring distance walked. *Medicine and Science in Sports and Exercise, 28,* 1071-1077.

Bassett, D.R., Ainsworth, B.E., Swartz, A.M., Strath, S.J., O'Brien, W.L., & King, G.A. (2000). Validity of four motion sensors in measuring moderate intensity physical activity. *Medicine and Science in Sports and Exercise, 32,* S471-S480.

Bassett, D.R., Cureton, A.L., & Ainsworth, B.E. (2000). Measurement of daily walking distance—questionnaire versus pedometer. *Medicine and Science in Sports and Exercise, 32,* 1018-1023.

Bassett, D.R., Giese, M.D., Nagle, F.J., Ward, A., Raab, D.M., & Balke, B. (1985). Aerobic requirements of overground versus treadmill running. *Medicine and Science in Sports and Exercise, 17,* 447-481.

Bassey, E.J., Bendall, M.J., & Pearson, M. (1988). Muscle strength in the triceps surae and objectively measured customary walking activity in men and women over 65 years of age. *Clinics in Science, 74,* 85-89.

Bassey, E.J., Patrick, J.M., Irving, J.M., Blecher, A., & Fentem, P.H. (1983). An unsupervised "aerobics" physical training programme in middle-aged factory workers: Feasibility, validation, and response. *European Journal of Applied Physiology, 52,* 120-125.

Belcher, G.L., Radomisli, T.E., Abate, J.A., Stabile, L.A., & Trafton, P.G. (1997). Functional outcome analysis of operatively treated malleolar fractures. *Journal of Orthopedic Trauma, 11,* 106-109.

Blair, S.N., Dowda, M., Pate, R.R., Kronenfeld, J., Howe, H.G., Jr., Parker, G., Blair, A., & Fridinger, F. (1991). Reliability of long-term recall of participation in physical activity by middle-aged men and women. *American Journal of Epidemiology, 133,* 266-275.

Crespo, C.J., Keteyian, S.J., Heath, G.W., & Sempos, C.T. (1996). Leisure-time physical activity among U.S. adults. *Archives of Internal Medicine, 156,* 93-98.

Dumbauld, E. (1946). *Thomas Jefferson: American tourist.* Norman, OK: University of Oklahoma Press.

Dunn, A.L., Marcus, B.H., Kampert, J.B., Garcia, M.E., Kohl, H.W., III, & Blair, S.N. (1999). Comparison of lifestyle and structured interventions to increase physical activity and cardiorespiratory fitness: A random trial. *JAMA, 281,* 327-334.

Fletcher, G.F. (1997). How to implement physical activity in primary and secondary prevention: A statement for healthcare professionals from the Task Force on Risk Reduction, American Heart Association. *Circulation, 96,* 355-357.

Fogelholm, M., Kukkonen-Harjula, K., Nenonen, A., & Pasanen, M. (2000). Effects of walking training on weight maintenance after a very-low-energy diet in premenopausal obese women. *Archives of Internal Medicine, 160,* 2177-2184.

Freedson, P.S., & Miller, K. (2000). Objective monitoring of physical activity using motion sensors. *Research Quarterly for Exercise and Sport, 71,* 21-29.

Gardner, A.W., Sieminski, D.J., & Killewick, L.A. (1997). The effect of cigarette smoking on free-living daily physical activity in older claudication patients. *Journal of Vascular Diseases, 48,* 947-955.

Gayle, R., Montoye, H.J., & Philpot, J. (1977). Accuracy of pedometers for measuring distance walked. *Research Quarterly, 48,* 632-636.

Gibbs-Smith, C. (1978). *The inventions of Leonardo da Vinci.* London: Phaidon Press.

Gretebeck, R., & Montoye, H. (1992). Variability of some objective measures of physical activity. *Medicine and Science in Sports and Exercise, 24,* 1167-1172.

Hakim, A.A., Curb, J.D., & Petrovitch, H. (1999). Effects of walking on coronary heart disease in elderly men: The Honolulu Heart Program. *Circulation, 100,* 9-13.

Hakim, A.A., Petrovitch, H., Burchfield, C.M., Ross, G.W., Rodriquez, B.L., White, L.R., Yano, K., Curb, J.D., & Abbott, R.D. (1998). Effects of walking on mortality among nonsmoking retired men. *New England Journal of Medicine, 338,* 94-99.

Hatano, Y. (1993). Use of the pedometer for promoting daily walking exercise. *International Council for Health, Physical Education, and Recreation, 29,* 4-8.

Hatano, Y. (1997). Prevalence and use of pedometer. *Research Journal of Walking, 1,* 45-54.

Hellmich, N. (1999). Journey to better fitness starts with 10,000 steps. *USA Today,* June 29, 8-9.

Hendelman, D., Miller, K., Bagget, C., Debold, E., & Freedson, P. (2000). Validity of accelerometry for the assessment of moderate intensity physical activity in the field. *Medicine and Science in Sports and Exercise, 32,* S442-S449.

Hoodless, D.J., Stainer, K., Savic, N., Batin, P., Hawkins, M., & Cowley, A.J. (1994). Reduced customary activity in chronic heart failure: Assessment with a new shoe-mounted pedometer. *International Journal of Cardiology, 43,* 39-42.

Ichihara, Y., Hattori, R., Anno, T., Okuma, K., Yokoi, M., Mizuno, Y., Iwatsuka, T., Ohta, T., & Kawamura, T. (1996). Oxygen uptake and its relation to physical activity and other coronary risk factors in asymptomatic middle-aged Japanese. *Journal of Cardiopulmonary Rehabilitation, 16,* 378-385.

King, G.A., McLaughlin, J.E., Howley, E.T., Bassett, D.R., & Ainsworth, B.E. (1999). Validation of Aerosport KB1-C portable metabolic system. *International Journal of Sports Medicine, 20,* 304-308.

Krucoff, C. (1999). 10,000 steps to better health. *The Washington Post,* November 23, 16.

Lee, C.J., Lawler, G.S., Panemangalore, M., & Street, D. (1987). Nutritional status of middle-aged and elderly females in Kentucky in two seasons. Part 1, Body weight and related factors. *Journal of the American College of Nutrition, 6,* 209-215.

Leenders, N.Y.J.M., Sherman, W.M., & Nagaraja, H.N. (2000). Comparison of four methods of estimating physical activity in adult women. *Medicine and Science in Sports and Exercise, 32,* 1320-1326.

Manson, J.E., Hu, F.B., Rich-Edwards, J.W., Colditz, G.A., Stampfer, M.J., Willett, W.C., Speizer, F.E., & Hennekens, C.H. (1999). A prospective study of walking as compared with vigorous exercise in the prevention of coronary heart disease in women. *New England Journal of Medicine, 341,* 650-658.

McClung, C.D., Zahiri, C.A., Higa, J.K., Amstutz, H.C., & Schmalzried, T.P. (2000). Relationship between body mass index and activity in hip or knee arthroplasty patients. *Journal of Bone and Joint Surgery, 18,* 35-39.

Meijer, G.A.L., Westerterp, K.R., Verhoeven, F.M.H., Koper, H.B.M., & Hoor, F.T. (1991). Methods to assess physical activity with special reference to motion sensors and accelerometers. *IEEE Transactions in Biomedical Engineering, 38,* 221-228.

Melanson, E.L., & Freedson, P.S. (1995). Validity of the Computer Science and Applications, Inc. (CSA) activity monitor. *Medicine and Science in Sports and Exercise, 27,* 934-940.

Montoye, H.J. (1988). Use of movement sensors in measuring physical activity. *Science and Sports, 3,* 223-236.

Montoye, H.J., Kemper, H.C.G., Saris, W.M., & Washburn, R.A. (1995). *Measuring physical activity and energy expenditure.* Champaign, IL: Human Kinetics.

Moreau, K.L., DeGarmo, R., Langley, J., McMahon, C., Howley, E.T., Bassett, D.R., & Thompson, D.L. (2001). Increasing daily walking lowers blood pressure in postmenopausal women. *Medicine and Science in Sports and Exercise, 33,* 1825-1831.

Nelson, T.E., Leenders, N.Y.J.M., & Sherman, W.M. (1998). Comparison of activity monitors worn during treadmill walking. *Medicine and Science in Sports and Exercise, 30,* S11 [Abstract].

Nock, A.J. (1926). *Jefferson.* New York: Harcourt, Brace, and Co.

Ohtsuka, T. (1997). Factors regulating the volitional walking velocity among various age and different sex groups. *Circular of Tests and Measurement Research Division, Japanese Society of PE, 52,* 109-113.

Paffenbarger, R.S., Wing, A.L., & Hyde, R.T. (1978). Physical activity as an index of heart attack risk in college alumni. *American Journal of Epidemiology, 108,* 161-175.

Palletta, L.A., & Worth, F. (Eds.) (1993). *World almanac of presidential facts.* New York: Pharos.

Pate, R.R., Pratt, M., Blair, S.N., Haskell, W.L., Macera, C.A., Bouchard, C., Buchner, D., Ettinger, W., Heath, G.W., King, A.C., Kriska, A., Leon, A.S., Marcus, B.H., Morris, J., Paffenbarger, R.S., Patrick, K., Pollock, M.L., Rippe, J.M., Sallis, J., & Wilmore, J.H. (1995). Physical activity and public health: A recommendation from the Centers for Disease Control and Prevention and the American College of Sports Medicine. *JAMA, 273,* 402-407.

Richardson, M., Ainsworth, B., Jacobs, D., & Leon, A. (1996). Validity of the Godin physical activity questionnaire in assessing leisure time physical activity. *Medicine and Science in Sports and Exercise, 28,* S32 [Abstract].

Richardson, M.T., Leon, A.S., Jacobs, D.R., Ainsworth, B.E., & Serfass, R. (1994). Comprehensive evaluation of the Minnesota Leisure-Time Physical Activity Questionnaire. *Journal of Clinical Epidemiology, 47,* 271-281.

Rowlands, A.V., Eston, R.G., & Ingledew, D.K. (1999). Relationships between activity levels, aerobic fitness, and body fat in 8- to 10-yr-old children. *Journal of Applied Physiology, 86,* 1428-1435.

Sallis, J.F., Haskell, W.L., Wood, P.D., Fortmann, S.P., Rogers, T., Blair, S.N., & Paffenbarger, R.S. (1985). Physical activity assessment methodology in the Five-City Project. *American Journal of Epidemiology, 121,* 91-106.

Schmalzried, T.P., Szuszczewicz, E.S., Northfield, M.R., Akizuki, K.H., Frankel, R.E., Belcher, G., & Amstutz, H.C. (1998). Quantitative assessment of walking activity after total hip or knee replacement. *Journal of Bone and Joint Surgery-America, 80A,* 54-59.

Schonhofer, B., Ardes, P., Geibel, M., Kohler, D., & Jones, P.W. (1997). Evaluation of a movement detector to measure daily activity in patients with chronic lung disease. *European Respiratory Journal, 10,* 2814-2819.

Sequeira, M.M., Richardson, M., Wietlisbach, V., Tullen, B., & Schutz, Y. (1995). Physical activity assessment using a pedometer and its comparison with a questionnaire in a large population study. *American Journal of Epidemiology, 142,* 989-999.

Shaukat, N., Douglas, J.T., Bennett, J.L., & Bono, D.P.D. (1995). Can physical activity explain the differences in insulin levels and fibrinolytic activity between young Indo-origin and European relatives of patients with coronary artery disease? *Fibrinolysis, 9,* 55-63.

Shepherd, E.F., Toloza, E., McClung, C.D., & Schmalzried, T.P. (1999). Step activity monitor: Increased accuracy in quantifying ambulatory activity. *Journal of Orthopaedic Research, 17,* 703-708.

Sieminski, D.J., Cowell, L.L., Montgomery, P.S., Pillai, S.B., & Gardner, A.W. (1997). Physical activity monitoring in patients with peripheral arterial occlusive disease. *Journal of Cardiopulmonary Rehabilitation, 17,* 43-47.

Smith, G., Boone, D.A., Joseph, A.W., & Aguila, M.A.D. (1999). Step activity monitor: Long-term, continuous recording of ambulatory function. *Journal of Rehabilitation Research and Development, 1,* 1-12.

Stunkard, A. (1960). A method of studying physical activity in man. *American Journal of Clinical Nutrition, 8*, 595-600.

Temes, W.C. (1994). Cardiac rehabilitation. In E.A. Hillegass & H.S. Sadowsky (Eds.), *Essentials of cardiopulmonary physical therapy* (pp. 633-675). Philadelphia: W.B. Saunders.

Tryon, W.W., Goldberg, J.L., & Morrison, D.F. (1992). Activity decreases as percent overweight increases. *International Journal of Obesity, 16*, 591-595.

Tryon, W.W., Pinto, L.P., & Morrison, D.F. (1991). Reliability assessment of pedometer activity measurements. *Journal of Psychopathology and Behavioral Assessment, 13*, 27-43.

Tudor-Locke, C.E., & Myers, A.M. (2001a). Challenges and opportunities for measuring physical activity in sedentary adults. *Sports Medicine, 31*, 91-100.

Tudor-Locke, C.E., & Myers, A.M. (2001b). Methodological considerations for researchers and practitioners using pedometers to measure physical (ambulatory) activity. *Research Quarterly for Exercise and Sport, 72*, 1-12.

U.S. Department of Health and Human Services. (1996). *Physical activity and health: A report of the Surgeon General.* Atlanta, GA: U.S. Department of Health and Human Services, Centers for Disease Control and Prevention.

van Ingen Schenau, G.J. (1980). Some fundamental aspects of the biomechanics of overground versus treadmill locomotion. *Medicine and Science in Sports and Exercise, 12*, 257-261.

Washburn, R., Chin, M.K., & Montoye, H.J. (1980). Accuracy of pedometer in walking and running. *Research Quarterly, 51*, 695-702.

Welk, G.J., Blair, S.N., Wood, K., Jones, S., & Thompson, R. (2000). A comparative evaluation of three accelerometry-based physical activity monitors. *Medicine and Science in Sports and Exercise, 32*, S489-S497.

Welk, G.J., Differding, J.A., Thompson, R.W., Blair, S.N., Dziura, J., & Hart, P. (2000). The utility of the Digi-Walker step counter to assess daily physical activity patterns. *Medicine and Science in Sports and Exercise, 32*, S481-S488.

Wilson, D.L., & Stanton, L. (Eds.) (1999). *Jefferson abroad.* New York: Modern Library.

Yamanouchi, K., Shinozaki, T., Chikada, K., Toshihiko, N., Katsunori, I., Shimizu, S., Ozawa, N., Suzuki, Y., Maeno, H., Kato, K., Oshida, Y., & Sato, Y. (1995). Daily walking combined with diet therapy is a useful means for obese NIDDM patients not only to reduce body weight but also to improve insulin sensitivity. *Diabetes Care, 18*, 775-778.

Yanagimoto, Y., Oshida, Y., & Sato, Y. (2000). Effects of walking on bone quality as determined by ultrasound in the elderly. *Scandinavian Journal of Medicine and Science in Sports, 10*, 103-108.

CHAPTER

11

Use of Direct Observation to Assess Physical Activity

Thomas L. McKenzie, PhD
San Diego State University

Direct (systematic) observation has a long history in the study of human behavior in natural settings (Ciminero et al., 1977; Hartman, 1982), including sport and physical education environments (Darst et al., 1983). While a common approach to behavioral research and essential in a wide range of applied disciplines (e.g., psychology, sociology, zoology), systematic observation has been viewed as labor intensive and tedious (Montoye et al., 1996) and has been historically overlooked by physical activity researchers. Because direct observation exceeds other measures of physical activity in providing contextually rich data, it has recently received increased attention by researchers using ecological and cognitive-behavioral approaches in their attempt to understand how physical activity is influenced by physical and social environments. Recent technological advances that permit observational codes to be entered, stored, and analyzed by handheld computers make this methodology of assessing physical activity much more useful and appealing.

Principles of Method

The essence of direct observation is to classify free living physical activity behaviors into distinct categories that can be quantified and analyzed in greater detail. This technique is particularly important in assessing physical activity in health-related research. Physical activity is contextual and always influenced by environmental factors. The importance of direct observation rests not only on its ability to measure physical activity but also on its capacity to identify the type of activity and when, where, and with whom it occurs.

Direct observation is particularly important in assessing the physical activity levels of children. Self-report methodology for physical activity has substantial limitations related to estimating time and recall bias, and accelerometer data do not provide information on activity contexts. Questionnaires and interviews are not very useful when used with children below the fourth-grade level because most children are inept at recalling the frequency, duration, and levels of their physical activity. The use of direct observation was supported by recent meta-analysis, which found that the size of the relationship between body fat and physical activity in children depended on the type of activity measure used (Rowlands et al., 2000). The authors recommended that direct methods of assessing physical activity, including observations and motion sensors, be used when investigating the relationship of activity levels with health instead of indirect methods (i.e., questionnaires and heart rate monitoring).

Instrumentation

Systematic observation is a procedure or method for generating data, not a single technique or instrument. Observation has been incorporated into numerous instruments or systems developed to study children's physical activity (e.g., Behaviors of Eating and Physical Activity for Children's Health: Evaluation System [BEACHES], Children's Activity Rating Scale [CARS], Fargo Activity Timesampling Survey [FATS], System for Observing Fitness Instruction Time [SOFIT]). Systems are designed for a specific purpose and include a definition of the behavioral categories of interest, protocols for use (e.g., pacing of observations, data entry, data summaries), and coding conventions (i.e., interpretations of common scenarios). Within a system, various observational recording techniques may be used to collect data, but they primarily measure the frequency (number of times), duration (length), and latency (time between a stimulus and onset of the behavior) of behaviors. Systems designed for observing physical activity also include various codes for scoring the context in which activity occurs. Contextual codes are usually intended for a particular environment (e.g., home, recess, physical education class); however, some instruments are designed for coding in diverse settings.

Review of Published Physical Activity Observation Instruments

Several years ago, a review of instruments that focused directly on measuring children's physical activity was published (McKenzie, 1991). This section updates that review and compares nine instruments on 11 characteristics. To be included in this review, the instrument's physical activity codes had to have been validated by an energy expenditure measure. Table 11.1 compares three instruments

designed specifically to observe children's physical activity in schools, and table 11.2 compares six instruments designed for more diverse settings. The categories presented in the tables are described as follows:

- Name: Identifies the name of the instrument. If it did not have a specific name, one was provided.
- Citation: Provides the initial reference used in the review of the system. Subsequent papers for some systems, including CARS (DuRant et al., 1993) and SOFIT, have also been published.
- Location: Identifies settings where the instrument has been used. "Diverse" indicates that the system may be used in any location.
- Observation strategy: Identifies the specific observation method the system uses (momentary time sampling = coding events that are occurring at the end of the observation interval; partial time sampling = coding events that have occurred during a specified time interval, usually 5 to 20 s in length). Even though observers might watch a subject continuously, a partial or momentary time sampling system could be used to record the data.
- Activity categories: Specifies the number and briefly identifies the activity categories coded.
- Validation: Identifies variables that have been correlated with the activity codes to permit energy expenditure to be estimated. Details on the validity of direct observation techniques are provided in a subsequent section.
- Data summary: Identifies the primary summary variables. Additional summary variables may be calculated for most instruments.
- Associated variables: Identifies associated variables typically coded simultaneously with the physical activity levels.
- Training time: Identifies the relative amount of time necessary for initially training observers to use the instrument reliably. Training time was reported for only a few instruments. Estimates of total hours, including classroom study and field practice, for initial training are low (10 to 15), medium (15 to 25), and high (25 to 40).
- Test site: Identifies the setting where participants were observed during field tests of the instrument.
- Subjects: Identifies the characteristics of participants in the field test.

- Reliability: Identifies the type(s) and results of reliabilities reported in field tests (reliability = the degree that two or more persons simultaneously viewing an activity using the same behavior definitions and coding conventions entered the same codes). When possible, the inter-observer agreement scores for the physical activity codes were isolated and are reported here (I-I = interval-by-interval reliabilities; ICC = intra-class correlations; and kappa is a statistic that takes chance agreement into consideration). I-I scores are typically higher than kappa scores. Additional information on reliability is provided in a subsequent section.

All nine reviewed instruments have been carefully constructed, field tested with children, report high reliabilities, and are generalizable for use in studying physical activity as it relates to health. Instruments validated for studying the physical activity levels of adults were not located.

The observation focus of eight instruments was on an individual child, while one (System

TABLE 11.1   Characteristics of Selected Instruments Designed Primarily for Observing Children's Physical Activity in School Settings

Instrument name:	CPAF (Children's Physical Activity Form)
Citation:	O'Hara et al. (1989)
Location:	Physical education classes
Observation strategy:	Partial time sampling; 1 min intervals
Activity categories:	4 (stationary, no movement; stationary, limb movement; slow trunk movement; rapid trunk movement)
Energy validation:	Heart rates
Summary variables:	Activity points; kcal
Recording method:	Paper/pencil
Associated variables:	None
Training time:	Moderate
Test site:	Physical education classes
Subjects:	3rd- through 5th-graders
Reliabilities:	I-I (96-98%)
Instrument name:	SOFIT (System for Observing Fitness Instruction Time)
Citation:	McKenzie, Sallis, & Nader (1991)
Location:	Physical education classes
Observation strategy:	Momentary time sampling; 10 s observe/record intervals
Activity categories:	5 (lying; sitting; standing; walking; very active)
Energy validation:	Heart rates; Caltrac
Summary variables:	% intervals/time
Recording method:	Paper/pencil; Scantron
Associated variables:	Lesson context; teacher behavior
Training time:	Low
Test site:	Physical education classes
Subjects:	3rd- through 5th-graders
Reliabilities:	I-I (92%)
Instrument name:	SOPLAY (System for Observing Play and Leisure Activity in Youth)
Citation:	McKenzie et al. (2000a)
Location:	All activity areas; before, during, and after school
Observation strategy:	Momentary time sampling (Placheck recording)
Activity categories:	3 (sedentary; walking; very active)
Energy validation:	Heart rates
Summary variables:	% attending; % in activity categories
Recording method:	Mechanical recorder; paper/pencil
Associated variables:	Area characteristics (e.g., usability, supervision, equipment)
Training time:	Low
Test site:	24 middle schools
Subjects:	6th- through 8th-graders
Reliabilities:	ICC (0.76-0.99); I-I (88-97% on contexts)

TABLE 11.2 Characteristics of Selected Instruments Designed Primarily for Observing Children's Physical Activity in Nonschool Settings

Instrument name: APEE (Activity Patterns and Energy Expenditure)
    Citation: Epstein et al. (1984)
    Location: Free play
    Observation strategy: Momentary time sampling; 15 s intervals
    Activity categories: 5 (sitting/lying quietly; standing quietly; sitting/lying while active; standing while active; very active/moving)

Energy validation: Heart rates
Summary variables: Mean activity score; kcal/kg/min
Recording method: Paper/pencil
Associated variables: None
Training time: Low
Test site: Free play in gym during day camp
Subjects: 5- to 8-year-old obese girls
Reliabilities: I-I (86-99%)

Instrument name: BEACHES (Behaviors of Eating and Physical Activity for Children's Health: Evaluation System)
    Citation: McKenzie, Sallis, Patterson et al. (1991)
    Location: Diverse
    Observation strategy: Momentary time sampling; 1 min intervals
    Activity categories: 5 (lying; sitting; standing; walking; very active)
    Energy validation: Heart rates
    Summary variables: % intervals/time; kcal/kg/min
    Recording method: Computer
    Associated variables: Location; persons present; television viewing; eating; interactors; prompts; consequences; child response

Training time: High
Test site: Home and school
Subjects: 4- to 9-year-olds
Reliabilities: Kappa (0.91)

Instrument name: CARS (Children's Activity Rating Scale)
    Citation: Puhl et al. (1990)
    Location: Diverse
    Observation strategy: Partial time sampling; 1 min intervals
    Activity categories: 5 (stationary, no movement; stationary, movement; translocation, easy; translocation, moderate; translocation, strenuous)
    Energy validation: Heart rates; $\dot{V}O_2$
    Summary variables: % intervals; kcal/kg/min
    Recording method: Computer
    Associated variables: Location; persons present; television; eating; interactors; prompts; consequences
    Training time: High
    Test site: Diverse
    Subjects: 3- to 6-year-olds
    Reliabilities: I-I (84.1%)

Instrument name: FATS (Fargo Activity Timesampling Survey)
    Citation: Klesges et al. (1984)
    Location: Home
    Observation strategy: Partial interval recording; 10 s observe/record intervals
    Activity categories: 8 (sleeping; lying down; sitting; crawling; climbing; standing; walking; running), each with 3 intensity levels (minimal; moderate; extreme)
    Energy validation: None; correlated with activity monitor (LSI) readings
    Summary variables: % intervals
    Recording method: Computer
    Associated variables: Location; persons present; interactors; interactions; child response
    Training time: High
    Test site: Home

Subjects:	20- to 48-month-old children
Reliabilities:	I-I (91-98%); kappa (0.90)
Instrument name:	LETO (Level and Tempo of Children's Activity)
Citation:	Bailey et al. (1995)
Location:	Natural settings, including home
Observation strategy:	Time sampling; 3 s intervals
Activity categories:	14 postures (each with 3 intensity levels)
Energy validation:	Heart rates; $\dot{V}O_2$
Summary variables:	Min, % time in categories
Recording method:	Paper/pencil
Associated variables:	Location; interactors; activity prompts; compliance
Training time:	Very high
Test site:	Natural settings, including home
Subjects:	6- to 10-year-olds
Reliabilities:	I-I (91%); kappa (0.90)
Instrument name:	SCAN CATS (Studies of Children's Activity and Nutrition: Children's Activity Timesampling Survey)
Citation:	Klesges et al. (1990)
Location:	Diverse
Observation strategy:	Momentary time sampling; 10 s observe/record intervals
Activity categories:	4 (stationary; minimal activity; slow movement; rapid movement)
Energy validation:	None
Summary variables:	% intervals/time
Recording method:	Computer
Associated variables:	Location; persons present; interactors; prompts
Training time:	High
Test site:	Home
Subjects	3- to 6-year-olds
Reliabilities:	Field scores not reported; kappa (0.91) on videotape calibration

for Observing Play and Leisure Activity in Youth [SOPLAY]) focused on group behavior. All nine used sampling procedures to estimate the amount of time children spent in various activity categories. Instruments using duration recording, which would provide a more precise measure of the length of time spent in specific categories, were not located.

Three instruments were designed specifically for researching children's physical activity levels at school (see table 11.1) and are appropriate for studying behaviors during recess, free play, and physical education classes. CPAF (Children's Physical Activity Form) and SOFIT were primarily designed for physical education classes, while SOPLAY was designed for leisure settings.

Six instruments (see table 11.2) were designed for observing children's physical activity in the home and can be used in a variety of other settings. They all permit the simultaneous coding of numerous events associated with physical activity and are useful for comparing physical activity levels in different locations.

The available instruments reflect the current interests of health-related researchers who primarily study the activity of children. All nine were de-

signed for studying the physical activities of the young, including one for 20- to 48-month-old children (Klesges et al., 1984). Most of the instruments were validated with energy expenditure estimations with young children, and none were tested for use with subjects beyond the 12th grade.

Commercial Software and Hardware Systems

Within the past decade, several computer packages for the collection, management, analysis, and presentation of observational data have become available commercially. The strengths and limitations of several packages have been reviewed (Kahng & Iwata, 1998). Codes designed for observing physical activity and related contexts can be inserted into many of them. The BEST (Behavior Evaluation Strategies and Taxonomies) (Sharpe & Koperwas, 2000) and The Observer (Noldus Information Technology, 2001; **www.noldus.com**) are two of the most flexible and comprehensive systems currently available.

Several packages permit the entry of data directly into a PC, Mac, or a handheld computer as well as allow events to be coded directly from videotape or digital media file. Entering data directly into a computer reduces error and speeds analyses and reporting. Handheld computers are advantageous for collecting physical activity data because they permit mobility in the field. Their small size, low weight, and long battery life make them preferable to laptops for data collection. Once data are collected in the field, they can be transferred to a stationary computer for further analysis.

Sophisticated commercial systems permit data analyses that were once not possible with older paper-and-pencil data entry. For example, during an observation session, events are logged by typing predefined key codes or by clicking items on the screen with a mouse or pen stylus. The computer automatically adds a time stamp to each event and writes it to a data file. In duration recording, behaviors (e.g., sitting) are scored by pressing a key at the start of the behavior and ended automatically when a key representing another behavior (e.g., walking) is pressed. Data analysis now can include time-event tables and plots, and generate reports with statistics on frequencies and durations, the sequential structure of the process, and the co-occurrence of events. Lag sequential analyses, which examine how often certain behaviors are preceded or followed by other events, are also possible. For example, it is possible to assess how often an activity prompt is followed by an actual increase in physical activity and the time it took for that increase to occur.

Assessment Procedures

Obtaining accurate information with direct observation requires considerable rigor and attention to detail. This section describes the general assessment procedures and highlights specific issues that are critical for the direct observation technique.

Classes of Physical Activity Behavior

In observational research, physical activity behaviors of interest (and associated elements) are arranged in one or more classes (i.e., groups of mutually exclusive categories). Within a class, only one state can be occurring (and thus recorded) at a time. Location, posture, social interactions, and physical activity are examples of independently scored classes of behavior. Systems for scoring physical activity typically use five or more mutually exclusive categories or codes (see tables 11.1 and 11.2). For example, BEACHES codes physical activity into five levels that are associated with body posture (i.e.,

lying down, sitting, standing, walking, and very active), while LETO (Level and Tempo of Children's Activity) (Bailey et al., 1995) codes 14 postures, each with 3 intensity levels. More codes and levels of codes (i.e., subcodes) allow greater precision during analyses, but there are some disadvantages. Additional codes make observer training more challenging, and coding in the field is more tedious, causing observers to become fatigued more quickly and reliability scores to be lowered. The increased precision gained by using more codes may not be necessary to answer some research questions or to be of real-world value for practitioners. For example, SOPLAY, which uses only three codes to categorize physical activity (i.e., sedentary, walking, and very active), was sufficiently complex to classify and interpret physical activity by time period and gender in 151 activity environments in 24 schools (McKenzie et al., 2000a).

Sampling Methods

In observational studies, the sampling method specifies which subjects to watch, when to watch them, and how to record their behavior. Continuous recording involving the observation of a subject continuously for a specified amount of time and recording the onset and ending of each behavior is the most informative method, but it is also the least feasible. When several behaviors of interest are to be recorded simultaneously (e.g., physical activity plus prompts and encouragement for activity), time sampling involving interval recording is often the best choice. In this case, observation periods are typically divided into short observe-record segments. For example, an assessor might observe a child for 10 seconds and then use the next 10 seconds to enter codes for the behaviors. The result, in this case, would be the entering of three data codes for the child's physical activity and each associated event during each minute. Observe and record intervals are typically the same length, and these have ranged from 3 to 35 seconds in length.

Several sampling techniques are available, including momentary time sampling (i.e., instantaneous or scan sampling), partial time sampling (i.e., recording the event if it occurs at any time during the observe interval), and whole-interval sampling (i.e., recording the event only if it occurs throughout the entire interval). Each sampling method has its advantages and limitations (Sulzer-Azaroff & Mayer, 1977), but a detailed assessment of each technique is beyond the scope of this chapter.

Pacing of Observations

When time sampling is used, the pacing of observations must be prompted. Stopwatches were initially used for this function, but they required observers to look away from the observational viewing area. Audio tape players with prerecorded signals to initiate and end recording are much preferred. The wearing of ear jacks allows only the observer to hear the pacing signals. During reliability assessments, a Y adapter is used so that two independent observers can simultaneous hear the signals. When data are entered directly into a computer, the computer's internal clock can be used to pace the intervals by making audible signals and/or by visual displays on the screen.

Time Frame for Observations

The determination of the amount of data to collect in a study depends on a number of factors. Because of the intensive nature of direct observation, the costs of data collection and analyses as well as observer and respondent (i.e., the observed subject) burden must be considered. Additional factors to consider in measuring physical activity are location, season, weather, temperature, and time of day.

Decisions on how long, when, and how often observations need to take place must be determined individually for each study. Decisions often depend on the research question and the setting of interest (e.g., physical education lesson, recess, before school, only when parents are present). For example, a pilot study found that the 94% of physical activity variance in 90-minute observations using the BEACHES instrument in children's homes was accounted for by observing for 60 minutes, and 73% was accounted for in 30 minutes of observing (McKenzie, Sallis, Patterson et al., 1991). In this case, 60 minutes was selected as the time frame for observations, and these were scheduled in the late afternoon and early evening when parents were most likely to be present.

Because physical activity is highly variable, there is interest in determining the number of measurements needed to correctly classify physical activity for generality purposes. Because few studies have been completed, that number is essentially unknown. It will likely depend, however, on numerous factors such as the age of the subject, location, physical activity measure (i.e., observation system), and whether specific activity levels or an activity summary score is used in the analysis (Levin et al., 2001).

With young children, minute-to-minute variability is extremely high (Bailey et al., 1995). Baranowski et al. (1993) showed, however, that gender, month, and location accounted for 75% of the variability in the overall activity levels of three- and four-year-old children. These researchers also showed that month-to-month variability may be related to temperature and humidity. Klesges et al. (1984) and McKenzie, Sallis, Patterson et al. (1991) have suggested that at least four observation sessions are needed to adequately estimate the physical activity of young children at home.

Regression models on data collected using the SOFIT system showed that schools, schools by semester, and weeks explained over 33% of the variability in physical activity provided in physical education lessons (Levin et al., 2001). The magnitude of variation was greater for vigorous physical activity than moderate-to-vigorous physical activity and energy expenditure. In physical education classes, variability in activity levels results from numerous factors, including lesson goals, content and placement within an instructional unit, teacher behavior, available equipment and facilities (e.g., size of indoor space), and individual differences among children. Each of these factors needs to be considered when determining the number of observations needed for studying physical activity in physical education settings.

Observer Training and Maintenance

Regardless of which observation system is used, the extent to which recorded data reflect the actual occurrence of behaviors is largely dependent on the skills of human observers. To ensure the accuracy of data, substantial initial observer training is necessary. As observer's skills may deteriorate or "drift" over time, it is also important to plan for observer monitoring and retraining during the data collection period. The following steps are recommended for developing and maintaining observer accuracy:

1. Observer orientation
2. Study of the observation manual and the memorization of categories
3. Direct practice using modeling and videotaped segments
4. Assessments and feedback using gold-standard videotapes
5. Practice in the observational setting with immediate feedback
6. Field practice with concurrent reliability assessments with a certified assessor

7. Monitoring assessments and booster training during the data collection period to detect and prevent "observer drift"

The length of initial training and the frequency of booster sessions depend on several factors, including the number and the complexity of the physical activity codes and associated variables. During the initial phase, observers need to be appropriately oriented to the scientific method, systematic observation, and the nature of the study. The content of this orientation will depend on the observer's experience and education. Before mastering the instrument itself, observers need to learn about ethical issues, the importance of objectivity and confidentiality, etiquette about how to behave in the observation environment, and how to reduce subject reactivity. Observers need to generally know about the study, but if at all possible, they should not know the study hypotheses or the exact intervention components being tested.

A clearly written observation manual is a necessity, and it should be kept up to date (e.g., with "coding conventions" that contain answers to observers' questions about novel instances they encounter in the field). The manual should contain operational definitions for all constructs being examined. Each construct should be clearly defined and be objectively identified through the use of both positive and negative instances.

Training should begin with relatively simple codes and sequences. Then, when these are mastered, more complex scenes should be introduced. Training sessions should be progressively lengthened until they match the time of the data collection sessions. Practicing in the observational setting before the recording of data for the study is critical. It not only provides observers an opportunity for live recording with feedback and debriefing from a certified trainer, but it also ensures that trainees become familiar with observational equipment and how to operate it when they are under some pressure.

Preparing Videotapes for Observer Training

Numerous videotaped samples are needed for training, certification, and reassessment. High-quality, precoded videotapes are particularly useful in longitudinal studies. For example, the SOFIT system is currently being used in the Child and Adolescent Trial for Cardiovascular Health (CATCH) study to assess the maintenance effects

of an intervention that took place in 96 schools up to 9 years ago. The same observers were not available for the follow-up study, but comparisons can be made among the data sets because all CATCH observers throughout the years were trained, certified, and recalibrated using the same videotapes.

Videotaped samples should be available for all study variables, including all physical activity levels and all contextual categories. For example, BEACHES was designed to assess family influences on children's activity and nutrition. Therefore, it was necessary not only to have videotaped segments to illustrate physical activity levels in various locations but also to show samples of eating behavior, prompts and consequences for physical activity, and children's reactions to those prompts and consequences.

Training and assessment tapes should be shot in the study environment or in settings that are similar to it. In preparation for a study in numerous schools, video samples from a variety of schools should be prepared. Getting samples of physical activity levels is relatively easy compared to recording viewable/audible examples of prompts and consequences for activity that occur intermittently and at low frequencies. With complex systems, in addition to having authentic samples, it may be necessary to fabricate some segments to ensure that observers can be trained and tested on all constructs.

Shooting videotapes that are useful for editing and coding requires careful planning. Appropriate school, child, and parent clearance must be obtained before tapes are made. If the observational target is an individual child, children with relevant characteristics (e.g., age, gender, ethnicity) who look and behave like the subjects should also be on the videotapes. Children targeted for videotaping should be those who are physically active and show variety. A tape of a child who just sits during an entire video sequence makes activity coding easy and it promotes high inter-observer agreement scores, but is not very useful for training observers.

To enable the analysis, the subject needs to be clearly visible on the tape (e.g., by wearing a brightly colored shirt or pants). Zooming in is often necessary to locate the subject in a crowd, but there should be sufficient background in view to enable observers to accurately record other categories, such as the location or interactors. The subject should wear a remote wireless microphone to ensure that verbal prompts can be heard. Panning and tilting of the camera should be done slowly and smoothly to avoid distracting viewers from the main observational focus.

Usable training tapes can be made with equipment found at most universities; however, top-quality tapes require an experienced producer with access to professional equipment. Digital video tapes provide better quality images, and there is little loss during editing. Editing long tapes on computer is still problematic for amateurs because of the skills, expensive software, and large computer storage capacity needed.

Using Videotapes for Data Collection

In addition to using videotapes or film in training observers and monitoring their accuracy, data can be coded directly from them. One of the first studies of physical activity to use systematic observation involved the motion picture analyses of obese and non-obese girls (Bullen et al., 1964). Now, small, portable, low-cost video systems are available. Videotapes provide permanent samples that can be viewed repeatedly, by many different observers, and in an environment removed from distractions. Videotapes are particularly useful when observation codes are complex, the activity setting is fixed (e.g., in a single gymnasium), and when only a few subjects are involved, such as in validating another physical activity measurement technique.

While videotapes provide advantages in obtaining accurate data, researchers need to consider that using this technology has several disadvantages, including time, inconvenience, and cost (e.g., camera, wireless microphone, tapes, playback machine, monitor). Data collection time is more than doubled because time is needed not only to analyze tapes but also to set up and record them. Some subjects and institutions are reluctant to provide consent for videotaping, and subject reactivity to being videotaped is greater than that of simply being observed.

Outcome Variables

Direct observation involves researching applied settings, so the data output should be useful to both practitioners and researchers. No matter how physical activity is classified, data summaries from most observation systems include both minutes and percentage of observed time spent at various activity levels (see figures 11.1, 11.2, and 11.3). Systems that have been validated for energy expenditure also typically report both total and rate of estimated energy expenditure (e.g., kcal/kg/min).

Measurement Properties: Validity

Validity refers to a measure assessing what it purports to measure. Systematic observation is a direct measure of behavior that requires little inference or interpretation and, therefore, has high internal validity, sometimes referred to as *face validity*. Direct observation is therefore sometimes used as a criterion method for validating other measures of physical activity, such as self-report and mechanical and electronic monitoring.

In considering the validity of a direct observation system, it is important to acknowledge that physical activity and energy expenditure are substantially different. Physical activity, which refers to body movement, can be assessed directly through observation methodology. Energy expenditure, on the other hand, results primarily from body movement and can only be inferred from observations. For example, a lean child and an obese child may be observed and accurately (i.e., validly) recorded as engaging in the same physical activity behavior (e.g., walking a mile in 15 continuous minutes), but their energy expenditure resulting from walking that mile would be substantially different. Because many health-related researchers are interested in caloric expenditure, many published instruments for studying physical activity have undergone external validations using measures of energy expenditure or heart rates (see tables 11.1 and 11.2). Other instruments have been validated with mechanical recording devices and accelerometers, such as the Large Scale Integrated (LSI), Caltrac, and Computer Science and Applications (CSA) monitors.

External validation is especially important for investigators interested in energy costs, but it may not be a priority for all physical activity researchers or for most practitioners. The cost of validating physical activity observation codes is substantial, and beginning investigators should consider selecting an already validated physical activity measure rather than developing their own. Most of the published instruments can be modified to permit investigators to modify contextual codes while keeping the validated physical activity codes intact.

Most energy expenditure validations with observation instruments have been established with children below the sixth grade, although some have included middle school and high school students. An instrument validated for observing the physical activity of young children may not necessarily be appropriate for use with older populations. Therefore, further validation is needed before some instruments can be used with adults, particularly if energy expenditure is to be estimated.

Measurement Properties: Reliability

Reliability refers to consistency, and it is used as a measure of the quality of collected data. In observational research, reliability usually refers to the level of agreement among trained, independent observers. It is expressed as an index (typically percentage of agreement) that is based on the degree of correspondence between codes in two or more data files. Reliability is usually assessed on the results obtained by different observers coding the same subjects on the same occasion (i.e., inter-observer reliability). It may also involve assessing the degree of correspondence between scores made when the same person codes the same videotape twice (i.e., intra-observer reliability).

Different statistics are used in reporting reliability measures, including I-I reliabilities, ICC, and kappa. I-I scores are the most commonly used because they are particularly useful in assessing consistency both during observer training and in the field. During training, the examination of individual observation intervals can identify specific problem areas that can be countered by additional instruction and practice. Reliability scores from the field observations are always reported in research papers. Tables 11.1 and 11.2 show that published systems for measuring physical activity generally have very high inter-observer agreement. Reliability scores vary with the complexity of the observation system and the method used to calculate agreement. The kappa statistic takes chance agreement into consideration, resulting in kappa scores being somewhat lower than I-I scores.

In addition to assessing the reliability of data collected in the field, reliability analyses are important in checking the consistency of observers over time in order to reduce observer drift. In this case, observers throughout a study periodically code gold-standard videotapes that were coded by the observer trainer before the start of the research. Reduced reliability scores indicate slippage or drift from the original coding definitions and should result in additional training via booster sessions.

Applications for Direct Observation

Direct observation has become an increasingly popular tool for health-related research. Few stud-ies, however, have used observational methods to assess physical activity in adults, and the only observation systems validated using energy expenditure included children and adolescents (see tables 11.1 and 11.2). Nonetheless, direct observation of adults engaging in physical activity could be useful, particularly when environmental factors are of interest. For example, a recent study used direct observation to assess adult behavior in exercise counseling sessions provided by physicians in community family practice (Podl et al., 1999).

As described previously, the selection of an appropriate direct observation tool depends on the setting and the type of data being sought. This author has been involved in the development, assessment, and use of three observation systems designed to assess physical activity in very different settings. The first, BEACHES, was designed specifically to assess the behaviors of young children at home and during preschool recess. The second, SOFIT, was designed for assessing structured physical education classes. The third, SOPLAY, is an innovative system for generating physical activity and contextual data on groups in nonstructured leisure-time settings. It is the only system currently published that measures the physical activity of groups.

All three systems use the same physical activity codes (i.e., lying down, sitting, standing, walking, and very active) that have been validated by both heart rate monitors (McKenzie, Sallis, & Nader 1991; Rowe, Schuldheisz et al., 1997; Rowe, van der Mars et al., 1997) and accelerometers (McKenzie et al., 1994). In SOPLAY, three codes (i.e., lying down, sitting, and standing) are collapsed into a "sedentary" category. This section describes the application of these tools and provides examples for how they have been used in specific research projects.

BEACHES

BEACHES was designed to simultaneously code the physical activity and eating behaviors of children and related environmental factors (McKenzie, Sallis, Patterson et al., 1991). It can be used at home (Elder et al., 1998), at school (McKenzie, Sallis, Elder et al. 1997), and in other settings where a targeted child might go. The system was developed from a behavior analytic (i.e., operant psychology/social learning theory) viewpoint and includes the coding of 10 categories simultaneously. Selected environmental factors (e.g., the presence of others, the availability of food, whether or not the child is watching television) and the child's physical activity level, eating behavior, and physical location are coded during each interval. Additional codes are entered whenever a physical activity or eating behavior is prompted or consequated, including the interactor, type of prompt or consequence, and the child's response to being prompted.

SOFIT

SOFIT was designed to measure student physical activity, lesson context, and teacher behavior during physical education classes (McKenzie, Sallis, & Nader, 1991). It has been used to assess the quality of physical education instruction in three large-scale intervention projects supported by the National Heart, Lung, and Blood Institute: CATCH (McKenzie et al., 1996), Middle School Physical Activity and Nutrition (M-SPAN) (McKenzie et al., 2000b), and Sports, Play, and Active Recreation for Kids (SPARK) (McKenzie, Sallis, Kolody et al., 1997; Sallis et al., 1997). While the main focus of SOFIT is the coding of student physical activity levels, selected environmental factors (i.e., lesson context and teacher behavior) associated with opportunities for students to be physically active and become physically fit are recorded simultaneously. Lesson context time is categorized as management, knowledge, fitness, skill drills, game play, and free play. Teacher behavior is coded as promotes fitness, demonstrates fitness, instructs generally, manages, observes, and other. SOFIT data can be entered by Scantron (McKenzie et al., 2000b) or computer (Keating et al., 1999), and the physical activity summary scores are useful in making comparisons to national recommendations for physical activity accrual.

Figure 11.1 illustrates the use of SOFIT during a descriptive study of 430 middle school coeducational physical education classes (McKenzie et al., 2000b). During the average lesson, 10 minutes were allocated to game play with only 1.8 minutes for skill drills. Moderate-to-vigorous physical activity (MVPA) varied by lesson context, and boys were significantly more active than girls in free play, game play, and skill drill situations.

SOPLAY

SOPLAY was more recently developed, and considerable detail is included in this chapter because it is the only instrument so far to assess the physical activity of groups of people (McKenzie et al., 2000a). Other validated methods, including observation systems such as BEACHES and SOFIT, heart rate monitoring, accelerometers, and questionnaires, were designed to measure the physical activity of individuals only. Without objective instruments to assess physical activity in open environments, such as recreational and leisure settings, limited research from an ecological perspective can occur in these areas. Measuring activity in open settings is complicated because both the number of participants and their activity levels change frequently.

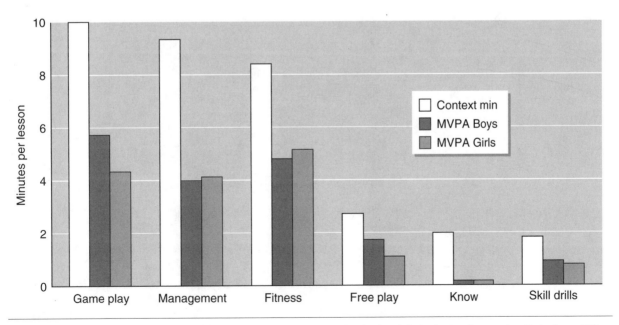

Figure 11.1 Lesson context and MVPA (moderate-to-vigorous physical activity) minutes by gender. Data from 430 lessons in 24 different schools.

SOPLAY assesses the number of people in a designated activity area and their activity levels using momentary time sampling. Environmental aspects of the settings that might influence physical activity are also recorded. Using SOPLAY to study an environment (e.g., recreation center, designated park, or school area) requires that the area for physical activity be identified and measured prior to data collection. The location, size, and boundaries of each target area are determined, and maps detailing this area are made to ensure that observers collect data from a consistent spot.

The reliability, validity, and feasibility of using SOPLAY to study physical activity and numerous environmental characteristics were tested by observations in 151 activity areas in 24 middle schools during 72 days (McKenzie et al., 2000a). Abbreviated definitions and the protocol used in that study follow on pages 192-194. Adaptations can be made to the independent characteristics (i.e., the nonphysical activity categories) so that SOPLAY can be used to study the physical activity of diverse participants in any designated area.

Figure 11.2 illustrates the use of SOPLAY to assess the percentage of students enrolled in 24 middle schools that were in physical activity areas before school, at lunchtime, and after school during a 72-day period. The data show that relatively few students visited the activity areas, but a significantly larger proportion of boys than girls attending school were in the areas before school and at lunchtime. Figure 11.3 shows that the boys in the activity areas were significantly more physically active than girls who were there before school and at lunchtime (McKenzie et al., 2000a). Those in the targeted areas after school were primarily engaged in interscholastic team practices. To facilitate additional research on patterns of physical activity in intact groups and settings, SOPLAY protocols and recording procedures are provided on pages 192-194.

Summary

Direct (systematic) observation is an important tool in the assessment of physical activity because it uses an objective method and it can provide contextually rich data on the setting in which activity occurs. It is particularly useful for researchers studying children and for those using ecological and cognitive-behavioral approaches to examine how physical activity is influenced by physical and social environments. This chapter provided a background and overview of systematic observation as it applies to studying physical activity. The classification of physical activity in observation systems, validity and reliability of observations, and techniques and systems for generating observational data were discussed. Practical information was provided for making decisions on behavior sampling, training and calibrating observers, using videotape

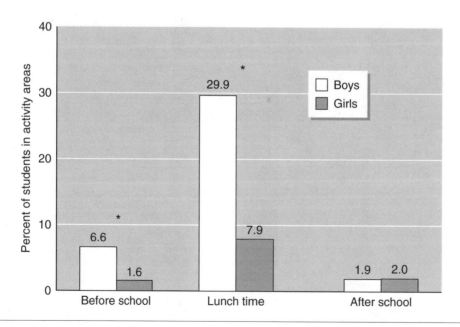

Figure 11.2 Percentage of students in activity areas. Data were collected from 151 areas in 24 M-SPAN schools (gender differences: * = p < 0.001).

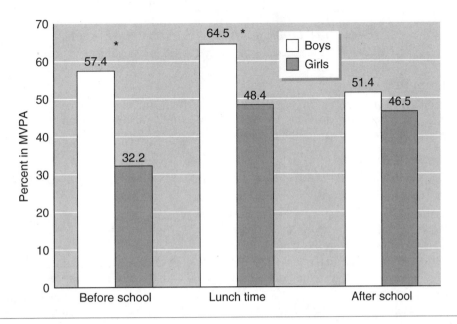

Figure 11.3 Percentage of MVPA by gender. Data from 151 activity areas in 24 M-SPAN schools (gender differences: * = p < 0.01).

technology, and selecting commercially prepared software for collecting, storing, and presenting observational data. The application of several direct observation tools for use in various research applications was also described.

References

Bailey, R.C., Olson, J., Pepper, S.L., Porszasz, J., Barstow, T.J., & Cooper, D.M. (1995). The level and tempo of children's physical activities: An observational study. *Medicine and Science in Sports and Exercise, 27,* 1033-1041.

Baranowski, T., Thompson, W., DuRant, R., Baranowski, J., & Puhl, J. (1993). Observations on physical activity in physical locations: Age, gender, ethnicity and month effects. *Research Quarterly for Exercise and Sport, 64,* 127-133.

Bullen, B.A., Reed, R.B., & Mayer, J. (1964). Physical activity of obese and non-obese adolescent girls. *American Journal of Clinical Nutrition, 14,* 211-223.

Ciminero, A.R., Calhoun, K.S., & Adams, H.E. (Eds.) (1977). *Handbook of behavioral assessment.* New York: Wiley.

Darst, P., Zakrajsek, D., & Mancini, V. (Eds.) (1983). *Analyzing physical education and sport instruction.* (2nd ed.). Champaign, IL: Human Kinetics.

DuRant, R.H., Baranowski, T., Puhl, J., Rhodes, T., Davis, H., Greaves, K.A., & Thompson, W.O. (1993). Evaluation of the Children's Activity Rating Scale (CARS) in young children. *Medicine and Science in Sports and Exercise, 25,* 1415-1421.

Elder, J.P., Broyles, S.L., McKenzie, T.L., Sallis, J.F., Berry, C.C., Davis, T.B., Hoy, P.L., & Nader, P.R. (1998). Direct home observations of the prompting of physical activity in sedentary and active Mexican- and Anglo-American children. *Journal of Developmental and Behavioral Pediatrics, 19,* 26-30.

Epstein, L., McGowan, C., & Woodall, K. (1984). A behavioral observation system for free play in young overweight female children. *Research Quarterly for Exercise and Sport, 55,* 180-183.

Hartman, D.P. (Ed.) (1982). *Using observers to study behavior.* San Francisco: Jossey-Bass.

Kahng, S.W., & Iwata, B.A. (1998). Computerized systems for collecting real-time observational data. *Journal of Applied Behavior Analysis, 31,* 253-262.

Keating, X.D., Kulinna, P.H., & Silverman, S. (1999). Measuring teaching behaviors, lesson context, and physical activity in school physical education programs: Comparing the SOFIT and the C-SOFIT instruments. *Measurement in Physical Education and Exercise Science, 3,* 207-220.

Klesges, R.C., Coates, T.J., Moldenhauer-Klesges, L.M., Holzer, B., Gustavson, J., & Barnes, B. (1984). The FATS: An observational system for assessing physical activity in children and associated parent behavior. *Behavioral Assessment, 6,* 333-345.

Klesges, R.C., Eck, L., Hanson, C., Haddock, K., & Klesges, L. (1990). Effects of obesity, social interactions, and physical environment on physical activity in preschoolers. *Health Psychology, 9,* 435-449.

Levin, S., McKenzie, T.L., Hussey, J., Kelder, S., & Lytle, L. (2001). Variability of physical activity in indoor elementary school physical education lessons taught by PE specialists. *Measurement in Physical Education and Exercise Science, 5,* 207-218.

McKenzie, T.L. (1991). Observational measures of children's physical activity. *Journal of School Health, 61,* 224-227.

McKenzie, T.L., Marshall, S.J., Sallis, J.F., & Conway, T.L. (2000a). Leisure-time physical activity in school environments: An observational study using SOPLAY. *Preventive Medicine, 30,* 70-77.

McKenzie, T.L., Marshall, S.J., Sallis, J.F, & Conway, T.L. (2000b). Student activity levels, lesson context, and teacher behavior during middle school physical education. *Research Quarterly for Exercise and Sport, 71,* 249-259.

SOPLAY

Instrument Purpose

To obtain data on the number of students and their physical activity levels during play and leisure opportunities in a specified area. During the M-SPAN study, observations were made before school (BS), during each lunch period (L), and after school (AS).

Definition of Terms

target area: A predetermined observation area in which students potentially engage in leisure-time physical activity.

scan space: A subdivision of a target area in which an assessor makes an observation sweep. Target areas are subdivided into scan spaces when the number of students is large and they are engaged actively.

sweep: A single observation movement from left to right across a target area or scan space. During a sweep, individual students are counted and coded for their activity level.

activity levels: The activity level of each student is coded into one of three categories using momentary time sampling (i.e., what the student was doing when viewed during the observational sweep):

1. **code S:** (sedentary, i.e., lying, sitting, or standing) unless the student is expending more energy than that required for an ordinary walk.

2. **code W:** (walking) when the student is walking.

3. **code V:** (very active) when the student is expending more energy than he or she would during ordinary walking (e.g., running, jogging, skipping, hopping). Do not consider body position only. For example, code V if the student is wrestling with a peer (even though lying on his or her back) or pedaling a stationary bike (even though sitting).

Target Areas

1. Observations are made in all locations likely to provide opportunities for students to be physically active. These areas are predetermined and measured. A map of areas is provided and a standard observation order is established for each school.

2. During occasions of high student density, target areas are subdivided into smaller scan spaces. Observers use court or field markings to determine appropriate scan spaces within a target area. Data from the smaller scan spaces are summed to provide an overall measure for a target area.

3. A decision to subdivide a target area depends upon the number of students in the area and the type of student activity. Fast moving activities with students clustered together and moving in diverse directions (e.g., during basketball) require smaller scan spaces.

Scanning Procedures

1. Scan from left to right, observing each student only once. If an observed student reappears in the scan area, do not record him or her a second time. If a new student appears in part of the area already scanned, do not backtrack to count that student.

2. First scan the entire target area for girls, enter their data, and reset the counter. Then scan the area for boys, enter their data, and reset the counter. Move to the next target area.

Recording Procedures and Directions

1. On the abbreviated SOPLAY data recording form (see form 11.1), enter school ID, date, observer ID, whether or not the observation is a reliability measure, the temperature, and observation period.

2. For each target area, record the start time, condition (i.e., description of area), and the number of students sedentary, walking, and very active—separately for girls and boys—and indicate the most prominent physical activity they were doing (Act).

reliability: Circle No unless you are the second observer.

temp.: Enter the temperature at the start of the observation period.

period: Circle the code to identify whether observations were made before school (BS), at lunch time (LU), or after school (AS).

start time: Enter the start time (2400 hours) of the sweep for that designated area.

area: Enter the number of the target area from the school map.

condition: Circle N or Y to describe specific conditions for each area.

A = Area is accessible to students (e.g., not locked or rented to others).

U = Area is usable for physical activity (e.g., is not excessively wet or windy).

S = Area is supervised by school or adjunct personnel (e.g., teachers, playground supervisors, registered volunteers). The supervisor must be in or adjacent to the area (i.e., available to direct students and respond to emergencies) but does not necessarily instruct, officiate, or organize activities.

O = Organized physical activity is occurring in the area (i.e., scheduled, with leadership provided by school or agency personnel; includes intramurals, directed fitness stations, and interscholastic and club practices).

E = Equipment (e.g., balls, jump ropes) is provided by the school or other agency. Do not code Y if the only equipment available is permanent (e.g., basketball hoops) or is owned by students themselves.

S W V: S = sedentary; **W** = walking; **V** = very active

act.: Enter the code for the most prominent physical activity in which girls and boys are participating.

Observer Preparation

1. Before arrival at school, prepare materials including synchronized wristwatch, thermometer, counter, clipboard, recording forms, and pencils.

2. Arrive at least 60 minutes before the official start of school. Review sequence for visiting target areas. Prepare mentally by scanning areas a few times.

Before-School Observations

The objective is to obtain a measure of the number of students engaged in physical activity before school starts. The last scan begins 15 minutes before school starts. Begin at school start – 40 minutes (with 6 target areas), – 30 minutes (with 4 target areas), or – 25 minutes (with 3 target areas).

Lunchtime Observations

The objective is to obtain a measure of the number of students engaged in physical activity at lunchtime (outside of required physical education). Begin observations at lunch start + 15 minutes.

After-School Observations

The objective is to obtain a measure of the number of students engaged in physical activity beginning at school end + 15, 45, and 75 minutes.

Sample School SOPLAY Schedule

9:00 A.M. school start; 4 target areas; 3 lunch periods

8:00 A.M.	Check target areas, prepare data forms
8:25 A.M.	Initiate scan in target area 1 and follow established sequence
9:00 A.M.	School start
11:30 A.M.	Lunch 1 starts; initiate scan in target area 1 at 11:45
12:00 P.M.	Lunch 2 starts; initiate scan in target area 1 at 12:15
12:30 P.M.	Lunch 3 starts; initiate scan in target area 1 at 12:45
15:00 P.M.	School end
15:15 P.M.	Initiate scan in target area 1; continue to all areas
15:45 P.M.	Initiate scan in target area 1; continue to all areas
16:15 P.M.	Initiate scan in target area 1; continue to all areas

SOPLAY

(System for Observing Play and Leisure Activity in Youth)

School ID: _____   Date: ____/____/____   Obs. ID #: _____

Reliability: (No) Yes   Temp.: _68_ °F   Period: (BS) LU AS

Start time	Area	Condition A U S O E	Girls S W V Act.	Boys S W V Act.
08:00	1	N N (N) (N) (N) (Y) (Y) Y Y Y	5 2 0 13	9 4 4 3
08:05	2	N N (N) (N) (N) (Y) (Y) Y Y Y	11 3 1 9	8 5 4 9

Form codes:

Obs. = Observer

Temp. = Fahrenheit temperature

BS = Before school

LU = Lunchtime

AS = After school

A = Accessible

U = Usable

S = Supervised

O = Organized activity

E = Equipment provided

S = Sedentary

W = Walking

V = Very active

Act. = Prominent activity

Activity codes:

0 = No identifiable activity

1 = Aerobics

2 = Baseball or softball

3 = Basketball

4 = Dance

5 = Football

6 = Gymnastics

7 = Martial arts

8 = Racket sports

9 = Soccer

10 = Swimming

11 = Volleyball

12 = Weight training

13 = Other playground games

14 = None of the above

McKenzie, T.L., Nader, P.R., Strikmiller, P.K., Yang, M., Stone, E.J., Perry, C.L., Taylor, W.C., Epping, J., Feldman, H., Luepker, R.V., & Kelder, S.H. (1996). School physical education: Effect of the Child and Adolescent Trial for Cardiovascular Health (CATCH). *Preventive Medicine, 25*, 423-431.

McKenzie, T.L., Sallis, J.F., & Armstrong, C.A. (1994). Association between direct observation and accelerometer measures of children's physical activity during physical education and recess. *Medicine and Science in Sports and Exercise, 26*, S143 [Abstract].

McKenzie, T.L., Sallis, J.F., Elder, J.P., Broyles, S.L., Berry, C.C., Hoy, P.L., Nader, P.R., Zive, M., & Broyles, S.L. (1997). Physical activity levels and prompts in young children at school recess: A two-year study of a bi-ethnic sample. *Research Quarterly for Exercise and Sport, 68*, 195-202.

McKenzie, T.L., Sallis, J.F., Kolody, B., & Faucette, N. (1997). Long-term effects of a physical education curriculum and staff development program: SPARK. *Research Quarterly for Exercise and Sport, 68*, 280-291.

McKenzie, T.L., Sallis, J.F., & Nader, P.R. (1991). SOFIT: System for Observing Fitness Instruction Time. *Journal of Teaching in Physical Education, 11*, 195-205.

McKenzie, T.L., Sallis, J.F., Patterson, T.L., Elder, J.P., Berry, C.C., Rupp, J.W., Atkins, C.J., Buono, M.J., & Nader, P.R. (1991). BEACHES: An observational system for assessing children's eating and physical activity behaviors and associated events. *Journal of Applied Behavior Analysis, 24*, 141-151.

Montoye, H.J., Kemper, H., Saris, W., & Washburn, R.A. (1996). *Measuring physical activity and energy expenditure.* Champaign, IL: Human Kinetics.

Noldus Information Technology. (2001). *The Observer.* Wageningen, The Netherlands: Noldus Information Technology.

O'Hara, N., Baranowski, T., Simons-Morton, B., Wilson, S., & Parcel, G. (1989). Validity of the observation of children's physical activity. *Research Quarterly for Exercise and Sport, 60*, 42-47.

Podl, T.R., Goodwin, M.A., Kikano, G.E., & Stange, K.C. (1999). Direct observation of exercise counseling in community family practice. *American Journal of Preventive Medicine, 17*, 207-210.

Puhl, J., Greaves, K., Hoyt, M., & Baranowski, T. (1990). Children's Activity Rating Scale (CARS): Description and evaluation. *Research Quarterly for Exercise and Sport, 61*, 26-36.

Rowe, P.J., Schuldheisz, J.M., & van der Mars, H. (1997). Measuring physical activity in physical education: Validation of the SOFIT direct observation instrument for use with first- to eighth-grade students. *Pediatric Exercise Science, 9*, 136-149.

Rowe, P., van der Mars, H., Schuldheisz, J.M., & Fox, S. (1997). *Measuring physical activity: Validating SOFIT for use with high school students.* Paper presented at the fourth Pacific Rim Conference on Exercise Science and Sports Medicine, Corvallis, OR.

Rowlands, A.V., Ingledew, D.K., & Eston, R.G. (2000). The effect of type of physical activity measure on the relationship between body fatness and habitual physical activity in children: A meta-analysis. *Annals of Human Biology, 27*, 479-497.

Sallis, J.F., McKenzie, T.L., Alcaraz, J.E., Kolody, B., Faucette, N., & Hovell, M.F. (1997). The effects of a 2-year physical education program (SPARK) on physical activity and fitness in elementary school students. *American Journal of Public Health, 87*, 1328-1334.

Sharpe, T.L., & Koperwas, J. (2000). *Software assist for education and social science settings: Behavior Evaluation Strategies and Taxonomies (BEST) and accompanying qualitative applications* (2nd ed.). Thousand Oaks, CA: Sage-Scolari.

Sulzer-Azaroff, B., & Mayer, G.R. (1977). *Applying behavior analysis procedures with children and youth.* New York: Holt, Rinehart, & Winston.

12

Use of Doubly Labeled Water and Indirect Calorimetry to Assess Physical Activity

Raymond D. Starling, PhD
Pfizer, Inc.

The various metabolic processes within the body provide the energy needed to perform all required functions. During periods of increased physical activity, the body speeds up these metabolic processes to meet the increased energy demand of movement. The doubly labeled water (DLW) technique is a biochemical procedure that essentially tracks the rate of these metabolic processes. Because the energetics of these metabolic reactions are well described, the procedure can provide a highly accurate assessment of overall energy expenditure.

Schoeller and van Santen (1982) first introduced DLW as a potential method to measure total daily energy expenditure in humans; however, Lifson began the original development in rodents during the late 1940s and early 1950s. The discovery by Lifson et al. (1949) that O_2 in respiratory CO_2 was in equilibrium with O_2 in body water (H_2O) became an important first step in the development of the

DLW method. The tracking of O_2 in body water became a potential means to determine CO_2 production and ultimately energy expenditure. This discovery and the development of early mass spectrometry equipment to measure isotopic enrichments led to the first measurement of CO_2 production using DLW. Lifson et al. (1955) enriched the body water of mice with DLW and tracked the elimination of ^2H and ^{18}O from the body by blood samples over several days. These data were compared to CO_2 production data simultaneously collected from a metabolic chamber.

Despite the successful introduction of the DLW method to measure daily energy expenditure, the exceptionally high cost of ^{18}O limited early use of this new technique. Before the early 1970s, the only application study completed was by LeFebvre (1964). This study examined energy expenditure in racing pigeons over a 300-mi race by utilizing DLW methodologies first proposed by Lifson and compared this to direct chemical analyses of pigeon carcasses. The high price of the isotopes and lack of precise analytical instrumentation to measure isotopic enrichments in bodily fluids limited the use of DLW to small animal studies for nearly 20 years, until the reintroduction of the method in humans by Schoeller and van Santen (1982).

Recently, there has been considerable interest in quantifying the specific contribution of physical activity to overall energy expenditure. To address this need, recent efforts have begun to incorporate indirect calorimetry procedures to allow the contribution from physical activity energy expenditure to be partitioned out. This is accomplished by measuring total daily energy expenditure by DLW while simultaneously measuring two components of total daily energy expenditure, resting metabolic rate (RMR) and thermic effect of food (TEF), by indirect calorimetry (Bray, 1997). The remaining component of total daily energy expenditure, physical activity energy expenditure, can be estimated by subtracting RMR and TEF from total daily energy expenditure (figure 12.1).

Principles of Method: Biochemical Basis of Energy Expenditure

Energy expenditure can be assessed accurately with both the DLW method and indirect calorimetry. Emphasis in this chapter is on the combined use (DLW/indirect calorimetry) because this provides a way to assess free-living physical activity.

With this method a person drinks a standardized amount of the two stable isotopes deuterium (^2H) and oxygen-18 (^{18}O), which compose DLW ($^2H_2$$^{18}O$), and is then allowed to return to a normal lifestyle with only urine samples collected at the beginning and end of the measurement period. By measuring the elimination kinetics of ^2H and ^{18}O from a person's body, total carbon dioxide (CO_2) production is determined for the measurement period and, ultimately, an estimation of total daily energy expenditure. Indirect calorimetry measurements of RMR and TEF are needed to complete the estimate of physical activity energy expenditure. RMR is the largest component of total daily energy expenditure (see figure 12.1) and is associated with the energy needs to sustain life. TEF, or dietary-induced thermogenesis, is the smallest component of daily energy expenditure and is associated with the energy needs to process nutrients. Both RMR and TEF can be estimated by determining the concentration

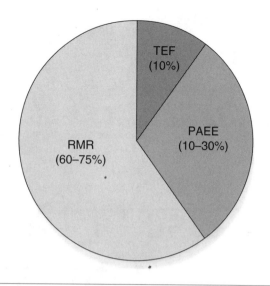

Figure 12.1 Total daily energy expenditure and its components: resting metabolic rate (RMR); thermic effect of food (TEF); physical activity energy expenditure (PAEE).

of CO_2 and O_2 in respiratory gases. Additional information on the principles of DLW and indirect calorimetry are included in this chapter to provide a better understanding of the biochemical basis of this approach.

Doubly Labeled Water

Isotopic tracers can be used to monitor metabolic events in the human body (Wolfe, 1992). An ideal isotopic tracer is very similar to the compound of interest or the tracee. In the case of measuring total daily energy expenditure, we are interested in tracing the elimination of hydrogen and oxygen from the human body. Specifically, the two stable isotopes, ^2H and ^{18}O, are used to measure total daily energy expenditure. An isotope has the same chemical identity as its parent element; however, its atomic mass is slightly different. This difference in mass allows the isotope to be a tracer and distinguishable from the parent element or tracee. A stable isotope is nonradioactive and poses no risk to the human body.

Isotopes of hydrogen include protium (^1H), deuterium (^2H), and tritium (^3H). Because ^1H is the most abundant mass of hydrogen and ^3H is radioactive, ^2H becomes the hydrogen isotope of choice in the DLW method. Isotopes of oxygen include ^{16}O, ^{17}O, and ^{18}O. ^{18}O is the tracer of choice in the DLW method because ^{16}O is the most abundant mass of oxygen and ^{17}O is much more expensive than ^{18}O. Both ^2H and ^{18}O are in trace amounts in nature and the human body, and must be produced via column distillation techniques. ^2H and ^{18}O are commercially available through chemical isotope companies. ^2H is very inexpensive, whereas 10% ^{18}O costs approximately $4.00 to $6.00 per g, when production and availability are high.

Indirect Calorimetry

Chemical energy, or food, is liberated in the presence of oxygen to produce mechanical energy and heat. This conversion allows the daily functions of life to occur, from walking to scratching your nose. The heat produced during this conversion is equal to the liberated chemical energy. That is, energy is neither created nor destroyed; rather, it is transformed from one form to another, as stated by the laws of thermodynamics.

Energy expenditure or the capacity of our bodies to perform work can be assessed by direct or indirect calorimetry methods. Direct calorimetry measures energy expenditure by directly assessing heat production, such as a bomb calorimeter to measure energy content of food items. Indirect calorimetry measures energy expenditure by assessing O_2 consumption and not heat production. As mentioned previously, heat production is equal to liberated chemical energy. Thus, indirect calorimetry can be used to assess energy expenditure by indirectly measuring O_2 consumption associated with metabolizing foodstuffs and not by directly measuring heat production.

Later sections of this chapter describe the assessment of energy expenditure by indirect calorimetry, but it is important here to describe the biochemical principle behind the conversion of O_2 consumption values to heat equivalents or kcal. This process is highly dependent on knowing the composition of foodstuffs being metabolized. It is known that the three primary foodstuffs are carbohydrate, fat, and protein. Each is metabolized in the presence of O_2 to produce CO_2, H_2O, and energy. The composition of the foodstuffs metabolized determines the amount of energy expended. Specifically, the ratio of CO_2 produced to O_2 consumed, referred to as the *respiratory exchange ratio* (RER), will determine the kcal equivalent value for each liter of O_2 consumed. A typical physiological RER ranges from 0.7 to 1.0. When 100% carbohydrate is being metabolized, the RER equals 1.0, and when 100% fat is being oxidized, the RER equals 0.7. A 50:50 mixture of both foodstuffs is equivalent to an RER of 0.85. It should be noted that protein contributes little to energy production, except in periods of prolonged starvation.

Because carbohydrate is a more O_2-efficient fuel (i.e., less O_2 is used to metabolize 1 g of carbohydrate versus 1 g of fat), more kcal are liberated per liter of O_2 consumed when metabolizing carbohydrate compared to fat. Using a person's measured RER, along with a published caloric equivalent table, energy expenditure can be determined in kcal. For example, an RER of 0.7 corresponds to a caloric equivalent of 4.686 kcal/L of O_2 consumed. An RER of 1.0 corresponds to a caloric equivalent of 5.047 kcal/L of O_2 consumed. Thus, a person metabolizing 100% carbohydrate will expend nearly 0.5 kcal more per liter of O_2 consumed compared to a person metabolizing 100% fat. Those in a fasted state typically utilize more fat than carbohydrate (0.75 < RER < 0.82), while those participating in strenuous aerobic exercise will utilize more carbohydrate (0.88 < RER < 0.95). Overall, an accurate assessment of energy expenditure using indirect calorimetry methods is highly dependent on measuring O_2 consumption and knowing the nature of foodstuffs being metabolized.

Assessment Procedures

Total daily energy expenditure includes all the energy expended by a person during a 24-hour period. This includes energy to sustain biological processes of the resting body, digestion, and daily activities (see figure 12.1). Although this chapter focuses on DLW, total daily energy expenditure can be assessed by whole-room calorimetry, which utilizes the principles of indirect calorimetry (Ravussin et al., 1986). For this procedure, an individual lives for several days within the controlled environment of a metabolic chamber. The chamber is built and set up to resemble typical living arrangements (i.e., a room about 10' × 10'), although the size of most calorimeters must be limited to assure proper mixing and to prevent trapping of respiratory gases.

Although a whole-room calorimeter is an excellent tool to measure RMR, TEF, and substrate oxidation rates over extended periods of time, it poses an artificial environment to measure daily energy expenditure. Total daily energy expenditure can be underestimated by as much as 15 to 20% in a whole-room calorimeter (Rosenbaum et al., 1996). This is primarily due to the underestimation of physical activity in this confined environment. A primary strength of the combined DLW/indirect calorimetry method is the ability to accurately measure physical activity energy expenditure under free-living conditions. The following sections provide details of these assessments.

Doubly Labeled Water Methodology

The basic principle behind the DLW method to measure total daily energy expenditure can be seen in figure 12.2. Specifically, an initial standardized dose of $^2H_2^{18}O$ is given orally to increase the abundance or enrichment of these isotopes in the human body. The new ^2H and ^{18}O enrichment levels are determined from urine samples and the person is then allowed to return to a normal environment for, generally, 7 to 14 days. After this time frame, the person returns to the laboratory so that final urine enrichment levels of ^2H and ^{18}O can be determined. As seen in figure 12.2, the decline in ^2H is a function of H_2O turnover and the loss of ^{18}O is a function of H_2O turnover and also CO_2 production. That is, ^{18}O is lost from the body both as H_2O and as CO_2. As Lifson discovered in the 1940s, oxygen in body water and in respiratory CO_2 are in equilibrium. Thus, the difference between the ^2H and ^{18}O water turnover curves is the ^{18}O that is lost in the form of CO_2. The difference between the ^2H and ^{18}O curves is integrated over the desired measurement period to determine the total CO_2 production for this period. These data are then used to determine total daily energy expenditure (see the section on calculations later in this chapter, p. 202).

Before the reintroduction of the DLW method, the assessment of total daily energy expenditure in free-living humans had been problematic. Tradition-ally, indirect assessments occurred by measuring energy intake from food records and diaries over several days to weeks. When a person's weight is stable for this period, the daily energy intake is a surrogate measure of daily energy expenditure. That is, a person is in energy balance when energy intake matches energy output. However, it is now well documented that the recording of accurate dietary food records is difficult and most people traditionally underreport what they consume (Sawaya et al., 1996). Thus, the reintroduction of the DLW method in humans provided a revolutionary assessment of energy expenditure.

Dosing and Urine Collections. Along with being a free-living measure of daily energy expenditure, another major strength of the DLW method is the unobtrusive nature of the measurement. The person being dosed with DLW can actually be blinded to the entire experiment. Aside from determining daily energy expenditure, a person's total body water is derived using DLW. This total body water data can be used to estimate fat and fat-free mass. Thus, if subjects are told that body composition is being tested instead of activity, they may be less likely to change their daily routine.

The actual dosing of a person is very simple and can be done easily with minimal supplies; however, attention to detail is critical. First, a large volume of DLW must be prepared to dose a group of people. This typically occurs by diluting a volume of 10% ^{18}O with 99.9% ^2H to produce a final mixed dose of $^2H_2^{18}O$ with an ^{18}O concentration between 5 and 9%. An aliquot of this mixed dose should be stored in, for example, a vacutainer tube and stored in the freezer for future analysis.

Second, the subject will orally ingest a standardized volume of this DLW. In general, a certain amount of DLW needs to be administered to significantly increase body levels of ^2H and ^{18}O above normally present background levels. It should be remembered that these two stable isotopes are in trace amounts in the human body. Specific propagation of error analyses have been completed to determine the optimal dose that will raise enrichments levels of both ^2H and ^{18}O to provide the best ratio of the two isotopes and minimize the cost of dosing a person because of the higher cost of ^{18}O. Traditionally, a dose containing approximately 0.25 g of ^2H and 0.12 g of ^{18}O per kg of estimated total body water is given to a person (Schoeller, 1983).

The length of the measurement period is also a consideration when dealing with populations at the extremes of the spectrum. Because the elimination

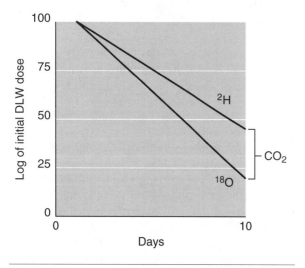

Figure 12.2 Decline of deuterium (^2H) and oxygen-18 (^{18}O) from the body water pool over a 10-day period. ^2H is lost completely from water turnover, whereas ^{18}O is lost from H_2O and CO_2 turnover.

of both isotopes from the body is directly related to water turnover, the measurement period should correspond to expected water turnover. For example, a person such as an ultra-endurance athlete has a much higher water turnover than a sedentary older adult. A general rule of thumb is that the measurement period should last 2 to 3 half-lives of each isotope, which would correspond to roughly 8 to 10 days for a very active person and approximately 14 to 18 days for a very sedentary older person. This will ensure that enrichment levels of ^{2}H and ^{18}O at the end of the measurement period will be well above normally present background levels. It is also imperative that the source of daily drinking water for the person does not change over the measurement period, as levels of ^{2}H and ^{18}O can differ from source to source.

The actual dose of DLW should be weighed out on an electronic balance within ±0.2%. A sterile cup with a lid can be used to weigh out the dose, and the exact weight of the dose should be recorded, which is critical for future isotope calculations. Keep the lid on the dose before administering it to prevent spillage of the DLW and any potential isotopic fractionation (i.e., isotope transfer from solid to gas phase).

Before dosing, a urine sample must be obtained to determine background enrichment of ^{2}H and ^{18}O in the body. It is imperative that the baseline urine be collected before dosing. Generally, a baseline urine sample is obtained between 16:00 and 21:00 hours in a 2- to 4-hour postprandial state. Two aliquots of urine (total of 10 to 20 ml) are then quickly placed in airtight containers such as vacutainers without additive. Excessive exposure to atmospheric air can increase the chance of isotopic fractionation. Thus, the storage tubes should be airtight, no airspace should be left in the tube above the sample, and processing of all urine should be done promptly. Urine samples are stored at –20° C until future analyses. After collecting the baseline urine, the DLW is orally ingested from the sterile cup via a straw to avoid spilling the dose. The sterile cup is then filled with 50 to 100 ml of tap water to rinse any residual DLW and to ensure that all DLW reaches the stomach. Additionally, the time and date of the baseline urine collections and DLW dosing are recorded.

Two additional urine samples are then obtained the following morning after the initial morning void. These samples should be at least 20 minutes apart. The overnight void is discarded because it is difficult to place an accurate time of collection on this sample. Again, the time and date of the baseline

urine collections are recorded. After these urine collections, this individual will return back to a normal environment to complete the measurement period, which typically lasts 10 to 14 days. After this period, two additional samples are obtained with a minimum of 20 minutes between samples.

Analyses. Over a 10- to 14-day measurement, a total of five urine samples will be collected and frozen for future analyses (i.e., one sample before dosing, two samples on day 1, and two at the end of the measurement period). The isotopic enrichment of these urine samples is measured by gas isotope ratio mass spectrometry (IRMS). Specifically, the IRMS will measure isotopic enrichment of elements in a gaseous form such as H_2, CO_2, or N_2 (Wolfe, 1992). With respect to the DLW analyses, both ^{2}H and ^{18}O in the urine samples must be processed into a gaseous form prior to IRMS analyses.

^{2}H in the urine samples is prepared for IRMS analysis during a reduction reaction with zinc under extreme heat to produce H_2 (Kendall & Copelan, 1985). The H_2 is in equilibrium with the ^{2}H in the urine, and the isotopic enrichment of the urine can therefore be determined by measuring H_2 using IRMS. This conventional processing method has been used for many years, but the method is labor intensive and requires a special off-line vacuumed processing system. Recent work suggests that a platinum reduction method may be as reliable as the zinc method and is less costly and labor intensive (Herd et al., 2000). In contrast, the preparation of ^{18}O urine samples for IRMS analyses is very simple. An aliquot of urine is placed in a vacutainer, which is then filled with low-pressure CO_2 and shaken overnight at room temperature. The oxygen in the CO_2 equilibrates with the oxygen in the urine, and the ^{18}O isotopic enrichment of the CO_2 is then measured by IRMS.

An IRMS unit has a variety of applications and can be found in chemistry or geology laboratories to identify chemical substance or date rock formation, or in research centers that complete human studies of metabolism (Wolfe, 1992). The use of an IRMS requires an individual with expert knowledge of isotope analyses. This analytical equipment requires regular maintenance and likely is not a piece of equipment that a laboratory will invest in unless it is completing regular DLW analyses. In fact, many laboratories around the country contract their IRMS and expertise to complete DLW urine sample analyses. Nonetheless, the basic principle behind the IRMS for DLW analyses is the determination of the mass-to-charge ratio of hydrogen and oxygen.

Specifically, the DLW-prepped gas samples are introduced into the IRMS and they are immediately ionized. The ionized elements are deflected based on their atomic mass by an electromagnet, and a special detector collects the ions and generates a signal proportional to their masses. In the case of the DLW method, we are examining the ratios of $^2H:^1H$ and $^{18}O:^{16}O$. These data along with reference sample data are used to determine the exact isotopic enrichment of ^2H and ^{18}O from baseline and measurement period urine samples. These enrichment data are used to calculate total daily energy expenditure as explained in the following section.

Calculations. As mentioned, the ability to calculate total daily energy expenditure is dependent on determining the rate of CO_2 production, which is related to the rate at which ^2H and ^{18}O leave the body water pool over the measurement period. The purpose of this section is to provide a basic understanding about calculating total daily energy expenditure. However, additional readings are suggested to understand the background and certain assumptions associated with these calculations (IDECG 1990; Schoeller, 1988). The following equation is used to estimate the rate of CO_2 production (rCO_2) in moles/day:

$$rCO_2 \text{ (moles/day)} = TBW/2[(1.041 \times k^2H) - (1.007 \times k^{18}O)]$$

TBW is total body water (in L) and k is the rate of loss of ^2H and ^{18}O from the body over the measurement period. The individual k values are the slopes of the elimination curves calculated from the individual enrichment data points. The factors of 1.041 and 1.007 are the hydrogen and oxygen dilution spaces, respectively (Racette et al., 1994). These factors account for the additional space in the body that hydrogen and oxygen are present, outside of the normal body water pool. The individual rCO_2 (in moles/day) can be converted to total daily energy expenditure (kcal/day) using the Weir equation (Weir, 1949):

$$\text{Total daily energy expenditure (kcal/day)} = (3.94 / RQ + 1.10) \times 22.41 \times rCO_2$$

The rCO_2 is converted to L/day by multiplying by 22.41 (1 L = 22.41 moles). RQ (respiratory quotient) is an index of the composition of the foodstuffs in a person's diet. A value of 0.85 is typically assumed instead of making a direct measurement from labor-intensive food records. However, it has been shown that a 24-hour dietary recall can pro-

vide accurate food quotient data (Surrao et al., 1998).

Indirect Calorimetry Methodology

The most common and simplest indirect calorimetry measurement of RMR or TEF is the dilution canopy, also referred to as the *ventilated hood indirect calorimetry technique* or simply the *ventilated hood technique*. An example of the ventilated hood technique is shown in figure 12.3. Respiratory gases are pulled from the plastic canopy through the plastic tubing to the O_2 and CO_2 analyzers by a low-speed pump. Room air is subsequently pulled into the plastic canopy. Thus, room air is diluted by the O_2 and CO_2 in the person's breath. Typical concentrations of room O_2 and CO_2 are 20.93 and 0.03%, respectively. The low-speed pump is set so that the dilution of room air in the canopy produces a CO_2 concentration of approximately 0.5 to 1.0%. The pump rate may be 15 to 35 L/min, depending on the indirect calorimetry system and the size of the adult. Overall, the fraction of inspired and expired O_2 and CO_2, pump rate, and ambient conditions are used in conjunction with the Weir equation (Weir, 1949) to calculate RMR or TEF in kcal/day.

Determination of Resting Metabolic Rate. The largest component of daily energy expenditure is RMR (see figure 12.1). RMR is all expended energy associated with the physiological processes to sustain life—that is, if a person were to lie awake in bed 24 hours a day, the energy expenditure associated with heart and lung function, conduction of nerve impulses, kidney filtration, and so on. Remarkably, this accounts for nearly 60 to 75% of a person's daily energy expenditure. Furthermore, nearly 40% of an individual's total RMR is a function of energy expended by the brain, liver, and kidneys, which constitute only 5% of a person's total body weight (Elia, 1992).

As would be expected, fat-free mass (i.e., metabolically active tissues) is the primary factor that predicts a person's RMR. Nearly 60 to 70% of the variation in RMR can be explained by a person's total amount of fat-free mass (Tataranni & Ravussin, 1995). However, other factors do influence RMR, including gender, race, age, and fitness. Women typically have a lower RMR than men after adjusting for differences in body weight (Arciero et al., 1993) and African-American women have a lower RMR than Caucasian women (Gannon et al., 2000). Aging is typically associated with a 2 to 3% decline in RMR (Tzankoff & Norris, 1978), which is prima-

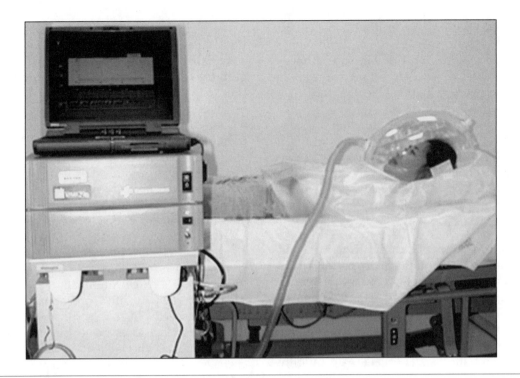

Figure 12.3 RMR measurement utilizing the ventilated hood indirect calorimetry technique.
Courtesy of Raymond D. Starling.

rily linked to loss of fat-free mass and potentially the metabolic activity of various organs.

Pre-RMR testing conditions are important to obtain accurate and reliable data. *Basal metabolic rate* (BMR) is the classical term used to characterize resting metabolism and is obtained under strict conditions. That is, a person is measured on waking and after a 12- to 18-hour fast in a quiet and thermoneutral environment (i.e., approximately 23° C), with the body and mind at complete rest. Nonetheless, this type of controlled environment is difficult to duplicate, particularly with the natural anxiety of a person going through a procedure. Thus, true BMR is typically unobtainable and the term *RMR* (or *resting metabolic rate*) is typically used in the field of energy metabolism. RMR may be slightly higher than a true BMR, but it is possible that no differences exist if tight control occurs with the RMR measurement.

Many research laboratories with overnight facilities will measure RMR after a subject awakens. This measurement should occur after a 10- to 12-hour overnight fast in a lightly lit and thermoneutral room with no noise or other mental distraction, as the person rests quietly in an awakened state. Additionally, individuals should abstain from structured exercise and alcohol for 24 hours before measurement because of the potential influence of these factors on RMR.

A typical RMR measurement can be completed in 15 to 30 minutes if the person reaches a steady state. A general rule of thumb is that a coefficient of variation of <10% for 5 minutes of consecutive energy expenditure measurements can be considered a steady state. In some people, this criterion may be obtained in 10 to 15 minutes; however, it may take 30 minutes in other individuals. The collection period should be limited because extending the time beyond 30 minutes may cause people to become restless, which will artificially increase their metabolism. Allowing participants to become familiar with the plastic canopy the night before the RMR measurement will minimize any potential anxiety.

It should be pointed out that RMR could be measured accurately if inpatient testing facilities are not present (Bullough & Melby, 1993). If one is collecting outpatient data, it is important that subjects minimize their morning activity before arriving at the testing facilities. Furthermore, they will need to rest quietly in bed for 30 to 45 minutes so that their metabolic rate can return back to resting levels. Additionally, the menstrual cycle should be considered when measuring RMR in women. RMR has been shown to be 4% higher during the luteal phase compared to the follicular phase (Matsuo et al., 1999). The higher concentrations of estradiol and catecholamine during the luteal phase may be responsible for the higher RMR. However, others have demonstrated no difference in RMR between the luteal and follicular phases (Piers et al., 1995). Nonetheless, care should be taken to measure all menstruating women in the same period of the menstrual cycle, preferably in the follicular phase (i.e., 6 to 10 days after beginning menses).

Overall, standardized procedures and a controlled environment are important to obtain an accurate RMR. The coefficient of variation for day-to-day RMR measurements is in the order of 3 to 6% for most laboratories. Thus, repeat RMR measurements may be advised, particularly if trying to detect small differences in RMR (i.e., 100 to 150 kcal/day).

Determination of the Thermic Effect of Food (TEF). The smallest component of a person's daily energy expenditure is that energy associated with dietary-induced thermogenesis. TEF is expended energy associated with digestion, absorption, transport, and storage of daily nutrients (i.e., the energy above RMR associated with daily meals, beverages, and snacks). TEF makes up roughly 10% of most people's daily energy expenditure. The TEF response is composed of the obligatory and facultative components. The obligatory component is associated with all phases of metabolism from digestion to storage of nutrients, while the facultative component is associated with sympathetic activity.

The measure of TEF follows the same standard procedures and control as the assessment of RMR. The only major differences are that a standardized meal is given before measurement and the measurement period is longer, lasting 4 to 8 hours. The premeasurement meal should resemble a typical meal in size and composition. The meal size should be standardized, either as a percentage of energy requirements or per kg of body weight or fat-free mass. A good rule of thumb is one-third of daily energy requirements (i.e., based on three meals a day) or 10 kcal/kg of body weight. For example, a very sedentary 70-kg male with a daily energy requirement of roughly 2,100 kcal would ingest a 700-kcal meal before the TEF measurement. The composition of the meal should resemble the usual diet of the population to be studied. Traditionally, a mixed nutrient diet is used, comprising 50% energy from carbohydrate, 35% from fat, and 15% from protein. It should be pointed out that the meal size and composition will influence the magnitude and duration of TEF, so care should be taken to standardize these factors.

Test meals should be consumed within 15 minutes of beginning the TEF measurement. As with measuring RMR, the plastic canopy is placed over the head as the subject rests in a supine and awake state. TEF is typically assessed for the first 10 to 20 minutes of every 30 minutes with the remaining time for a break from the canopy, while the person remains in bed. TEF is generally measured for 4 to 6 hours. These data are plotted over the measurement period, and TEF in kcal is calculated as the area under the response curve minus RMR. These kcal data can be multiplied by 3 (i.e., three meals a day) to calculate a total daily TEF.

The major limitation with assessing TEF is the long measurement period. The person being measured can become very restless, resulting in elevated energy expenditure from fidgeting. The TEF response persists for 6 hours after ingestion of a meal (Reed & Hill, 1996). Additionally, 40% and 22% of TEF may be missed with 3- or 4-hour measurement periods, respectively. Thus, a true assessment of TEF requires a longer assessment period to capture all dietary-induced thermogenesis. Furthermore, the coefficient of variation for day-to-day TEF measurements can be as high as 30 to 35%.

Outcome Variables

The main outcome from the DLW method is energy expenditure. Frequency, intensity, and duration of activity cannot be determined, but the incorporation of indirect calorimetry data allows the energy expenditure of physical activity to be determined. This section describes the calculations needed to make these estimations.

Physical Activity Energy Expenditure

Physical activity energy expenditure is all expended energy not associated with RMR and TEF. Physical activity energy expenditure includes both volitional and nonvolitional activities. Volitional activities range from tasks on the job and at home to exercise and leisure activities. Nonvolitional activities include shivering, fidgeting, and maintenance of postural muscles. Physical activity energy expenditure may comprise anywhere from 10 to 50% of a person's daily energy expenditure. Very sedentary adults may expend no more than several hundred physical activity kcal/day, whereas cyclists participating in the Tour de France may expend as much as 6,000 to 7,000 kcal/day.

Calculation of physical activity energy expenditure using DLW/indirect calorimetry data is very simplistic and involves basic mathematical principles. The following equation is typically used to calculate physical activity energy expenditure in kcal/day:

Physical activity energy expenditure (kcal/day) = (total daily energy expenditure × 0.9) – RMR

Total daily energy expenditure in kcal/day is derived from the DLW. This value is multiplied by the factor 0.9 to account for the kcal associated with TEF. As mentioned in the previous section, measurement of TEF is a long procedure with high day-to-day variability. Thus, most laboratories will estimate TEF because it is such a small fraction and consistently 10% of a person's daily energy expenditure. Finally, RMR is subtracted to derive physical activity energy expenditure.

Normalization of Physical Activity Energy Expenditure

An additional factor to consider when utilizing DLW/indirect calorimetry is the correction of physical activity energy expenditure data for differences in body weight. Because the majority of daily activities are weight dependent (i.e., involve movement of a person's body weight), a correction for this may be appropriate, particularly since heavier individuals expend more energy for the same activity compared to a smaller person. Several approaches have been suggested to normalize physical activity energy expenditure (Prentice et al., 1996). One is to report a physical activity level (PAL) ratio. PAL is the ratio of total daily energy expenditure to RMR, and the correction for body size is related to the fact that RMR is dependent on body size. Typical PAL ratios can range from 1.5 for sedentary individuals to >2.0 for highly active individuals. This approach provides a uniform method to compare data across studies but provides no information about the total number of kcal expended during daily activity.

To alleviate this problem, a correction using body weight can be completed. Originally, protocols called for physical activity energy expenditure to be divided by body weight to some exponent. The exponent is used because daily activities are a mix of weight-bearing and non-weight-bearing activities. However, it is difficult to ascertain the appropriate exponent; typically a factor of 0.5 is used. Others suggest that directly dividing physical activity by body weight (i.e., ratio method) is inappropriate because the relationship between energy expenditure and body weight has a non-zero inter-

cept when plotted (Carpenter et al., 1995). Using the ratio method may cause spurious results, particularly for those with lower body weights. It is suggested that an analysis of covariance be completed with body weight as the covariate, which will account for the non-zero intercept resulting in appropriate adjustments for both larger and smaller people. Despite the presence of several methods to adjust physical activity energy expenditure, body weight does influence expenditure and should be accounted for when comparing obese and lean people.

As mentioned previously, physical activity energy expenditure includes all energy expended during daily activities. Thus, the total amount of daily physical activity energy expenditure can be derived, although no information can be obtained about the duration and intensity of these activities. If conducting studies examining the influence of total daily physical activity on energy balance and subsequently the prevalence of obesity, DLW/indirect calorimetry provides the most accurate assessment of all physical activity. However, DLW and indirect calorimetry will not allow one to examine, for example, the influence of the intensity of daily activities on rates of morbidity and mortality. Therefore, DLW/indirect calorimetry should only be used when the dimension of physical activity to be assessed is the total amount of physical activity energy expenditure.

Partitioning of Volitional and Nonvolitional Activity

A novel use of DLW/indirect calorimetry has been the recent attempt to separate nonvolitional from volitional activities. As mentioned previously, nonvolitional activities are shivering, fidgeting, and maintenance of postural muscle, while all other activities are considered volitional. Previous research using infrared sensors in a whole-room calorimeter suggests that some people may expend 100 to 700 kcal/day fidgeting (Ravussin et al., 1986). To further assess the impact of nonvolitional activity, a recent study utilized DLW/indirect calorimetry in conjunction with an activity recall and pedometer to separate physical activity energy expenditure into volitional and nonvolitional activities (Levine et al., 1999). Energy expenditure from daily work, exercise, and leisure activities assessed by recall and pedometer was subtracted from physical activity energy expenditure assessed by DLW/indirect calorimetry. The difference was energy expenditure most likely associated with nonvolitional activities. Data from this study showed that those who are most resistant to weight gain during overfeeding might have high levels of nonvolitional activity. Although these are preliminary data, DLW/indirect calorimetry may be a valuable tool to separate volitional and nonvolitional activities.

Applications of the DLW and Indirect Calorimetry Method

Because the DLW/indirect calorimetry method provides a precise measure of energy expenditure, it is most commonly used as a criterion measure to validate other techniques. It is also used as an outcome measure in some mechanistic studies on weight control. This section describes these applications in more detail.

Validation Studies

DLW/indirect calorimetry has traditionally been the gold standard for assessing physical activity. Studies have attempted to validate other physical activity assessment methods against DLW/indirect calorimetry. Many of these studies have reported significantly low to moderate correlations between their methods and DLW/indirect calorimetry. Most concluded, therefore, that their methods are a proxy measure of physical activity energy expenditure. However, this may be an inappropriate conclusion for several reasons.

First, some of these other physical activity tools do not assess the same dimension of physical activity. For example, some recall questionnaires may assess only leisure-time activities and, therefore, not all daily activities, as does DLW/indirect calorimetry. The assessment method of choice must also measure physical activity energy expenditure in order to effectively use DLW/indirect calorimetry for validation purposes. In addition, most studies correlate the assessment method of choice with DLW/indirect calorimetry. Even though the assessment method in question may measure the same dimension of physical activity as DLW/indirect calorimetry, a strong correlation or relationship between the two methods does not mean they are in total agreement. That is, data points from both methods may lie *across* a straight line, giving a high correlation, but unless the points lie *on* the line of identity, there is not complete agreement between the methods. Rather than complete simple correlation analyses, it is suggested that data

between two methods be examined using Bland-Altman (Bland & Altman, 1986) plots, which are simple computations and provide information about agreement between two methods. (For further details on these analyses, see Bland & Altman, 1986.)

Finally, it is important that validation studies be completed in an adequate number of people with equal gender distribution. If a physical activity assessment method is to be used in large-scale epidemiological studies, it is important that the method has been validated against a gold standard in equal numbers of females and males.

To date, a handful of well-controlled validation studies in adults have been completed comparing other physical activity assessment methods to DLW/indirect calorimetry. Although the purpose of this section is not to assess previous validation studies, one DLW/indirect calorimetry study is presented here to highlight conclusions that can be drawn from a well-controlled study.

Figure 12.4 presents physical activity energy expenditure data for 35 elderly women and 32 elderly men. The data were collected using DLW/indirect calorimetry as the gold standard and several other assessment methods (Starling et al., 1999). The methods to be validated included the Minnesota Leisure-Time Physical Activity Questionnaire (LTPAQ) developed in the late 1970s, the Yale Physical Activity Survey (YPAS) developed in the early 1990s, and the Caltrac uniaxial accelerometer, commercially available since the late 1980s. Physical activity energy expenditure was underestimated by approximately 55 to 60% in both women and men using the Minnesota LTPAQ questionnaire or the Caltrac. Underestimation with the Minnesota LTPAQ was most likely because this recall questionnaire only measures exercise and leisure-time activities. Also, the questionnaire was developed in younger adults and does not encompass the types of activities that elderly adults participate in. The difference between the Caltrac and DLW/indirect calorimetry measurements was most likely because the Caltrac accelerometer only measures activity in one plane and does not account for water activities because it cannot be submerged. In contrast, physical activity energy expenditure was measured accurately when utilizing the YPAS (see figure 12.4). This may be because the YPAS was developed in older adults and includes the types of activities present in their normal work, exercise, and leisure routines.

These data demonstrate that physical activity energy expenditure can be assessed accurately us-

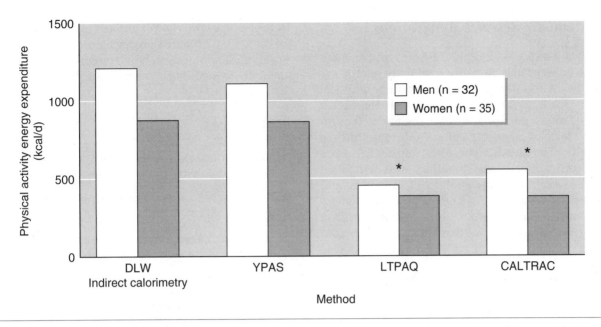

Figure 12.4 Physical activity energy expenditure data comparing DLW/indirect calorimetry to the Minnesota Leisure-Time Physical Activity Questionnaire (LTPAQ), the Yale Physical Activity Survey (YPAS), and a Caltrac uniaxial accelerometer (* = different than DLW/indirect calorimetry [p < 0.05]).

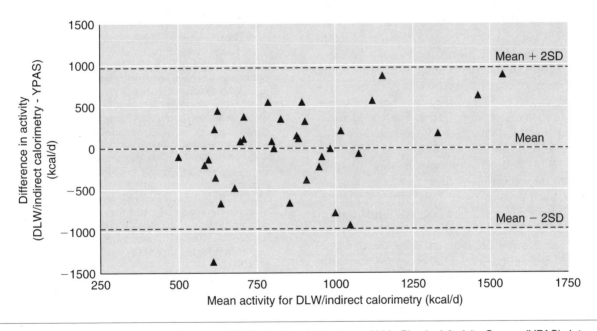

Figure 12.5 A Bland-Altman plot between DLW/indirect calorimetry and Yale Physical Activity Survey (YPAS) data for 35 elderly women.

ing well-developed recall questionnaires. However, despite the accuracy on a group basis, the YPAS did not accurately assess physical activity energy expenditure when examining individual data for the group. This can be seen in figure 12.5, which presents a Bland-Altman plot of agreement for the women. The limits of agreement ranged from –963

to +981 kcal/day, demonstrating a wide range of over- and underestimation when individual data are considered outside of the group data. Thus, this study suggests that the YPAS will provide excellent accuracy when examining physical activity energy expenditure in a group; however, accuracy for an individual is more variable.

Mechanistic Research Applications

DLW and indirect calorimetry are powerful research tools to accurately measure total daily energy expenditure and its components, including physical activity energy expenditure. This has important implications, particularly when the research of interest focuses on energy balance (e.g., the study of obesity). The causes of obesity are multifaceted and not well understood. Therefore, well-controlled studies examining the etiology of obesity would benefit from the DLW/indirect calorimetry method. For example, recent work utilizing DLW/indirect calorimetry demonstrate that higher than expected levels of daily physical activity may be needed to prevent weight regain after completing a weight-loss program (Schoeller et al., 1997).

Understanding the energy requirements of today's society, including children, adults, the elderly, and individuals with disease, has important public health implications. Current worldwide energy requirement recommendations are primarily based on the summation of estimated energy expenditure values for a variety of daily activities, referred to as the *factorial method*. This energy requirement value is reported as a multiple of RMR, or a PAL ratio, as mentioned previously. PAL is a ratio of total daily energy expenditure to RMR. DLW/indirect calorimetry provides an objective method to assess energy requirements of various populations without the need to use the factorial method, which is labor intensive and can be highly subjective.

The DLW/indirect calorimetry method also provides an important research tool to assess the impact of various factors on daily physical activity energy expenditure, such as aging, different exercise prescriptions, and genetics. Overall, DLW/indirect calorimetry provides an accurate and free-living assessment of total daily energy expenditure and its components, including physical activity energy expenditure. Thus, any research question in which the accurate assessment of energy expenditure is important may benefit from the use of DLW/indirect calorimetry.

Summary

DLW/indirect calorimetry is the most accurate method to assess physical activity energy expenditure. The primary strengths of this method are the free-living assessment and the ability to capture all volitional and nonvolitional daily activities. Because of the isotope costs, DLW/indirect calorimetry may be difficult to utilize in large-scale studies. Smaller-scale validation studies should be conducted before using these methods in larger-scale projects.

References

Arciero, P.J., Goran, M.I., & Poehlman, E.T. (1993). Resting metabolic rate is lower in women than in men. *Journal of Applied Physiology, 75*, 2514-2520.

Bland, J.M., & Altman, D.G. (1986). Statistical methods for assessing agreement between two methods of clinical measurement. *The Lancet, 1*, 307-310.

Bray, G.A. (1997). Energy expenditure using doubly labeled water: The unveiling of objective truth. *Obesity Research, 5*, 71-77.

Bullough, R.C., & Melby, C.L. (1993). Effect of inpatient versus outpatient measurement protocol on resting metabolic rate and respiratory exchange ratio. *Annals of Nutrition and Metabolism, 37*, 24-32.

Carpenter, W.H., Poehlman, E.T., O'Connell, M., & Goran, M.I. (1995). Influence of body composition and resting metabolic rate on variation in total energy expenditure: A meta-analysis. *American Journal of Clinical Nutrition, 61*, 4-10.

Elia, M. (1992). Organ and tissue contribution to metabolic rate. In J.M. Kinney & H.N. Tucker (Eds.), *Energy metabolism: Determinants and cellular corollaries* (pp. 1-79). New York: Raven Press.

Gannon, B., DiPietro, L., & Poehlman, E.T. (2000). Do African Americans have lower energy expenditure than Caucasians? *International Journal of Obesity, 24*, 4-13.

Herd, S.L., Vaughn, W.H., & Goran, M.I. (2000). Comparison of zinc reduction with platinum reduction for analysis of deuterium-enriched water samples for the doubly labeled water technique. *Obesity Research, 8*, 302-308.

IDECG. (1990). *The doubly-labelled water method for measuring energy expenditure*. Vienna: International Atomic Energy Agency.

Kendall, C., & Copelan, T.B. (1985). Multisample conversion of water to hydrogen by zinc for stable isotope determination. *Analytical Chemistry, 57*, 1437-1440.

LeFebvre, E.A. (1964). The use of D2O18 for measuring energy metabolism in *Colomba livier* at rest and in flight. *Auk, 81*, 403-416.

Levine, J.A., Eberhardt, N.L., & Jensen, M.D. (1999). Role of nonexercise activity thermogenesis in resistance to fat gain in humans. *Science, 283*, 212-214.

Lifson, N., Gordon, G.B., & McClintock, R. (1955). Measurement of total carbon dioxide production by means of D2O18. *Journal of Applied Physiology, 7*, 704-710.

Lifson, N., Gordon, G.B., Visscher, M.B., & Nier, A.O. (1949). The fate of utilized molecular oxygen and the source of heavy oxygen of respiratory carbon dioxide, studied with the aid of heavy oxygen. *Journal of Biological Chemistry, 180*, 803-811.

Matsuo, T., Saitoh, S., & Suzuki, M. (1999). Effects of menstrual cycle on excessive postexercise consumption in healthy women. *Metabolism, 48*, 275-277.

Piers, L.S., Diggavi, S.N., Rijskamp, J., van Raaij, J., Shetty, P.S., & Hautvast, J. (1995). Resting metabolic rate and thermic effect of a meal in the follicular and luteal phases of the menstrual cycle in well-nourished Indian women. *American Journal of Clinical Nutrition, 61,* 296-302.

Prentice, A.M., Goldberg, G.R., Murgatroyd, P.R., & Cole, T.J. (1996). Physical activity and obesity: Problems in correcting expenditure for body size. *International Journal of Obesity, 20,* 688-691.

Racette, S.B., Schoeller, D.A., Luke, A.H., Shay, K., Hnilicka, J., & Kushner, R.F. (1994). Relative dilution spaces of 2H to 18O-labeled water in humans. *American Journal of Physiology, 267,* E585-E590.

Ravussin, E., Lillioja, S., Anderson, T.E., Christin, L., & Bogardus, C. (1986). Determinants of 24-hour energy expenditure in man. *Journal of Clinical Investigation, 78,* 1568-1578.

Reed, G.W., & Hill, J.O. (1996). Measuring thermic effect of food. *American Journal of Clinical Nutrition, 63,* 164-169.

Rosenbaum, M., Ravussin, E., Matthews, D.E., Gilker, C., Ferraro, R., Heymsfield, S.B., Hirsch, J., & Leibel, R.L. (1996). A comparative study of different means of assessing long-term energy expenditure in humans. *American Journal of Physiology, 270,* R496-R504.

Sawaya, A.L., Tucker, K., Willet, W., Saltzman, E., Dallal, G.E., & Roberts, S.B. (1996). Evaluation of four methods for determining energy intake in young and older women: Comparison with doubly labeled water measurements of total energy expenditure. *American Journal of Clinical Nutrition, 63,* 491-499.

Schoeller, D.A. (1983). Energy expenditure from doubly labeled water: Some fundamental considerations in humans. *American Journal of Clinical Nutrition, 38,* 999-1005.

Schoeller, D.A. (1988). Measurement of energy expenditure in free-living humans by using doubly labeled water. *Journal of Nutrition, 118,* 1278-1289.

Schoeller, D.A., Shay, K., & Kushner, R.F. (1997). How much physical activity is needed to minimize weight gain in previously obese women? *American Journal of Clinical Nutrition, 66,* 551-556.

Schoeller, D.A., & van Santen, E. (1982). Measurement of energy expenditure in humans by doubly labeled water. *Journal of Applied Physiology, 53,* 955-959.

Starling, R.D., Matthews, D.E., Ades, P.A., & Poehlman, E.T. (1999). Assessment of physical activity in older individuals: A doubly labeled water study. *Journal of Applied Physiology, 86,* 2090-2096.

Surrao, J., Sawaya, A.L., Dallal, G.E., Tsay, R., & Roberts, S.B. (1998). Use of food quotients in human doubly labeled water studies: Comparable results obtained with 4 widely used food intake methods. *Journal of the American Dietetic Association, 98,* 1015-1020.

Tataranni, P.A., & Ravussin, E. (1995). Variability in metabolic rate: Biological sites of regulation. *International Journal of Obesity, 19,* S102-S106.

Tzankoff, S.P., & Norris, A.H. (1978). Longitudinal changes in basal metabolism in man. *Journal of Applied Physiology, 45,* 536-539.

Weir, J.B. de V. (1949). New methods for calculating metabolic rate with special reference to protein metabolism. *Journal of Physiology, 109,* 1-9.

Wolfe, R.R. (1992). *Radioactive and stable isotope tracers in biomedicine.* New York: Wiley-Liss.

Innovative Approaches for Physical Activity Assessments

13

Applying Multiple Methods to Improve the Accuracy of Activity Assessments

Margarita S. Treuth, PhD
Johns Hopkins University

Many methods exist to assess physical activity, including self-report instruments, electronic monitoring devices, heart rate monitors, pedometers, observation, and doubly labeled water. These methods have been discussed in detail in previous chapters of this book.

All methods to assess physical activity have their own strengths and weaknesses, and the choice of which method to use is dependent on many factors. In some ways, the combination of methods might provide the best possible information. However, there has been little research concerning the use of multiple measures. This may be because administration of many methods can be burdensome to the participant, costly, and possibly more difficult to interpret. While field applications pose some challenges, several studies have described the potential benefits of combining different methods to increase the accuracy of measuring physical activ-

ity. These studies are discussed in detail in subsequent sections. The focus of this chapter is to present information regarding the application of multiple methods to improve the accuracy of activity assessments.

The combination of physical activity methods is a fairly new and innovative way to measure physical activity. Six studies were identified that tested the potential for combining measures of physical activity. All investigations that have combined methods have been published within the last 10 years. An overall hypothesis running through all the studies is that the combination of measures would provide a better indication of physical activity than a single measure alone. For each of these studies, the methods that have been combined, how to combine and analyze these methods, and the extent of the improvement in the data are discussed.

History of Multiple Methods

Numerous studies in the literature have measured physical activity using more than one instrument (Aaron et al., 1995; Emons et al., 1992; Jacobs et al., 1993; Janz et al., 1992; Johnson et al., 1998; Meijer et al., 1989; Weston et al., 1997). Several of these are very recent (Ainsworth et al., 2000; Bassett et al., 2000; Handelman et al., 2000; Welk et al., 2000), as there has been a surge of interest in the field. Most of these studies were designed to test the validity or reliability of two methods. (Several excellent reviews have also focused on these issues: Baranowski et al., 1992; Melanson & Freedson, 1996; Pate, 1996; Rowlands et al., 1997; Wareham & Rennie, 1998.) Other chapters in this book provide background and results from studies that have compared self-report questionnaires, accelerometers, doubly labeled water, pedometers, and heart rate monitors. A review of these methods is beyond the scope of this chapter. Because they provide good examples on the background and reasoning behind combining methods, however, two articles (Meijer et al., 1989; Emons et al., 1992) are briefly discussed here.

Meijer et al. (1989)

Meijer et al. (1989) estimated energy expenditure by recording both heart rate and body acceleration, but the estimates were done separately for the two methods. Adults (N = 16) underwent measurements of heart rate and oxygen consumption while sitting, standing, actively standing, and walking and running on a treadmill. In four subjects, the accelerometer was also worn for 7 days of free living. Energy intake was assessed using a food intake diary during the same period.

Two separate regression equations were developed in which energy expenditure was linearly related to the accelerometer output, with a small standard error of the estimate (SEE). Linearity was also observed between heart rate and energy expenditure. It should be noted that 5 of 11 subjects were excluded because of erroneous heart rate data. The mean difference between energy intake and the accelerometer was 30.4 ± 10.7%, similar to the mean difference (i.e., 33.1 ± 28.6%) between energy intake and energy expenditure predicted from heart rate. The correlation between the total number of activity counts by the accelerometer and energy intake for a week was 0.99. As acknowledged by the authors, a more objective method such as doubly labeled water may be needed in free-living situations.

Emons et al. (1992)

Emons et al. (1992) compared three different estimates of 24-hour energy expenditure for boys (N = 9) and girls (N = 10). Energy expenditure was quantified by three methods: a 24-hour period in a room respiration calorimeter, a 14-day measurement of free-living energy expenditure, and an estimation of energy expenditure based on 24-hour heart rate monitoring. Two individual regression equations were determined for each subject. One was based on quiet activities (i.e., sleeping and standing), and the other on dynamic activities (i.e., five submaximal exercise intensities on the treadmill). Using the Flex HR method (see chapter 8), the transition point between the two regression lines was set at the heart rate while standing.

The results showed that the heart rate procedure overestimated energy expenditure compared to the measured energy expenditure by 24-hour calorimetry and doubly labeled water. These errors were 10.4 and 12.3%, respectively. The range was also quite large, from 6.3 to 16.2%. When using both regression equations (one based on the quiet activities and the other on dynamic activities), there were much larger overestimations of energy expenditure than when using only one equation. This study showed how, in this case, two different equations for the same method did not improve the accuracy of estimating energy expenditure. The large differences in some cases between methods points to the need to combine methods to accurately determine physical activity. Accurate and valid information on physical activity across different ages, ethnicities, and health status are necessary in many fields/disciplines.

Previous Recommendations on Which Methods to Combine

Several ways to combine methods have been suggested by Saris (1985). The recommendations were specifically for children, but the same suggestions could be applied to adults or other special populations. The suggested recommendations (Saris, 1985) for multiple methods to measure daily physical activity in children are as follows:

- for > 100 children, a simple questionnaire about regular activities (child/parents/ teacher) and a movement sensor;
- for 20 to 100 children, heart rate monitoring and a movement counter; and

- for < 20 children, doubly labeled water technique, heart rate monitoring, indirect calorimetry, and observation.

In general, then, with fewer number of subjects to be studied (N < 20), more precise and labor-intensive measures can be employed. In contrast, larger epidemiological studies (N > 100) would require a low cost and an easily administered method.

Available Instrumentation for the Combination of Methods

The primary method that has been combined with other measures has been heart rate monitoring. All the studies reviewed here (Eston et al., 1998; Haskell et al., 1993; Luke et al., 1997; Moon & Butte, 1996; Rennie et al., 2000; Treuth et al., 1998) combined heart rate with activity/motion sensors. The method of validation also varies across studies, but it involved some form of measurement of oxygen consumption ($\dot{V}O_2$) either during specific laboratory tests or 24-hour room calorimetry.

The reasons heart rate monitoring is the most common method to be combined with other methods include the low cost, relatively low participant burden, and the heart rate-$\dot{V}O_2$ (HR-$\dot{V}O_2$) relationship. In addition, the newer heart rate devices have the capability of storing data for several days. Newer electronic monitoring devices can also measure physical activity for several days, and these data can therefore be combined using computer-based analyses.

The heart rate method is based on the assumption of a close linear relationship between heart rate and oxygen consumption. However, there are limitations to this method. The HR-$\dot{V}O_2$ relationship, which reflects energy expenditure, varies between individuals depending on their endurance capacity (Haskell et al., 1993; Livingstone et al., 1990). This therefore requires the generation of individual calibration curves. This is especially problematic at low energy expenditure values when the slope is flat and can lead to substantial error (Livingstone et al., 1990). The HR-$\dot{V}O_2$ relationship also varies for different activities (Haskell et al., 1993; Li et al., 1993)—for instance, arm vs. leg exercise or static vs. dynamic exercise (Haskell et al., 1993; Maas et al., 1989). These limits of agreement between individual curves can vary widely (−2,399 to 1,817 kJ over 16 hours) for 18 activities. Finally, other factors can substantially affect heart rate, such as posture, emotional status, and environmental conditions (Haskell

et al., 1993). Despite these limitations, the heart rate method has been the method of choice to combine with other methods.

There are several devices that can measure both heart rate and activity simultaneously, storing the data in one unit. The Vitalog (Vitalog Corp., Redwood City, CA) measures heart rate by an electrocardiogram (ECG) signal, and body movement can be calculated from motion sensors. The Mini-logger 2000 (Minimitter Inc., Eugene, OR) is another unit that uses a heart rate band and a leg/arm sensor. It has been on the market for several years, yet it is prohibitively expensive and realistically can be used only for small research studies. The HR+M is another one-unit monitor recently developed and is worn on the chest (Rennie et al., 2000). No mention of cost or possible use by other researchers was reported in the study, perhaps because this instrument was developed by and for use in their laboratory (Rennie et al., 2000). These instruments provide a strong advantage over using two instruments because the individual is only required to wear a single instrument. Using more than one instrument may be especially difficult in certain populations (e.g., children). With further technological advances in the activity units themselves and an increase in their demand, the costs may decrease. Therefore, to maximize the likelihood of obtaining good data and subject compliance, all of these types of concerns must be addressed when designing and implementing a research study.

Recent Applications: Review of Heart Rate and Activity Monitor Studies

The following sections review several studies on heart rate and activity sensors. The procedures, outcome variables, validity and reliability, and acceptability of the combined methods are discussed, if available.

Haskell et al. (1993)

Haskell et al. (1993) used simultaneous recording of heart rate and body motion via arm/leg movement sensors to combine analysis of the data to more accurately measure physical activity. In the laboratory, 19 men performed a variety of exercises while heart rate, leg and arm motion, and oxygen uptake were recorded. The activities included walking and running at various speeds and grades, arm cranking, bicycling, Air-Dyne ergometer riding, and bench step. The Vitalog PMS-8 monitor (Vitalog Corp., Redwood City, CA) simultaneously recorded heart rate (using disposable electrodes) and activity from two motion sensors (placed on the wrist and thigh).

Multiple regression analyses were performed to predict O_2 uptake (outcome variable) from heart rate, leg motion, and arm motion during all activities. In 15 subjects with complete data, the R^2 for predicting the O_2 uptake during all of the activities ranged from 0.86 to 0.93. The mean was 0.89 with a SEE of 2.3 ml/kg/min. When data for all activities and all subjects were combined, the R^2 was 0.73 (a decrease) and the SEE was 5.2 ml/kg/min (an increase). Haskell et al. (1993) provided an example of a case where the addition of the data from the motion sensor substantially increased the R^2 over what was obtained for heart rate alone. When riding the Air-Dyne ergometer, the R^2 increased from 0.69 to 0.82 when heart rate and the arm motion sensor were combined. The authors concluded that the accuracy of estimating oxygen uptake during a wide range of activities was improved when individualized heart rate–oxygen uptake regressions were used, and heart rate and body movements were analyzed simultaneously rather than separately (Haskell et al., 1993).

This study was the first to provide an interesting way to examine physical activity (e.g., using multiple methods and two activity sensors). Several key findings were noted: The location of the monitor influences the improvement in accuracy resulting from combining methods; however, the optimal site varies for different activities (Haskell et al., 1993). In other words, the accuracy for estimating energy expenditure from heart rate can be improved by using separate regression lines for arm vs. leg exercise.

Luke et al. (1997)

Luke et al. (1997) evaluated the use of heart rate monitoring and motion to assess energy expenditure (outcome variable). In contrast to the Haskell et al. (1993) study, Luke et al. (1997) used motion sensor data and heart rate only at the low end of energy expenditure. In this study, 10 healthy adults (8 women, 2 men) participated in a submaximal treadmill test and the activities of the daily living circuit test, consisting of sitting, simulated grocery shopping, vacuuming, walking, and climbing stairs. Motion was recorded by the Ambulatory Monitoring System 1000 (AMS-1000; Consumer Sensory

Products, Palo Alto, CA). The monitor was worn at the waist and was capable of storing motion for several weeks. This instrument could also store heart rate, but because of inaccurate data and the loss of data, the authors chose to record heart rate by telemetry. Oxygen consumption was measured during both tests.

The motion counts were calculated for 30-second segments, with heart rate determined manually every minute. Data were analyzed using individual and group general linear models for the prediction of $\dot{V}O_2$ from heart rate and/or motion data. The data showed that heart rate alone ($R^2 = 0.81$, SEE = 3.25 ml/kg/min) was a better predictor of $\dot{V}O_2$ for the activities of the daily living circuit test than the motion sensor ($r^2 = 0.53$). The same was true for the treadmill test. The predicted oxygen consumption was greater than measured during the treadmill test and the activities of the daily living circuit test by 4.5 ± 8.3% and by 1.0 ± 3.7%, respectively.

For the combination of heart rate and motion, the data was pooled for both activity tests. The difference between the baseline heart rate (defined as the lowest heart rate observed while sitting) and the heart rate measured during each sampling period was calculated. This was done so as to observe whether heart rate above rest could be predictive of $\dot{V}O_2$ in a single group equation. The mean R^2 was 0.85 (SEE = 2.95 ml/kg/min). For the pooled data, the $R^2 = 0.74$ (SEE = 4.16 ml/kg/min). The authors concluded that for the group as a whole, the combination of methods appeared to be reasonably good. This was not true for individuals, however. The $\dot{V}O_2$ ranges for individuals for the activities of the daily living circuit test were –25.6% to +14.1%. For the treadmill test, the ranges were –22.5% to +27.9%.

A key finding from this study, similar to the study by Haskell et al. (1993), was that the ability to improve estimation of oxygen consumption (energy expenditure) is dependent on the type of activity and the subjects' response to that activity (Luke et al., 1997). Luke et al. (1997) suggested adding measurement of motion during low-intensity activities because the majority of an individual's day is spent at this level. Overall, these two studies provided excellent background and awareness of the issues involving the use of multiple assessments. One limitation of the Haskell et al. (1993) study was that when the article was published, there were no activity monitors that could give minute-by-minute activity, whereas the heart rate monitors were able to do so. Newer activity monitors are now able to overcome this difficulty by recording minute-by-minute activity counts. Luke et al. (1997) had to manually calculate the heart rate each minute from the ECG strip because of erroneous data from their instrument. This would be both time consuming and labor intensive. Newer heart rate monitors are quite reliable, can easily store the heart rate each minute for several days, and can be downloaded by custom software. This not only saves tremendous research time, but it also allows for free-living heart rate to be assessed for longer periods of time. Thus, these studies gave direction and helpful hints for the more recent studies to be implemented.

Eston et al. (1998)

Eston et al. (1998) examined the accuracy of heart rate monitoring, triaxial accelerometry, uniaxial accelerometry, and pedometry to estimate energy expenditure (outcome variable). The instruments were evaluated both separately and in combination to determine the potential benefits of combining assessments. Thirty children, 15 boys and 15 girls with a mean age of 9.2 years, participated in the study. Oxygen uptake was measured while walking at two speeds, running at two speeds, playing hopscotch, playing catch, and sitting and crayoning. Each child wore three pedometers (Yamax Digi-Walker SW-200, Yamasa Corp., Tokyo) placed on the wrist, ankle, and waist. The pedometer provided total activity counts. The uniaxial accelerometer (WAM, model 7164; Computer Science and Applications [CSA], Shalimar, FL) was set to collect data for each minute. The Tritrac-R3D accelerometer (model T303) was also set for 1 minute and had the additional capability of assessing activity in three dimensions. Both accelerometers were worn on the hips. Data from the triaxial accelerometer were also examined by each axis (i.e., x, y, and z). Simple linear regressions were computed to predict $\dot{V}O_2$ from each measure, whereas multiple regression equations were used to predict $\dot{V}O_2$ from pairs of measures.

Table 13.1 shows the regression analysis for predicting $\dot{V}O_2$ for each measure separately and then when combined with another method. The overall vector magnitude, expressed as Tritrac$_{xyz}$, was the best single predictor, accounting for 82.5% of the variance. The separate axes for the Tritrac provided the next three best predictors, followed by the pedometer at the hip. For the multiple methods, the best model contained Tritrac$_{xyz}$ and heart rate ($R^2 = 0.849$) in which heart rate added 2% more to the variance than the Tritrac$_{xyz}$ alone. Conversely, when the Tritrac$_{xyz}$ was added to heart rate, the variance increased by 21.1%. For the pedometer on the hip

added to heart rate, a 16.4% increase in the variance was observed ($R^2 = 0.802$).

The authors attributed the low ability of heart rate to add much to the variance of the Tritrac$_{xyz}$ to the fact that the activities measured were of low intensity. It is also known that heart rate does not correlate well with oxygen consumption at low-intensity levels (Eston et al., 1998). This finding has two implications:

1. The additional burden and slight cost of the heart rate monitor to the Tritrac$_{xyz}$ is probably not warranted.
2. If one had to choose between the Tritrac$_{xyz}$ and heart rate or the pedometer and heart rate based on cost, the pedometer and heart rate would be recommended.

The difference between the total amount of variance explained between these sets of methods is only 2.3%. Thus, this study provided a very thorough examination of many activities and several different methods of assessment in a fairly large sample of children. However, individual regression equations for the HR-$\dot{V}O_2$ relationship were not developed. In summary, this study found that the combination of methods (Tritrac$_{xyz}$ and heart rate) would not be justified in much larger studies because of the extra cost and labor of adding a second method.

Rennie et al. (2000)

A new instrument (HR+M) has recently been developed and tested. Rennie et al. (2000) hypothesized that the combination of heart rate monitoring and a movement sensor might have theoretical advantage compared to either instrument alone for measuring energy expenditure. This single instrument piece, capable of storing movement and heart rate for 5 days, was worn around the chest and recorded heart rate and movement minute by minute. As described by Rennie et al. (2000), the HR+M recorder weighs only 100 g and attaches to a standard heart rate monitoring band. The monitor starts when heart rate is detected. The movement sensor is within the recorder and gives a binary output corresponding to 7.5° tilts of the monitor from the horizontal, which provides an indication that the subject moved the chest. The output from the heart signal (similar to recording heart rate from an ECG signal) and movement sensor is stored each minute. A custom-designed interface is used to download the data.

Subjects (N = 8, mean age of 31.3 years) underwent individual calibration in which energy expenditure

TABLE 13.1   Regression Analyses Predicting $\dot{V}O_2$

Predictor variables	Increment in R^2	Intercept	Unstandardized regression coefficient	Standardized regression coefficient	SEE	Percentage of mean $\dot{V}O_2$
			Final equation			
Single predictor variable						
Tritrac$_{xyz}$	0.825	25.640	0.012	0.908	10.31	18.23
Tritrac$_z$	0.794	28.237	0.015	0.891	11.19	19.79
Tritrac$_y$	0.767	23.659	0.026	0.876	11.92	21.08
Tritrac$_x$	0.718	29.616	0.031	0.847	13.11	23.18
Heart rate	0.638	−32.640	0.627	0.799	14.91	26.36
WAM	0.609	36.917	0.004	0.780	15.71	27.78
Hip pedometer	0.650	26.572	0.254	0.806	14.60	25.81
Ankle pedometer	0.623	27.046	0.246	0.789	15.15	26.79
Wrist pedometer	0.442	29.951	0.274	0.665	18.43	32.59
Two predictor variables (in order of forced entry)						
Tritrac$_{xyz}$	0.829	6.209	0.010	0.735	9.66	17.08
Heart rate	0.020		0.176	0.225		
Heart rate	0.638	6.209	0.010	0.735	9.66	17.08
Tritrac$_{xyz}$	0.211		0.176	0.225		
Tritrac$_{xyz}$	0.829	25.490	0.010	0.755	9.86	17.43
WAM	0.017		0.001	0.204		
WAM	0.609	25.490	0.010	0.755	9.86	17.43
Tritrac$_{xyz}$	0.238		0.001	0.204		
Tritrac$_{xyz}$	0.825	26.281	0.014	1.000	10.29	18.19
Hip pedometer	0.002		−0.032	−0.101		
Hip pedometer	0.650	26.281	0.014	1.000	10.29	18.19
Tritrac$_{xyz}$	0.177		−0.032	−0.101		
Heart rate	0.655	−12.277	0.416	0.518	13.11	23.18
WAM	0.077		0.002	0.403		
WAM	0.604	−12.277	0.416	0.518	13.11	23.18
Heart rate	0.128		0.002	0.403		
Heart rate	0.638	−17.211	0.385	0.491	11.05	19.53
Hip pedometer	0.164		0.161	0.509		
Hip pedometer	0.649	−17.211	0.385	0.491	11.05	19.53
Heart rate	0.153		0.161	0.509		
Hip pedometer	0.653	25.651	0.168	0.525	11.96	21.14
WAM	0.122		0.003	0.450		
WAM	0.609	25.651	0.168	0.525	11.96	21.14
Hip pedometer	0.166		0.003	0.450		

$\dot{V}O_2$ was measured in ml $\times$ kg$^{-0.75}$ $\times$ min^{-1} SEE. N varies from 159 to 177 because of listwise deletion for missing values. Significance at 0.01 level.

Reprinted from Eston, 1998.

(oxygen uptake) was measured at rest (i.e., lying prone and seated) and during a submaximal bicycle ergometer test (four workloads). Each subject then spent 24 hours in a whole-room respiration calorimeter, following a protocol with periods of inactivity and activity (rest, exercise, and sedentary activities such as reading and watching television).

Modifications to the original Flex HR method (Spurr et al., 1988) were made. Two flex points were

defined: Flex1 heart rate was calculated as the lowest resting heart rate and Flex2 as the lowest heart rate recorded during cycling at 37.5 watts. This produced two slopes at low and moderately high activity, as shown in figure 13.1.

Minute-by-minute heart rate was converted to energy expenditure (outcome variable) using individual calibration curves. The motion sensor data were used to discriminate between periods of inac-

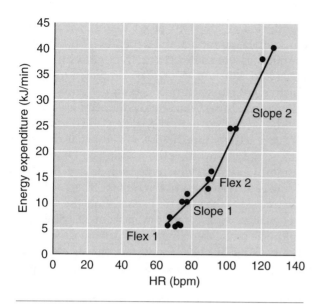

Figure 13.1 Individual calibration of energy expenditure against heart rate for the HR+M method.

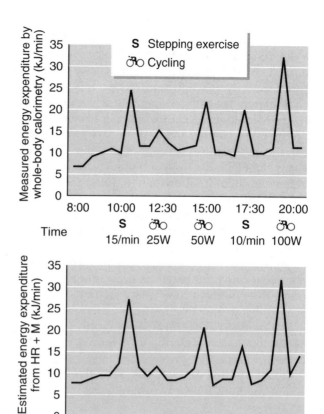

Figure 13.2 Simultaneous 30-minute energy expenditure (kJ/min) calculated in whole-body calorimetry and from HR+M in subject 4 over a 12-hour study period.

tivity and activity at these low heart rate levels. (For the details for computing energy expenditure for different heart rates, see Rennie et al., 2000.) Energy expenditure was also calculated using the Flex HR method that relies on heart rate alone. These estimates were compared to the measured energy expenditure in the calorimeter.

The results were promising. Illustrated in figure 13.2 is the estimated energy expenditure by indirect calorimetry and by HR+M. Note the similarity between the two graphs in terms of the magnitude of the peaks.

The mean percentage error of the HR+M method calculating 24-hour energy expenditure vs. the measured 24-hour energy expenditure was 0.00% (95% confidence interval [CI] of the mean error: –0.25 to 1.25) with a standard deviation (SD) of 12.5%. The range was from –22 to +19% (equivalent to –1,522 kJ to +1,180 kJ). The Flex HR method using heart rate alone resulted in a mean percentage error of 16.5% (95% CI of the mean error: –0.57 to 1.76) with an SD of 30.2%. The range was from –15 to +73% (equivalent to –1,046 kJ to +4,809 kJ). The authors concluded that the combined heart rate and motion sensor method was able to estimate energy expenditure as well as the pattern of energy expenditure and activity throughout the day (Rennie et al., 2000). Using only one instrument is also more acceptable to the participant. The validity and reliability of this new instrument was not reported. For this new instrument to be widely used in large research studies, however, future validation work needs to be done. Rennie et al. (2000) suggested that validation

needs to be done using doubly labeled water in the free-living state. This would allow for validation during an individual's usual activities rather than during a fixed activity protocol.

Moon and Butte (1996) and Treuth et al. (1998)

Two studies (Moon & Butte, 1996; Treuth et al., 1998) combined heart rate with activity sensors and validated the estimation of energy expenditure (outcome variable) by the same highly accurate, 24-hour room respiration calorimeters. The subjects in the Moon and Butte (1996) study included adult men and women, whereas Treuth et al. (1998) studied children 8 to 12 years of age. Both studies measured oxygen consumption ($\dot{V}O_2$) and carbon dioxide production ($\dot{V}CO_2$) by 24-hour room respiration calorimetry while simultaneously electronically recording heart rate and physical activity by activity sensors. The predicted 24-hour $\dot{V}O_2$ and $\dot{V}CO_2$ were validated by the calorimetry. Each study included a day of calibration in the calorimeter, free-living days at home, followed by a second day of unstructured activity (except required exercise) in the calorimeter. The subjects' activity was monitored by the Mini-logger 2000, including a sensor placed on the dominant leg and a wristwatch/wristband for the heart rate (Vantage XL; Polar Electro, Kempele, Finland). These monitors were worn for either two (children) or three (adults) free-living days.

The data processing included three methods of estimating 24-hour $\dot{V}O_2$ and $\dot{V}CO_2$ and estimation of energy expenditure. In the study by Moon and Butte (1996), the first series of models evaluated linear and nonlinear functions based on heart rate alone. Because the nonlinear functions were not shown to be promising, only the linear functions were tested in the Treuth et al. (1998) study and only these linear functions are described. Method 2 included heart rate and activity, whereas method 3 separated the awake period into active and inactive periods.

The methods for the two studies are as follows.

• Method 1 consisted of using heart rate alone (combining awake and asleep). HR, $\dot{V}O_2$, $\dot{V}CO_2$, and energy expenditure measured for the entire 24 hours in the calorimeter for day 1 were used to generate three regression equations:

1. $\dot{V}O_2 = a(HR) + b$;
2. $\dot{V}O_2 = a(HR)^2 + b$; and
3. $\dot{V}O_2 = a(HR)^3 + b$, where a is the slope and b is the intercept.

• The same regression equations were completed, substituting $\dot{V}CO_2$ and energy expenditure for $\dot{V}O_2$. The heart rate values used in the regression equations were defined as the mean of the heart rates recorded every minute.

• Method 2 (awake and asleep) divided the 24-hour period into awake and asleep portions. With the use of graphs of energy expenditure or $\dot{V}O_2$ plotted against time, sleep was determined by the drop in energy expenditure and minimal movement (<50 counts) detected by the activity sensor. Three regression equations using HR, HR^2, and HR^3 for $\dot{V}O_2$, $\dot{V}CO_2$, and energy expenditure were generated separately for the awake and asleep data. The results were combined for the 24-hour period.

• Method 3 (heart rate and activity) divided the awake period into active and inactive segments. Prediction equations were then developed for $\dot{V}O_2$ and $\dot{V}CO_2$ for both the active and inactive data. The equations (Moon & Butte, 1996: $\dot{V}O_2 = [a \times HR^3 + b]$;

Treuth et al., 1998: $\dot{V}O_2 = [a \times HR + b]$), with different coefficients for a and b depending on whether the minutes were defined as active or inactive, were generated. $\dot{V}O_2$ and $\dot{V}CO_2$ were predicted from the inactive HR equation unless both the activity and HR exceeded fixed thresholds defined for each individual. For the Moon and Butte (1996) study, the physical activity threshold was defined as the median activity from either the walking or the slowest treadmill exercise, whichever was lower. For the Treuth et al. (1998) study, the activity threshold was determined as the level (registered from the leg sensor) elicited during random movement while the child was in the calorimeter. At this time, the child was packing his/her belongings or was standing and getting ready to exercise on day 1. For both studies, the heart rate threshold was determined by examining the intersection of the active and inactive curves. To assign a heart rate to the active equation, the heart rate for the current minute had to exceed the HR threshold, and the activity for the current of either of the two previous minutes had to exceed the activity threshold.

Figure 13.3 illustrates a plot for one child's 24-hour calorimetry of $\dot{V}O_2$ versus HR, with the data separated into inactive and active periods. Figure 13.4 shows similar data for an adult. Note the differences in the slopes between the inactive and active curves.

Moon and Butte (1996) studied 10 men and 10 women. Day 1 was for calibration of heart rate and physical activity and included defined periods of

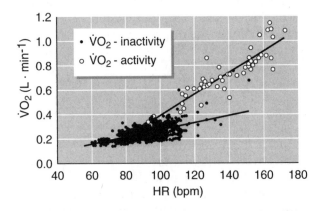

Figure 13.3 Oxygen consumption ($\dot{V}O_2$) and heart rate (HR) relationship in the room calorimeter for one subject, with separate curves for active vs. inactive data for awake portion of the day (method 3).

Reprinted, by permission, from M.S. Treuth, 1998, "Energy expenditure in children predicted from heart rate and activity calibrated against respiration calorimetry," *American Journal of Physiology* 275 (38): E15.

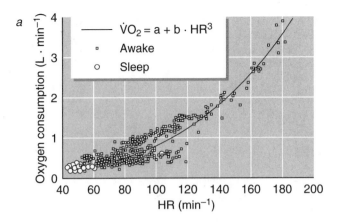

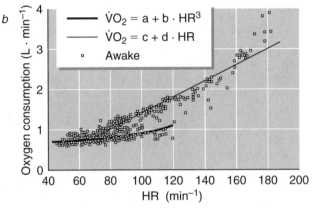

Figure 13.4 Relationship between oxygen consumption ($\dot{V}O_2$) rate and heart rate (HR) in a room calorimeter: *(a)* 1-minute values while subject was awake and asleep over 24 hours. Curve is a least squares fit of the equation $\dot{V}O_2 = a + b \times HR^3$, where *a* and *b* are coefficients, to 24-hour data; *(b)* separate curve fits to active (linear) and inactive (HR^3) portions of awake $\dot{V}O_2$-HR relationship.

Reprinted, by permission, from J.K. Moon and N.J. Butte, 1996, "Combined heart rate and activity improve estimates of oxygen consumption and carbon dioxide production rate," *Journal of Applied Physiology* 81 (4): 1754-1761.

rest, sitting, standing, and two levels of exercise, followed by three days of free living and then a second day of unstructured activity in the calorimeter. All data collection occurred within a period of 1 week.

Sleeping and basal metabolic rates could be accurately estimated from HR alone (with better estimates from the nonlinear models). The respective mean errors were -0.2 ± 0.8 (SD) and $-0.4 \pm 0.6\%$. By examining the data and separating the heart rates into active and inactive minutes, the smallest prediction errors for 24-hour $\dot{V}O_2$ and $\dot{V}CO_2$ were determined. During active periods, errors in estimated $\dot{V}O_2$ and $\dot{V}CO_2$ were -3.4 ± 4.5 and $-4.6 \pm 3.6\%$, respectively.

Figure 13.5 is a Bland-Altman plot that illustrates the difference between oxygen consumption and mean oxygen consumption. Moon and Butte (1996) concluded that combined heart rate and physical activity measured 24-hour $\dot{V}O_2$ and $\dot{V}CO_2$ with a precision similar to alternative methods.

Treuth et al. (1998) predicted energy expenditure from heart rate and activity calibrated against 24-hour respiration calorimetry in 20 children (boys and girls). These children also wore the monitors for two free-living days (one complete day required for data analysis) to demonstrate the utility of the heart rate and activity monitor method. The children followed a set schedule in the calorimeter including periods of rest (30 minutes); three 10-minute inactive periods dedicated to lying still, sitting, and standing; exercise (two 20-minute sessions), and sleep. The child was sent home with a heart rate watch, transmitter belt, and activity monitor. The

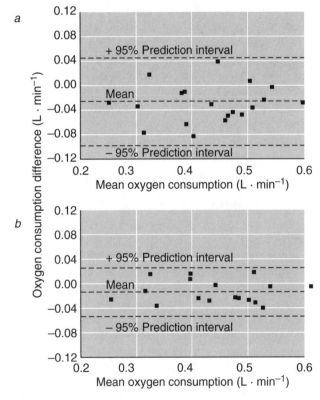

Figure 13.5 *(a)* Difference between awake $\dot{V}O_2$ measured in a calorimeter to $\dot{V}O_2$ predicted from HR^3; *(b)* HR combined with physical activity. Thick dashed lines represent group mean difference; thin dashed lines represent approximately 95% prediction interval ($\approx r_{0.975} \times$ SD) about mean.

Reprinted, by permission, from J.K. Moon and N.J. Butte, 1996, "Combined heart rate and activity improve estimates of oxygen consumption and carbon dioxide production rate," *Journal of Applied Physiology* 81 (4): 1754-1761.

child returned to the laboratory an average of 3 weeks later and spent an additional 24 hours in the calorimeter during which freedom of activity was allowed.

The errors between the predicted and actual data for day 1 were computed for the three equations. The best equation was equation 1, or linear heart rate. It had lower or similar errors when the data were compared between actual and predicted measures for day 1. Thus, linear heart rate was then used for all subsequent analysis. Table 13.2 illustrates the errors in predicting 24-hour $\dot{V}O_2$, $\dot{V}CO_2$, and energy expenditure using linear HR for the three methods.

A reduction in both the mean errors and standard deviations (for $\dot{V}O_2$) was seen with method 3. This combined activity and heart rate method gave individual errors ranging from −12.0 to +5.8% (mean ± SD = −2.6 ± 5.2) for 24-hour $\dot{V}O_2$ and −11.3 to +5.5% (mean ± SD = −2.9 ± 5.1%) for 24-hour energy expenditure. Thus, the awake period was further refined for method 3 by separating the data into active and inactive, and activity was then combined with heart rate to predict energy expenditure. The authors concluded that the combination of heart rate and physical activity is an acceptable method for determining energy expenditure, not only for groups of children but for individuals as well.

Comparisons of Studies

These last three studies applied the most sophisticated instrumentation and tested reasonably large sample sizes compared to the other studies. However, there were differences among them. The Rennie et al. (2000) study used a short calibration test in comparison to the other two studies (Moon & Butte, 1996; Treuth et al., 1998), but the test did not reflect usual activities (only resting and cycling). Moon and Butte (1996) and Treuth et al. (1998) simulated many activities with a wide range of intensity (e.g., reading, watching television, seated writing, cycling, and walking). Yet, three studies used calibrations based on confined calorimetry studies. This is not feasible in large epidemiological studies.

The range in errors was larger for the study by Rennie et al. (2000) than either of the other two (Moon & Butte, 1996; Treuth et al., 1998). However, the overall mean error (0.00%) was lower for Rennie et al. (2000). This indicates that their new instrument may be ideal to use in many research studies,

TABLE 13.2   Errors (%) in Predicting 24-Hour, Basal, and Sleep $\dot{V}O_2$, $\dot{V}CO_2$, and Energy Expenditure (EE) for Day 2 in the Calorimeter in Children

	$\dot{V}O_2$	$\dot{V}CO_2$	EE
Method 1: Combined awake and asleep			
Eq. 1 (HR)			
24 hours	−5.7 ± 8.0	−7.7 ± 8.8	−6.1 ± 7.9
Basal	4.4 ± 15.3	6.5 ± 18.2	4.8 ± 15.7
Sleep	−6.6 ± 14.0	−9.8 ± 15.5	−7.2 ± 14.2
Method 2: Separate awake and asleep			
Eq. 1 (HR)			
24 hours	−4.9 ± 7.0	−6.8 ± 8.4	−5.3 ± 7.0
Basal	−8.6 ± 6.3	−7.2 ± 5.9	−8.4 ± 6.0
Sleep	−1.0 ± 5.8	−2.3 ± 4.7	−1.2 ± 5.4
Method 3: HR and activity			
Eq. 1 (HR)			
24 hours	−2.6 ± 5.2	−4.1 ± 5.9	−2.9 ± 5.1
Basal	−8.6 ± 6.3	−7.2 ± 5.9	−8.4 ± 6.0
Sleep	−1.0 ± 5.8	−2.3 ± 4.7	−1.2 ± 5.4

Values are means ± SD.

Reprinted from Treuth, 1998.

although this appears to be a group effect because the range was quite large. In the Treuth et al. (1998) study, the conclusion was that the combination of methods does appear to work well for individual children and for groups. The drawback is that the data analysis for all studies appears complex because of the additional step of separating the data into two curves (active from inactive time) and other additional data processing. Thus, all three studies utilize a similar idea of using two regression lines, rather than a single line, to predict energy expenditure. This takes into account the influence of low levels of activity on the estimation of energy expenditure and allows for the differentiation between sedentary to low levels of activity and moderate to high levels of activity.

Future Directions

The use of combining multiple methods to improve estimations of physical activity is an exciting and innovative area in physical activity measurement. Because accurate measures of physical activity are needed in so many areas of research, there is the potential for further development. Some ideas for future studies are presented here.

There is no information on multiple methods of assessing physical activity in certain groups (e.g., preschool children, adolescents, older adults, the elderly) or any special populations (e.g., people with diabetes, cardiovascular patients, athletes, individuals at risk of developing obesity, obese or post-obese individuals).

Testing the various combinations of methods for larger sample sizes is also needed. For instance, none of the multiple methods discussed in this chapter could be used for epidemiological research in which large sample sizes are being studied. A study using a combination of a self-report questionnaire along with some other motion sensor or heart rate monitor is warranted.

Summary

The common theme evident among all the studies reviewed in this chapter is that use of multiple physical activity methods will improve the estimation of physical activity compared to only one measure. The studies provide a common combination of methods, namely, heart rate monitoring and activity sensors. In children and adults, the mean errors ranged from approximately –5% to 13%. Individual error ranges among some of the studies were quite large, pointing to the fact that future studies should be implemented to attempt to improve the estimations among individuals. The combined method could be applied in a variety of situations (e.g., for measurement of physical activity and/or energy expenditure during growth, during or following exercise training interventions, during weight loss) and also in special populations. The combined methods might be particularly useful when small changes are expected, and the combination of methods might reduce the error.

To implement the use of two methods in a research study, the following steps can be followed:

• Choice of methods: Table 13.3 suggests possible ways to combine physical activity methods. This is similar to that reviewed by Saris (1985) but with several modifications and additions. Other considerations, such as cost, participant burden, the number of subjects, precision, ease of downloading data, computer software analysis for speed and calculation of regression equations, and the extent of data analysis must also be considered, although table 13.3 can be used as a guide. The minute-by-minute data collection is recommended for heart rate and activity/motion if they are the chosen methods. It must be kept in mind that another appropriate method must be used to validate the new combined method if that is the purpose of the study. If, however, the study purpose is to use two methods that have previously been combined, the previously mentioned studies should serve as a guide.

• An additional issue when considering which methods to combine is the primary outcome variable the researcher is most interested in. If energy expenditure is the primary outcome, then very precise measures may be needed and one may need to consider doubly labeled water as one of the methods. If physical activity level is of interest, then a self-report questionnaire would be appropriate as one of the methods.

• Protocol development: The study design needs to be developed, as one would for any research study. The appropriate sample size needs to be determined, along with considerations of feasibility, cost, and other issues that would impact the study's success.

TABLE 13.3 Ways to Combine Physical Activity Methods

Number of subjects	Method 1	Method 2
10-20	Doubly labeled water technique	Heart rate monitoring
	24-hour indirect calorimetry	Heart rate monitoring
	24-hour indirect calorimetry	Accelerometer
	Observation	Heart rate monitoring
	Indirect calorimetry	Motion sensor
	Heart rate monitoring	Pedometer
	Heart rate monitoring	Accelerometer
100-200	Self-report questionnaire	Pedometer
	Self-report questionnaire	Accelerometer
>200	Self-report questionnaire	Heart rate monitoring

- Subject testing: A sample of individuals should have $\dot{V}O_2$ measured during activities of low and high intensity (a range). These activities need to be common to the population to be studied. Care should be taken in having sufficient (i.e., complete) data collection.

- Data downloading and storage: Most of the newer instruments come with fairly easy ways to download the data directly onto personal computers. Programmers can develop ways to speed up the processing of the data in batch form.

- Data analysis: The analysis of the data will depend on the methods chosen. However, if heart rate monitoring is one of the methods, the analysis will involve running individual curves/regressions. It is recommended that the data be separated into active and inactive regressions based on the differences between low-intensity and high-intensity type activities. The studies discussed in this chapter can be used as guides on the specific details of how to analyze the data.

References

Aaron, D.J., Kriska, A.M., Dearwater, S.R., Cauley, J.A., Metz, K.F., & LaPorte, R.E. (1995). Reproducibility and validity of an epidemiologic questionnaire to assess past year physical activity in adolescents. *American Journal of Epidemiology, 142*, 191-201.

Ainsworth, B.E., Bassett, D.R., Strath, S.J., Swartz, A.M., O'Brien, W.L., Thompson, R.W., Jones, D.A., Macera, C.A., & Kimsey, C.D. (2000). Comparison of three methods for measuring the time spent in physical activity. *Medicine and Science in Sports and Exercise, 32*, S457-S464.

Baranowski, T., Bouchard, C., Bar-Or, O., Bricker, T., Heath, G., Kimm, S.Y.S., Malina, R., Obarzanek, E., Pate, R., Strong, W.B., Truman, B., & Washington, R. (1992). Assessment, prevalence, and cardiovascular benefits of physical activity and fitness in youth. *Medicine and Science in Sports and Exercise, 24*, S237-S247.

Bassett, D.R., Ainsworth, B.E., Swartz, A.M., Strath, S.J., O'Brien, W.L., & King, G.A. (2000). Validity of four motion sensors in measuring moderate intensity physical activity. *Medicine and Science in Sports and Exercise, 32*, S471-S480.

Emons, H.J.G., Groenenboom, D.C., Westerterp, K.R., & Saris, W.H.M. (1992). Comparison of heart rate monitoring combined with indirect calorimetry and the doubly-labelled water ($^2H_2^{18}O$) method for the measurement of energy expenditure in children. *European Journal of Applied Physiology, 65*, 99-103.

Eston, R.G., Rowlands, A.V., & Ingledew, D.K. (1998). Validity of heart rate, pedometry, and accelerometry for predicting the energy cost of children's activities. *Journal of Applied Physiology, 84*, 362-371.

Handelman, D., Miller, K., Baggett, C., Debold, E., & Freedson, P. (2000). Validity of acccelerometry for the assessment of moderate intensity physical activity in the field. *Medicine and Science in Sports and Exercise, 32*, S442-S449.

Haskell, W.L., Yee, M.C., Evans, A., & Irby, P.J. (1993). Simultaneous measurement of heart rate and body motion to quantitate physical activity. *Medicine and Science in Sports and Exercise, 25*, 109-115.

Jacobs, D.R., Ainsworth, B.E., Hartman, T.J., & Leon, A.S. (1993). A simultaneous evaluation of ten commonly used physical activity questionnaires. *Medicine and Science in Sports and Exercise, 25*, 81-91.

Janz, K.F., Golden, J.C., Hansen, J.R., & Mahoney, L.T. (1992). Heart rate monitoring of physical activity in children and adolescents: The Muscatine study. *Pediatrics, 89*, 256-261.

Johnson, R.K., Russ, J., & Goran, M.I. (1998). Physical activity related energy expenditure in children by doubly labeled water as compared with the Caltrac accelerometer. *International Journal of Obesity, 22*, 1046-1052.

Li, R., Deurenberg, P., & Hautvast, J.G. (1993). A critical evaluation of heart rate monitoring to assess energy expenditure in individuals. *American Journal of Clinical Nutrition, 58*, 602-607.

Livingstone, M.B., Prentice, A.M., Coward, W.A., Ceesay, S.M., Strain, J.J., McKenna, P.G., Nevin, G.B., Barker, M.E., & Hickey, R.J. (1990). Simultaneous measurement of free-living energy expenditure by the doubly labeled water method and heart-rate monitoring. *American Journal of Clinical Nutrition, 52*, 59-65.

Luke, A., Maki, K.C., Barkey, N., Cooper, R., & McGee, D. (1997). Simultaneous monitoring of heart rate and motion to assess energy expenditure. *Medicine and Science in Sports and Exercise, 29*, 144-148.

Maas, S., Kok, M.L.J., Westra, G.H., & Kemper, H.C.G. (1989). The validity of the use of heart rate in estimating oxygen consumption in static and in combined static/dynamic exercise. *Ergonomics, 32*, 141-148.

Meijer, G.W., Westerterp, K.R., Koper, H., & Ten Hoor, F. (1989). Assessment of energy expenditure by recording heart rate and body acceleration. *Medicine and Science in Sports and Exercise, 21*, 343-347.

Melanson, E.L., & Freedson, P.S. (1996). Physical activity assessment: A review of methods. *Critical Reviews in Food Science and Nutrition, 36*, 385-396.

Moon, J.K., & Butte, N.F. (1996). Combined heart rate and activity improve estimates of oxygen consumption and carbon dioxide production rates. *Journal of Applied Physiology, 81*, 1754-1761.

Pate, R.R. (1996). Physical activity assessment in children and adolescents. *Critical Reviews in Food Science and Nutrition, 36*, 385-396.

Rennie, K., Roswell, T., Jebb, S.A., Holburn, D., & Wareham, N.J. (2000). A combined heart rate and movement sensor: Proof of concept and preliminary testing study. *European Journal of Clinical Nutrition, 54*, 409-414.

Rowlands, A.V., Eston, R.G., & Ingledew, D.K. (1997). Measurement of physical activity in children with particular reference to the use of heart rate and pedometry. *Sports Medicine, 24*, 258-272.

Saris, W.H.M. (1985). The assessment and evaluation of daily physical activity in children. A review. *Acta Paediatrica Scandinavian, 318*, 37-48.

Spurr, G.B., Prentice, A.M., Murgatroyd, P.R., Goldberg, G.R., Reina, J.C., & Christman, N.T. (1988). Energy expenditure from minute-by-minute heart-rate recording: Comparison with indirect calorimetry. *American Journal of Clinical Nutrition, 48*, 552-559.

Treuth, M.S., Adolph, A.L., & Butte, N.F. (1998). Energy expenditure in children predicted from heart rate and activity calibrated against respiration calorimetry. *American Journal of Physiology, 275,* E12-E18.

Wareham, N.J., & Rennie, K.L. (1998). The assessment of physical activity in individuals and populations: Why try to be more precise about how physical activity is assessed? *International Journal of Obesity, 22,* S30-S38.

Welk, G.J., Blair, S.N., Wood, K., Jones, S., & Thompson, R.W. (2000). A comparative evaluation of three accelerometry-based physical activity monitors. *Medicine and Science in Sports and Exercise, 32,* S489-S497.

Weston, A.T., Petosa, R., & Pate, R.R. (1997). Validation of an instrument for measurement of physical activity in youth. *Medicine and Science in Sports and Exercise, 29,* 138-143.

14

Physical Activity Assessment Issues in Population-Based Interventions: A Stage Approach

Claudio R. Nigg, PhD
University of Hawaii

The statement "some is better than none and more is better than some" sums up the current recommendations put forth by various professional groups about physical activity and exercise behaviors (American College of Sports Medicine, 1990, 1998; Pate et al., 1995; U.S. Department of Health and Human Services, 1996). However, to be able to determine if people are doing "some" or "more," we need to be able to measure physical activity. There is considerable evidence that physical activity can be changed. In 1996, a meta-analysis was published that examined the effects of interventions for increasing physical activity (Dishman & Buckworth, 1996). This analysis of 127 research articles and 14 dissertations published between 1965 and 1995 indicated an overall 0.75 standard deviation effect size for interventions influencing level of physical activity. One of the major limitations noted was the variation in the quantification of physical activity (Dishman & Buckworth, 1996).

Because of challenges associated with assessing physical activity, interventions have frequently used fitness indicators or other health-related variables as the primary outcome measure. For example, in reviewing the effects of physical activity interventions on youth's physical activity, several studies were excluded because they did not include measures of physical activity (Stone et al., 1998). However, the reliance on fitness indicators poses problems in many studies since physical activity and physical fitness are not always highly correlated. As behavioral research has progressed, it has become increasingly important to have more direct measures of the specific behaviors that are being promoted.

Assessment of Physical Activity for Intervention Research: The Nature of the Problem

The measurement of physical activity is inherently difficult because activity is a highly variable behavior. Individuals may be active on a given day or week but be completely inactive at other periods of time. Even with detailed information about what people do, it is difficult to quantify the actual amount of activity. There is considerable variability between individuals in the energy expenditure of various activities and in types of physical activities performed (Kriska & Casperson, 1997). In addition, there is the impact of physical fitness and health status, which can affect the levels and types of physical activity performed. These factors make it difficult to capture levels of physical activity to evaluate health-related interventions.

A variety of methods are available to assess physical activity, but the lack of a true gold standard makes it impossible to determine the actual validity of any one method of measurement (Montoye et al., 1996). The units of measurement of the various methods are different (e.g., METS, calories, total minutes, step counts), causing difficulty in using intercorrelations between methods as a measure of validity. In the laboratory, energy expenditure can be measured by direct or indirect calorimetry, and physical fitness can be measured (i.e., maximal oxygen uptake) as a criterion measure reflecting habitual physical activity. Field methods include doubly labeled water, physical activity diaries and questionnaires, motion accelerometers, and heart rate monitoring. All of the methods of assessment have error associated with them, and the advantages and disadvantages of using one or more of these approaches to measurement depend on the research questions and the population being studied (Kriska & Casperson, 1997; Montoye et al., 1996).

Considering the imperfect relationship between fitness indicators and physical activity behavior, which is magnified by the measurement error of some fitness indicators and the problems in assessing some of these indicators on a population basis,

it is prudent to look toward self-reported physical activity behavior assessments.

Physical activity questionnaires are the most commonly used method in population intervention studies. Questionnaires are ideal for this application because they are less likely to alter physical activity behavior, are cost effective, can be adapted to a variety of settings and populations, and have reasonable reliability and validity when compared against other objective measures.

As described in chapter 7, physical activity questionnaires differ on a number of attributes, including complexity, time frame, types of activities surveyed, and scoring. The choice of a physical activity questionnaire for a particular application must take these factors of complexity, time frame, types of activities surveyed, and scoring into consideration in relationship to the study hypotheses, population (i.e., gender, age, ethnicity), time and financial constraints, and other factors. For example, to evaluate the effectiveness of an intervention to increase physical activity in older adults, a questionnaire that is valid and reliable in older adults is necessary, and it needs to be sensitive to changes in the outcome variable (physical activity).

The remainder of this chapter describes a stage approach to physical activity assessment and presents a brief, valid, and reliable instrument that is appropriate to evaluate population-based physical activity interventions. Various analytical strategies are also presented to facilitate the use of stage-based assessments in behavioral research.

Application of Stage Models for Intervention Research

Stage models are based on the assumption that people move through a series of stages in their attempt to change a behavior. Stages can be both dynamic and stable. In other words, an individual can be in any one stage for a considerable period of time, but if (when) the conditions are right, stage transitions will occur.

Stage models have been used to understand motivational readiness for a variety of health behaviors, and a number of stage models have been proposed in the exercise domain (Dishman, 1990; Prochaska & DiClemente, 1983; Sallis & Hovell, 1990). Of these, the most widely validated is the Prochaska and DiClemente stages of change construct within the transtheoretical model (TTM) of behavior change (Prochaska & DiClemente, 1983, 1985; Prochaska & Velicer, 1997).

Transtheoretical Model (TTM) of Behavior Change

The TTM was developed as a general explanatory model of intentional behavioral change and has been used to describe motivational readiness to engage in exercise behavior across age groups (e.g., Gorely & Gordon, 1995; Marcus & Simkin, 1993; Nigg & Courneya, 1998). The TTM is useful for understanding the process of intentional behavioral change because it incorporates both experiential cognitive processes and behavioral markers related to the change process (Prochaska & Marcus, 1994). The TTM posits that transition among stages results via stage-specific experiential and behavioral processes. The experiential processes employed most in the early stages include consciousness raising, dramatic relief, environmental reevaluation, self-reevaluation, and social liberation. The behavioral processes, more frequently used during the later stages, are counterconditioning, helping relationships, reinforcement management, self-liberation, and stimulus control. Further, each of the stages is characterized by changes in the cognitive variables of decisional balance, the balance between benefits and costs associated with engaging in a particular behavior (Janis & Mann, 1977), and self-efficacy, the perceived ability that one can engage in a particular behavior under specific circumstances (Bandura, 1986). According to this model, interventions fail, in part, because the strategies applied do not match the stage the participants are in. (For a more detailed description of the TTM, see Prochaska & Velicer, 1997.)

Stages of Change Construct of the Transtheoretical Model (TTM)

The central construct integrating the variables of the TTM is the stages of readiness to change. The stages of change construct incorporates a temporal framework, which allows it to capture the change process. The TTM uses five stages of change to understand the dynamic process of motivational readiness through which an individual transits when making an intentional behavioral change (e.g., Prochaska & Marcus, 1994). This can be represented in a spiral progression through the different stages, which are precontemplation, contemplation, preparation, action, and maintenance (see figure 14.1).

1. Precontemplation is the stage when the individual is not doing the behavior and is not intending to make changes toward the new behavior within the foreseeable future (i.e., within 6 months). People in this stage are uninformed about the long-term effects of their behavior and/or are demoralized about their ability. They also do not want to think about change and may be defensive, partly because of social pressures to change. Precontemplators may wish to change, but this seems quite different from a serious consideration of change (Prochaska & DiClemente, 1983, 1985; Prochaska & Velicer, 1997).

2. Contemplation is the stage when the individual is considering change. People in the contemplation stage openly state their intent to change within the next 6 months. This is about as far in advance as an individual can plan decisions (Prochaska & Marcus, 1994). However, individuals may stay in this stage for more than 2 years, despite the intention to change. When contemplators substitute thinking for acting, they are labeled as chronic contemplators. Contemplators are ambivalent about changing behavior and typically see the pros and cons of sustaining the risk behavior as approximately equal. The central element of this stage is serious consideration of problem resolution (Prochaska & DiClemente, 1983, 1985; Prochaska & Velicer, 1997).

3. Preparation is the stage at which individuals intend to take steps to change, usually within the next 30 days. This time frame reflects the immediacy of the intention to act. Preparers may have already made minor adjustments in their thought patterns and behaviors. Typically, preparers have a plan of

Figure 14.1 A spiral model of the stages of change.
Reprinted with Permission, LifeScan, Inc.

action or have made some behavioral changes, but they do not reach the predetermined action criteria. Immediate intention to engage in the behavior is thought to be the most important part in the definition of the preparation stage. The preparation stage is not very stable and contains people who are more likely than precontemplators or contemplators to progress (Prochaska & DiClemente, 1983, 1985; Prochaska & Velicer, 1997).

4. The action stage is when the individual actively engages in the new behavior. At this stage, overt behavioral changes have occurred within the last 6 months. Individuals in the action stage change their behavior, experiences, or environment to overcome their problems. Modifications tend to be highly visible and receive the greatest recognition from others. This is the least stable stage and corresponds with the highest risk for relapse. For physical activity, the highest likelihood for relapse occurs within the first 6 months of starting a regular program (Dishman, 1994). For this stage, the usual time criteria given is 0 to 6 months. The hallmarks of the action stage are changing the target behavior to a preset criterion and making significant efforts to change (Prochaska & DiClemente, 1983, 1985; Prochaska & Velicer, 1997).

5. Maintenance is the stage when the individual is sustaining the change over time. This is a period of continued change where the individual is working to prevent relapse and to consolidate the gains attained during the action stage (Prochaska & DiClemente, 1983, 1985; Prochaska & Velicer, 1997). These individuals have successfully continued regular physical activity beyond 6 months (e.g., longer than a sport season).

Some conceptions of the TTM have postulated a sixth stage known as the termination stage. The definition of *termination* is "an absolute absence of temptation to engage in an old behavior and 100% self-efficacy in engaging in the new healthy behavior" (Prochaska & Velicer, 1997). It has generally been applied to cessation behaviors, and there has been some question about the relevance of this stage for adoption behaviors (Prochaska & Marcus, 1994). Cardinal (1999, 2000), for example, found preliminary evidence that a sixth stage for exercise could exist and labeled it the *transformed stage*. However, in the area of exercise adoption (e.g., formerly sedentary individuals), it would seem, individuals are always at a risk for relapse and must continue to work to stay in the maintenance stage. The reason "termination" is not typically included in stage-based models of physical activity is because an individual must make a conscious decision to be physically active; it is not the path of least resistance. In cessation or quitting behaviors, doing nothing is the path of least resistance and termination may be more likely.

Psychometric Characteristics of Stage Assessments

To provide a useful assessment for behavioral and intervention research, an instrument must demonstrate acceptable psychometric properties. This section reviews research on the validity, reliability, and sensitivity to detect change.

Validity. Numerous studies have documented the construct validity of the stages of change for exercise behavior. Past research has indicated that the stages of exercise behavior and level of physical activity appear to be related. In general, physical activity behavior increases from the stages of precontemplation to maintenance (Barké & Nicholas, 1990; Marcus & Simkin, 1993). Individuals who were classified in the action and maintenance groups reported significantly more moderate to vigorous exercise than those in precontemplation and contemplation in a sample of middle-aged adults (Marcus & Simkin, 1993) and older adults (Gorely & Gordon, 1995; Hellman, 1997). Illustrating further consistency with theory, active groups of older adults (Barké & Nicholas, 1990) and middle-aged police officers (Hausenblas et al., 1999) placed themselves farther along the stages of exercise compared to inactive groups. It is important to note that Marcus and Simkin (1993) found that moderate physical activity did not differentiate all categories of exercise behavior.

Cardinal (1995a) and Hausenblas et al. (1999) developed their own stage of exercise behavior scale and replicated these results. Cardinal (1995a) reported an increasing linear gradient across the stages of change in terms of self-reported exercise and physical activity as well as $\dot{V}O_2$ peak in middle-aged females. Hausenblas et al. (1999) reported an increase in leisure-time exercise and $\dot{V}O_2$max with each increase in stage. In addition, Emmons et al. (1994) investigated physical activity and other behaviors/constructs in predominantly blue-collar workers and found a linear relationship between stages of exercise, minutes per week spent exercising, and other physical activities (e.g., blocks walked per day, flights of stairs climbed per day).

Cardinal (1997) showed both physiological and self-reported validation of the stages of exercise in adults. Body mass index, cardiorespiratory fitness, exercise behavior, barriers, and self-efficacy all increased from precontemplation to maintenance. Further, a relapse indicator differentiated the inactive stages from the active stages.

Wyse et al. (1995) evaluated the concurrent validity of the stages of exercise behavior using self-reported exercise level based on the Godin's Leisure-Time Exercise Questionnaire for young British adults. Significant differences were found between the stages and self-reported levels of exercise behavior in a theoretically consistent manner. Cardinal (1995b) further validated the stages of change for the same age group (i.e., American undergraduate students). Leisure-time exercise, frequency of sweating, physical activity rating, and $\dot{V}O_2$max significantly increased with each increase in stage, while having relapsed significantly decreased with increases in stage.

To summarize, numerous studies have investigated the construct validity of the stages of change applied to exercise behavior. Studies have found that stage is associated with both self-reported exercise behavior (Calfas et al., 1994; Emmons et al., 1994; Hellman, 1997; Lee et al., 2001; Marcus & Simkin, 1993; Pinto & Marcus, 1995; Wyse et al., 1995) and measures of fitness level (Cardinal, 1995a, 1995b, 1997) in adolescent, college, adult, and older adult samples. The relationships are stronger if exercise stage is related to exercise behaviors and physical activity stage is related to physical activity behaviors. In general, the more progressed an individual is in the stages of change, the greater the self-reported physical activity, exercise, and objective fitness level.

Reliability. The stages of exercise behavior change have also been found to be reliable in different samples. Marcus, Selby et al. (1992) developed and tested a stages of change measure as part of a larger study and found a kappa index of reliability for the stages of change measure of 0.78 measured over a 2-week period using 20 work-site employees. Cardinal developed the stages of change exercise behavior scale and found the 3-day test-retest reliability of the scale to be 1.00 using Spearman's rho (p < 0.0001) for 12 middle-aged females (1995a) and 12 undergraduate students (1995b). Courneya (1995) reported a 2-week test-retest reliability of 0.79 for the stages of change for an elderly Canadian sample.

Ability of the Stages to Detect Change. To be useful as an outcome measure, the stages of change should be sensitive to changes in behavior. Two types of support can be found in the literature: longitudinal naturalistic studies and intervention studies. Six naturalistic longitudinal studies were identified that all illustrate changes in stage over time. Three studies investigated middle-aged American samples (Armstrong et al., 1993; Marcus et al., 1994; Marcus et al., 1996), two investigated older Canadian adults (Courneya et al., 1997; Courneya et al., 1998), and one study looked at adolescents (Nigg, 2001). All studies described movement across the stages of change over time.

An increasing number of studies have begun using the TTM for exercise-related interventions (Cardinal & Sachs, 1995, 1996; Cole et al., 1998; Jones et al., 2001; Marcus, Banspach et al., 1992; Marcus, Bock et al., 1998; Marcus, Emmons et al., 1998). Participants across age groups appear to be significantly more active after interventions (Jones et al., 2001; Marcus, Banspach et al., 1992), and most individuals have been found to increase their stage (i.e., 43.2%) or at least maintain their stage (i.e., 42%), whereas only 14.8% regressed (Cardinal & Sachs, 1995). While evaluating the relatively short-term effects of an intervention, Cole et al. (1998) further validated these findings, revealing that only a small percentage of participants regressed one or more stages (i.e., 6.5%), maintained their stage (i.e., 30.3%), and progressed one (i.e., 21.1%) or more (i.e., 14.3%) stages. In addition, recent research suggests that individualized, motivationally tailored interventions based on the TTM in comparison to standard interventions are more efficacious at increasing physical activity participation (Marcus, Bock et al., 1998) and result in more stage progression as well as decreased stage regression (Marcus, Emmons et al., 1998).

Additional Considerations

The stages of change should not be seen as a pure physical activity or exercise measure per se. Rather, they may be conceptualized as an outcome variable that increases the outcome space from pure behavior only to distinct gradations ranging from lack of intention to behavioral maintenance. With this conceptualization, interventions are able to detect a broader outcome of the behavioral change process than would normally be apparent with other approaches. Moving someone from precontemplation to contemplation (i.e., from not intending to change to intending to change) is an important step in the behavioral change process that does not involve overt behavioral change. Traditional activity measures cannot detect this kind of shift and may incorrectly perceive the intervention as unsuccessful. Relatedly, the stages of change allow the researcher to identify the process of behavioral change and which variables (mediators) are important at which specific stage to lead to stage progression. This will have direct implications for future intervention development.

As an outcome measure, the stages of change may be utilized in various ways. As the pre-action stages (i.e., precontemplation, contemplation, and preparation) are below a specified criterion, and the action and maintenance stages are meeting the criterion, a dichotomous outcome of percentage at criterion is feasible. This allows for comparison to previous research, as it is one of the traditional outcome measures in the physical activity literature (Courneya, 1995). Further, the metric of the stages of change is common across behaviors. For example, a precontemplator is someone who is not intending to engage in the health behavior at criteria. The criteria can apply to physical activity, fruit and vegetable consumption, smoking cessation, safe sex behaviors, sun-protection behaviors, seatbelt use, or any other health behavior with a defined criterion. This presents the opportunity for comparing whether treatment is as effective for one behavior as it is for another.

Using the same metric across behaviors also facilitates the identification of gateway behaviors. A gateway behavior can be visualized as a behavior that, when intervened, has positive effects on other behavioral changes. Basically, a few behaviors may be related to general health of a specific population. There is preliminary evidence that points toward this possibility, as stages of change for a large number of behaviors are somewhat related (Nigg et al., 1999). Examination of the effect of single behavioral change interventions on other health behavioral changes is needed to further develop knowledge regarding potential gateway behaviors. When using stages of change as a common metric across behaviors, the issue of similarity of criteria may need to be addressed. For example, is being physically active for 30 minutes or more on most days of the week on the same level of difficulty as quitting smoking or eating five servings of fruits and vegetables a day?

Assessing the Stage of Physical Activity Behavior

Reed et al. (1997) and Norman et al. (1998) investigated different staging algorithms that have been published. Based on criteria of non- or misclassification, theoretical consistency, and ease of under-

standing, the following recommendations were put forth.

1. The stage questions need to be introduced with a clear, concise definition of the criterion behavior. For physical activity behaviors, the definition needs to include frequency, duration, and intensity.
2. A variety of examples reflecting the definition, and appropriateness for the population, need to be included.
3. The structure of the response categories can either take a five-choice format (i.e., one for each stage; see form 14.1) or a four-item yes or no format (see form 14.2). Both formats take minimal time to complete (roughly 30 to 60 seconds) and are suited for questionnaire administration, whereas the four-item yes or no format is recommended for interviews. Form 14.1 uses the regular exercise behavior criteria as defined by the American College of Sports Medicine (1998), and form 14.2 uses physical activity as defined in Pate et al. (1995).

Analyzing Stages of Change Data

The analyses of stages of change data have proven challenging because of the multiple categories that can characterize an individual and because some individuals will progress in stages while others will regress. Cross-sectional data analysis of the stages of change is fairly straightforward because a stage can be viewed as an ordered categorical variable. Therefore, analyses illustrating stage differences (e.g., analysis of variance [ANOVA]) are appropriate. Classification analyses (e.g., discriminant function analysis) are also appropriate. To facilitate comparisons with other studies, it is recommended to always present the raw and percentage distribution.

A variety of statistical approaches have been used in analyzing stages of change as a primary or secondary intervention outcome. Some approaches are very well known, while others are more obscure. Each approach has specific advantages and disadvantages. Two methods commonly used in the published literature are described in the following section and a third method is recommended for research projects to enable study comparisons. There are more sophisticated techniques (e.g., latent transition analysis [LTA] and generalized estimating equations [GEE]) that are also appropriate for stages of change analyses. Readers interested in these approaches should consult more

Form 14.1   Stage Instrument for Questionnaires (Using Regular Exercise As an Example)

Regular Exercise: Any *planned* physical activity (e.g., brisk walking, jogging, bicycling, swimming, line dancing, tennis, etc.) performed to increase physical fitness. Such activity should be performed *3 or more times* per week for *20 or more minutes* per session at a level that increases your breathing rate and causes you to break a sweat.*

Do you exercise regularly according to the definition above?

Mark the one statement that applies to you.

SCORING

___Yes, I have been exercising regularly for more than 6 months.   **Maintenance**

___Yes, I have been exercising regularly, but for less than 6 months.   **Action**

___No, but I intend to exercise regularly in the next 30 days.   **Preparation**

___No, but I intend to exercise regularly in the next 6 months.   **Contemplation**

___No, and I do not intend to exercise regularly in the next 6 months.   **Precontemplation**

*Criterion from American College of Sports Medicine, 1998.

From *Physical Activity Assessments for Health-Related Research,* edited by Gregory J. Welk, 2002, Champaign, IL: Human Kinetics.

Form 14.2   Stage Instrument for Interviews (Using Regular Physical Activity As an Example)

Regular Physical Activity: For physical activity to be regular it must be done for *30 minutes at a time* (or more) per day, and be done *at least* 4 days per week. For example, you could take a 30-minute brisk walk or ride a bicycle for 30 minutes. Physical activity includes such activities as walking briskly, biking, swimming, line dancing, and aerobics classes or any other activities where the exertion is similar to these activities. Your heart rate and/or breathing should increase, but there is no need to exhaust yourself.*

Please answer all questions with either Yes or No.

According to the definition above:

1. Do you currently engage in regular physical activity?   **YES   NO**
   SKIP PATTERN: If YES go to #4, if NO go to #2

2. Do you intend to engage in regular physical activity in the next 6 months?   **YES   NO**
   SKIP PATTERN: If YES go to #3, if NO finish

3. Do you intend to engage in regular physical activity in the next 30 days?   **YES   NO**
   SKIP PATTERN: If YES or NO finish

4. Have you been regularly physically active for the past 6 months?   **YES   NO**
   SKIP PATTERN: If YES or NO finish

SCORING

If item 1 = NO and item 2 = NO   **Precontemplation**

If item 1 = NO and item 2 = YES and item 3 = NO   **Contemplation**

If item 1 = NO and item 2 = YES and item 3 = YES   **Preparation**

If item 1 = YES and item 4 = NO   **Action**

If item 1 = YES and item 4 = YES   **Maintenance**

*Criterion from Pate et al., 1995.

From *Physical Activity Assessments for Health-Related Research*, edited by Gregory J. Welk, 2002, Champaign, IL: Human Kinetics.

234

specific reviews of LTA (Collins & Wugalter, 1992; Graham et al., 1991; Martin et al., 1996) and GEE (Greenland, 1995a, 1995b; Zeger & Liang, 1986).

Traditional Analytical Approaches

As a dependent outcome variable, stages of change is most commonly conceptualized as the proportion of individuals moving to the action or maintenance stages of change. This may also be thought of as the proportion of individuals at a prespecified behavioral criterion that defines study outcome success. This approach is advantageous for several reasons:

1. It cuts across theoretical models.
2. It is generally equivalent to a bottom line, is easy to understand, and is an easily agreed-on measure of outcome success.
3. It results in an essentially dichotomous outcome measure that can be analyzed using well-known and widely available techniques.

The simplest analytical approaches are probably still the most commonly used, including chi-square tests and tests of proportions. Assuming all study participants are pre-action or "at risk" at baseline and have been randomized to treatment and control groups, the proportion reaching action or maintenance at follow-up represents an assessment of intervention outcome success. This approach assumes that the treatment and control group stage distributions are effectively equal at baseline. Stratified random assignment is often used to ensure equivalency.

While this analytical approach is very straightforward and easy to understand, there are serious disadvantages. One frequently overlooked is that it is usually not possible to include covariates in the analysis, at least not without difficulty and complication of what is, after all, supposed to be a relatively simple procedure. More often mentioned as the chief disadvantage of this approach is the dichotomization of a theoretically defined variable and related statistical power issues. Dichotomization of the stage variable results in significant loss of information and leads to a decrease in statistical power (Rossi, 1990). Furthermore, the use of nominal-level analytical procedures such as the chi-square test makes no assumptions concerning the underlying nature of the variable (e.g., normality, order). However, even when dichotomized, the stage variable is at least ordinal so that additional analytical sensitivity is lost when employing nominal-level techniques. More appropriate would be the use of certain nonparametric techniques sensitive to

ordinal-level change over time, such as the McNemar test or the Stuart-Bowker procedure. Unfortunately, these procedures are obscure at best and not readily available in most statistical computer packages.

Another approach to enhancing statistical sensitivity within the confines of the more traditional analytical approaches is to employ a normalizing and variance-stabilizing transformation procedure. The most appropriate procedure for proportions is the arcsine transformation (Rossi, 1985, 1990). This approach permits the use of more sensitive analytical procedures, such as the ANOVA. For multiple group outcomes, the Levy (1975) test for pair-wise comparisons among proportions—an analog of the Tukey test—is an effective follow-up procedure. The Levy procedure is also fairly obscure, but well worth seeking out.

Stage Progression Analysis

No matter what analytical approach is adopted, one disadvantage of dichotomizing stage into action vs. pre-action categories is that it fails to preserve all of the information that may be of interest in understanding how interventions are working (or not). In addition, intervention follow-up may be insufficient to capture much action, giving the impression that not much is happening. Depending on how recruitment is conducted and on the particular target behavior, it is likely that the majority of study participants will be in the precontemplation and contemplation stages of change at baseline. Movement to the action stage is therefore expected to take some time. A substitute for assessing movement to the action stage is to assess progress through the stages of change as an outcome variable. Typically, this is considered a secondary outcome measure. An advantage of the use of stage progression is that it nicely mirrors the primary intervention goal of most stage-based, tailored interventions (i.e., to enhance motivation or accelerate progress through the stages of change).

Analysis of stage progression raises the issue of what constitutes progress. Probably the most common approach is to count all forward stage movement as progress. Again, the result is usually the formation of a dichotomous variable with any forward stage movement counting as progress and no movement or regression counting as no progress. A trichotomous variable is also possible (i.e., progression vs. regression vs. no movement) to better reflect overall movement patterns. In addition to the previously described limitations associated with dichotomization, another problem with this approach is that it blurs the distinction between stages since all progression counts the same (i.e., there is no difference between advancing a single stage or multiple stages).

Because progression is typically constructed as a dichotomized variable, analysis typically follows that for dichotomous variables (e.g., chi-square test), as described earlier. A frequent misconception concerning the analysis of stage progression is that it will be more powerful than analyses utilizing progress to action as the outcome measure. This is not necessarily so. In fact, stage progression analyses are often likely to be less powerful in many circumstances. This is because the proportions of individuals progressing will usually be substantially greater than the proportions reaching action for both the treatment and control groups. Thus, study outcomes are more likely to be in the vicinity of 50% when movement to action is used as the outcome criterion. For example, let us assume that the proportion reaching action (from pre-action stages) for physical activity in the control and treatment groups is around the 10 to 25% range. Adding *any* plausible constant proportion to these numbers to simulate the results of stage progression will result in outcomes closer to 50%, where statistical analyses for proportions are least sensitive (Rossi, 1985). Thus, proposing stage progression as an outcome measure will typically have the effect of requiring *more* subjects for any specified level of statistical power than proposing movement to action as an outcome measure.

Recommended Stages of Change Analysis

A recommended alternative approach to conceptualizing stage progression as a dichotomous (or trichotomous) variable would be to count the number of stages progressed as the outcome variable. For example, an individual progressing a single stage would receive a score of 1, while an individual progressing two stages would receive a score of 2, and so on. Regression to an earlier stage would be assigned negative scores. This approach will illustrate comprehensive stage changes across all stages. Indeed, the direction of change and the magnitude of that change are given. If the interest is in investigating an intervention's effect on a particular stage, then the approach may be applied to individuals who started out at that specific stage. This approach would avoid the problems associated with dichotomous variable analysis and should be amenable to continuous variable techniques, such as the ANOVA. However, this approach does not yet seem to be commonly used.

Example of Different Stage Analyses

An example is presented here to describe how the outcome variable would vary depending on the types of analyses that were performed.

Study Background

These data stem from a 3-year longitudinal naturalistic design. The baseline and follow-up have been published (see Nigg, 2001; Nigg & Courneya, 1998). For the purposes of this chapter, the stage variable will be investigated in more detail.

Participants. Participants with both baseline and follow-up data (i.e., completers; n = 400, baseline age = 14.89, SD = 1.15; follow-up age = 17.62 years, SD = 1.18) were not different compared to follow-up noncompleters (n = 419) with regards to exercise staging (p < 0.001). At baseline, 43.0% of the completers were in grade nine (mean age = 13.92, SD = 0.50); 26.4% in grade 10 (mean age = 14.94, SD = 0.48); 20.3% in grade 11 (mean age = 15.94, SD = 0.55); and 10.3% in grade 12 (mean age = 16.99, SD = 0.71). At follow-up, completers identified themselves as employed full-time (19.5%; mean age = 18.72, SD = 1.03); employed part-time (12.0%; mean age = 17.71, SD = 1.05), unemployed (1.5%; mean age = 18.11, SD = 1.36), high school students (grade 12; 48%; mean age = 16.62, SD = 0.55), college students (16.8%; mean age = 17.61, SD = 0.49), and other (2.3%; mean age = 18.58, SD = 0.85).

Instruments. Marcus, Selby et al. (1992) adapted the stages of change questionnaire from the smoking literature to apply to exercise. Courneya (1995) further adapted this instrument, and this version is used for the current study. Five statements were given that represent each stage, and participants were asked to mark the one statement that applied to their current exercise status. The items were arranged so that if the first item was endorsed, the individual was placed in the precontemplation stage; if the second item was endorsed, the individual was categorized in the contemplation stage; and so on up to the maintenance stage (statement 5). The instrument has been validated for adolescents (Lee et al., 2001).

Procedures. The appropriate institutional and ethics review boards approved both time points. For the follow-up (summer 1998), questionnaires were mailed with up to seven telephone prompts. A short introduction and description of the project preceded the consent to participate. Participants under 18 years of age needed parental or guardian consent. If the questionnaire was completed and returned, the participant was entered into a drawing (twice if returned within a month) for one prize of $200 and two prizes of $50. In September, individuals who did not return questionnaires were reminded by phone at least twice. The procedure resulted in 400 individuals returning questionnaires, which was 48.84% of the original sample and 60.98% of individuals who could be contacted.

Analysis of the Example

The different stage outcome conceptualizations are presented in the following sections.

Traditional Analytical Approach. Categories were created for progressing to action/maintenance or not. This outcome variable was defined as any participants who were in precontemplation (PC), contemplation (C), or preparation (P) at baseline and progressed to action (A) or maintenance (M) at follow-up (n = 45) or any other stage combination (n = 349).

Stage Progression Analysis. Categories were created for stage progressors, regressors, stable nonexercisers, and stable exercisers, which were conceptualized as the outcome variable. Progressors were defined as any individuals who were at an earlier stage at baseline than at follow-up (n = 84). Regressors were defined by being in a later stage at baseline than at follow-up (n = 117). Stable non-exercisers were defined as individuals who were in a pre-criterion stage at baseline and in the same pre-criterion stage at follow-up (i.e., PC at baseline and PC at follow-up, or C at baseline and C at follow-up, or P at baseline and P at follow-up; n = 63). Participants at criterion at baseline and in the same stage at follow-up were the stable exercisers (i.e., A at baseline and A at follow-up, or M at baseline and M at follow-up; n = 130).

Recommended Stage Change Analysis

Five separate analyses were run, one for each baseline stage.

- Baseline precontemplation: The follow-up stage distribution was PC = 1; C = 1; P = 4; A = 2; M = 1.
- Baseline contemplation: The follow-up stage distribution was PC = 1; C = 2; P = 12; A = 2; M = 1.

- Baseline preparation: The follow-up stage distribution was PC = 5; C = 10; P = 60; A = 15; M = 26.
- Baseline action: The follow-up stage distribution was PC = 2; C = 2; P = 30; A = 10; M = 23.
- Baseline maintenance: The follow-up stage distribution was PC = 5; C = 10; P = 60; A = 15; M = 26.

As can be seen from the results of the different outcome conceptualizations, different conclusions and implications are possible. The traditional analytic result indicates the number of individuals adopting regular exercise. However, the group representing all other stage combinations is not very informative. The stage progression analysis is more informative in that results can be interpreted in terms of individuals who progressed in stage, regressed in stage, stayed inactive, or stayed active on a regular basis. However, the inherent benefit of a stage approach is that the important constructs related to each stage can be identified, which the stage progression does not allow. The recommended analysis method does allow information to be gathered specific to each stage and gives a complete picture of the stage changes that occurred for each stage. This allows conclusions to be drawn as to the pattern of change from each stage and what constructs or variables led to each stage change. This information is the most useful for how and why behavior change occurs and is maintained, and the results have direct implications for further stage-specific intervention designs.

Summary

In response to some of the issues with physical activity and exercise measurement for large-population projects, this chapter outlined a stage framework for physical activity and exercise behavior assessment. Supporting data, the actual instrument, and how to analyze the stages of change from the transtheoretical model of behavior change were presented. The stages of change instrument is short, valid, and reliable and is able to detect changes in physical activity and exercise behavior. Thus, this instrument is well suited for large-population interventions.

Acknowledgments

This chapter was written when the author was supported by Grant #1 RO1 AG16588 from the National

Institute on Aging. This chapter was prepared in partial collaboration with the Behavior Change Consortium Physical Activity Workgroup. The author would especially like to acknowledge the contributions of Drs. Joseph Rossi and Gregory Welk.

References

American College of Sports Medicine. (1990). Position stand: The recommended quantity and quality of exercise for developing and maintaining cardiorespiratory and muscular fitness in healthy adults. *Medicine and Science in Sports and Exercise, 22,* 265-274.

American College of Sports Medicine. (1998). ACSM position stand: The recommended quantity and quality of exercise for developing and maintaining cardiorespiratory and muscular fitness, and flexibility in healthy adults. *Medicine and Science in Sports and Exercise, 30,* 975-999.

Armstrong, C.A., Sallis, J.F., Hovell, M.F., & Hofstetter, C.R. (1993). Stages of change, self-efficacy and adoption of vigorous exercise: A prospective analysis. *Journal of Sport and Exercise Psychology, 15,* 390-402.

Bandura, A. (1986). *Social foundations of thought and action.* Englewood Cliffs, NJ: Prentice-Hall.

Barké, C.R., & Nicholas, P.R. (1990). Physical activity in older adults: The stages of change. *The Journal of Applied Gerontology, 9,* 216-223.

Calfas, K.J., Sallis, J.F., Lovato, C.Y., & Campbell, J. (1994). Physical activity and its determinants before and after college graduation. *Medicine, Exercise, Nutrition, and Health, 3,* 323-334.

Cardinal, B.J. (1995a). The stages of exercise scale and stages of exercise behavior in female adults. *Journal of Sport Medicine and Physical Fitness, 35,* 87-92.

Cardinal, B.J. (1995b). Behavioral and biometric comparisons of the preparation, action, and maintenance stages of exercise. *Wellness Perspectives: Research, Theory, and Practice, 11,* 36-43.

Cardinal, B.J. (1997). Construct validity of stages of change for exercise behavior. *American Journal of Health Promotion, 12,* 68-74.

Cardinal, B.J. (1999). Extended stage model of physical activity behavior. *Journal of Human Movement Sciences, 37,* 37-54.

Cardinal, B.J. (2000). Are sedentary behaviors terminable? *Journal of Human Movement Sciences, 38,* 137-150.

Cardinal, B.J., & Sachs, M.L. (1995). Prospective analysis of stage-of-exercise movement following mail-delivered, self-instructional exercise packets. *American Journal of Health Promotion, 9,* 430-432.

Cardinal, B.J., & Sachs, M.L. (1996). Effects of mail-mediated, stage-matched exercise behavior change strategies on female adults' leisure-time exercise behavior. *Journal of Sports Medicine and Physical Fitness, 36,* 100-107.

Cole, G., Leonard, B., Hammond, S., & Fridinger, F. (1998). Using the "stages of behavioral change" constructs to measure the short-term effects of a worksite-based intervention to increase moderate physical activity. *Psychological Reports, 82,* 615-618.

Collins, L.M., & Wugalter, S.E. (1992). Latent class models for stage-sequential dynamic latent variables. *Multivariate Behavioral Research, 27,* 131-137.

Courneya, K.S. (1995). Understanding readiness for regular physical activity in older individuals: An application of the theory of planned behavior. *Health Psychology, 14,* 80-87.

Courneya, K.S., Estabrooks, P.A., & Nigg, C.R. (1997). Predicting change in exercise over a three-year period: An application of the theory of planned behavior. *Avante, 3,* 1-13.

Courneya, K.S., Nigg, C.R., & Estabrooks, P.A. (1998). Relationships among the theory of planned behavior, stages of change, and exercise behavior in older persons over a three-year period. *Psychology and Health, 13,* 355-367.

Dishman, R.K. (1990). Determinants of participation in physical activity. In C. Bouchard, R.J. Shepard, T. Stephens, J.R. Sutton, & B.D. McPherson (Eds.), *Exercise, fitness, and health: A consensus of current knowledge* (pp. 75-101). Champaign, IL: Human Kinetics.

Dishman, R.K. (1994). *Advances in exercise adherence.* Champaign, IL: Human Kinetics.

Dishman, R.K., & Buckworth, J. (1996). Increasing physical activity: A quantitative synthesis. *Medicine and Science in Sports and Exercise, 28,* 706-719.

Emmons, K.M., Marcus, B.H., Linnan, L., Rossi, J.S., & Abrams, D.B. (1994). Mechanisms in multiple risk factor interventions: Smoking, physical activity, and dietary fat intake among manufacturing workers. *Preventive Medicine, 23,* 481-489.

Gorely, T., & Gordon, S. (1995). An examination of the transtheoretical model and exercise behavior in older adults. *Journal of Sport and Exercise Psychology, 17,* 312-324.

Graham, J.W., Collins, L.M., Wugalter, S.E., Chung, N.K., & Hansen, W.B. (1991). Modeling transitions in latent stage-sequential processes: A substance use prevention example. *Journal of Consulting and Clinical Psychology, 59,* 48-57.

Greenland, S. (1995a). Dose repose and trend analysis in epidemiology: Alternatives to categorical analysis. *Epidemiology, 6,* 356-365.

Greenland, S. (1995b). Avoiding power loss associated with categorization and ordinal scores in dose-repose and trend analysis. *Epidemiology, 6,* 450-454.

Hausenblas, H.A., Dannecker, E.A., Connaughton, D.P., & Lovins, T.R. (1999). Examining the validity of stages of exercise change algorithm. *American Journal of Health Studies, 15,* 94-99.

Hellman, E.A. (1997). Use of the stages of change in exercise adherence model among older adults with a cardiac diagnosis. *Journal of Cardiopulmonary Rehabilitation, 17,* 145-155.

Janis, I.L., & Mann, L. (1977). *Decision making: A psychological analysis of conflict, choice and commitment.* New York: Free Press.

Jones, N.D., DellaCorte, M.R., Nigg, C.R., Clark, P.G., Burbank, P.M., Garber, C.E., & Padula, C. (2001). Seniorcise: A print exercise intervention in older adults. *Educational Gerontology, 27,* 717-728.

Kriska, A.M., & Casperson, C.J. (1997). Introduction to a collection of physical activity questionnaires. *Medicine and Science in Sports and Exercise, 29,* S5-S9.

Lee, R.E., Nigg, C.R., DiClemente, C.C., & Courneya, K.S. (2001). Validating the stages of change for exercise behavior with adolescents. *Research Quarterly of Exercise and Sports, 72,* 401-410.

Levy, K.J. (1975). Large-sample pair-wise comparisons involving correlations, proportions, or variances. *Psychological Bulletin, 82,* 174-176.

Marcus, B.H., Banspach, S.W., Lefebvre, R.C., Rossi, J.S., Carleton, R.A., & Abrams, D.B. (1992). Using the stage of change model to increase the adoption of physical activity among community participants. *American Journal of Health Promotion, 6,* 424-429.

Marcus, B.H., Bock, B.C., Pinto, B.M., Forsyth, L.H., Roberts, M.B., & Traficante, R.M. (1998). Efficacy of an individualized, motivationally tailored physical activity intervention. *Annals of Behavioral Medicine, 20,* 174-180.

Marcus, B.H., Eaton, C.A., Rossi, J.S., & Harlow, L.L. (1994). Self-efficacy, decision-making and the stages of change: An integrative model of physical exercise. *Journal of Applied Social Psychology, 24,* 489-508.

Marcus, B.H., Emmons, K.M., Simkin-Silverman, L.R., Linnan, L.A., Taylor, E.R., Bock, B.C., Roberts, M.B., Rossi, J.S., & Abrams, D.B. (1998). Evaluation of motivationally tailored vs. standard self-help physical activity interventions at the workplace. *American Journal of Health Promotion, 12,* 246-253.

Marcus, B.H., Selby, V.C., Niaura, R.S., & Rossi, J.S. (1992). Self-efficacy and the stages of exercise behavior change. *Research Quarterly for Exercise and Sport, 63,* 60-66.

Marcus, B.H., & Simkin, L.R. (1993). The stages of exercise behavior. *Journal of Sports Medicine and Physical Fitness, 33,* 83-88.

Marcus, B.H., Simkin, L.R., Rossi, J.S., & Pinto, B.M. (1996). Longitudinal shifts in employees' stages and processes of exercise behavior change. *American Journal of Health Promotion, 10,* 195-200.

Martin, R.A., Velicer, W.F., & Fava, J.L. (1996) Latent transition analysis applied to the stages of change for smoking cessation. *Addictive Behaviors, 21,* 67-80.

Montoye, H.J., Kemper, H.C.G., Saris, W.H.M., & Washburn, R.A. (1996). *Measuring physical activity and energy expenditure.* Champaign, IL: Human Kinetics.

Nigg, C.R. (2001). Explaining adolescent exercise behavior change: A longitudinal application of the transtheoretical model. *Annals of Behavioral Medicine, 23,* 11-20.

Nigg, C.R., Burbank, P., Padula, C., Dufresne, R., Rossi, J.S., Velicer, W.F., Laforge, R.G., & Prochaska, J.O. (1999). Stages of change across ten health risk behaviors for older adults. *The Gerontologist, 39,* 473-482.

Nigg, C.R., & Courneya, K.S. (1998). Transtheoretical model: Examining adolescent exercise behavior. *Journal of Adolescent Health, 22,* 214-224.

Norman, G.J., Benisovich, S.V., Nigg, C.R., & Rossi, J.S. (1998). Examining three exercise staging algorithms in two samples. *Annals of Behavioral Medicine, 20,* S211.

Pate, R.R., Pratt, M., Blair, S.N., Haskell, W.L., Macera, C.A., Bouchard, C., Buchner, D., Ettinger, W., Heath, G.W., King, A.C., Kriska, A., Leon, A.S., Marcus, B.H., Morris, J., Paffenbarger, R.S., Patrick, K., Pollock, M.L., Rippe, J.M., Sallis, J., & Wilmore, J.H. (1995). Physical activity and public health. A recommendation from the Centers for Disease Control and Prevention and the American College of Sports Medicine. *Journal of the American Medical Association, 273,* 402-407.

Pinto, B.M., & Marcus, B.H. (1995). A stages of change approach to understanding college students' physical activity. *Journal of American College Health, 44,* 27-29.

Prochaska, J.O., & DiClemente, C.C. (1983). Stages and processes of self-change in smoking: Towards an integrative model of change. *Journal of Consulting and Clinical Psychology, 51,* 390-395.

Prochaska, J.O., & DiClemente, C.C. (1985). Common processes of self-change in smoking, weight control and psychological distress. In S. Shiffman & T.A. Willis (Eds.), *Coping and substance abuse* (pp. 345-363). New York: Academic Press.

Prochaska, J.O., & Marcus, B.H. (1994). The transtheoretical model: Applications to exercise. In R.K. Dishman (Ed.), *Advances in exercise adherence* (pp. 161-180). Champaign, IL: Human Kinetics.

Prochaska, J.O., & Velicer, W.F. (1997). The transtheoretical model of behavior change. *American Journal of Health Promotion, 12,* 38-48.

Reed, G.R., Velicer, W.F., Prochaska, J.O., Rossi, J.S., & Marcus, B.H. (1997). What makes a good staging algorithm: Examples from regular exercise. *American Journal of Health Promotion, 12,* 57-66.

Rossi, J.S. (1985). Tables of effect size for z score tests of differences between proportions and between correlation coefficients. *Educational and Psychological Measurement, 45,* 737-743.

Rossi, J.S. (1990). Statistical power of psychological research: What have we gained in 20 years? *Journal of Consulting and Clinical Psychology, 58,* 646-656.

Sallis, J.F., & Hovell, M.F. (1990). Determinants of exercise behavior. In K.B. Pandolph & J.O. Holloszy (Eds.), *Exercise and sport sciences reviews* (vol. 18, pp. 307-330). Baltimore, MD: Williams & Wilkins.

Stone, E.J., McKenzie, T.L., Welk, G.J., & Booth, M.L. (1998). Effects of physical activity interventions in youth: Review and synthesis. *American Journal of Preventive Medicine, 15,* 298-315.

U.S. Department of Health and Human Services. (1996). *Physical activity and health: A report of the Surgeon General.* Atlanta, GA: U.S. Department of Health and Human Services, Centers for Disease Control and Prevention.

Wyse, J., Mercer, T., Ashford, B., Buxton, K., & Gleeson, N. (1995). Evidence for the validity and utility of the stages of exercise behavior change scale in young adults. *Health Education Research, 10,* 365-377.

Zeger, S.L., & Liang, K.Y. (1986). Longitudinal data analysis for discrete and continuous outcomes. *Biometrics, 42,* 121-130.

15

Environmental and Policy Measurement in Physical Activity Research

Adrian Bauman, PhD
Liverpool Hospital, Sydney, Australia

James F. Sallis, PhD
San Diego State University

Neville Owen
University of Queensland

There have been almost five decades of purposive attempts to measure physical activity and exercise in humans, with the scientific goal of relating physical activity to health outcomes. There is now a substantial literature that describes the measurement of activity in laboratory and field situations as well (Montoye et al., 1996). The measurement properties of diverse assessments of physical activity are well described, as are the relationships between physical activity and health outcomes (U.S. Department of Health and Human Services, 1996).

Over the past decade, there has been increasing epidemiological evidence that some of the ascribed health benefits may occur at lower intensities of physical activity than was previously thought, lead-

ing to the public health recommendations to "accumulate at least half an hour of at least moderate intensity physical activity on most days of the week" (U.S. Department of Health and Human Services, 1996). The epidemiological evidence points to decreased total energy expenditure, rather than just specific or vigorous exercise, as contributing to a number of health outcomes. It is assumed that population-wide declines in energy expenditure are contributing to the substantial increases in obesity in many nations (World Health Organization, 1997). It is believed that introduction of labor-saving devices at work and at home, reliance on automobiles, rapidly evolving electronic forms of entertainment, and changes in the design of cities and suburbs are having major effects on producing declines in physical activity.

Understanding the impact of these societal changes on physical activity requires further development in two areas of measurement. First, in terms of characterizing population levels of activity, new dimensions of physical activity are being explored with increasing interest. These include occupational activity, domestic activities and gardening, activity related to transportation, and incidental physical activity, loosely defined as opportunistic energy expenditure in everyday life. There has been a resurgence of interest in the measurement and contribution of occupational energy expenditure, and in the measurement of sedentary or inactive time. These dimensions require new ways of assessing total physical activity and specific health-related activity components. Their assessment will allow us to track, and hopefully to understand better, changes in physical activity. Second, there is the need to understand large-scale influences on physical activity and sedentary living in whole populations. In this context, there is a need for measures that accurately characterize those environmental and policy variables that may be responsible for population declines in overall physical activity.

This chapter focuses specifically on the assessment of environmental and policy variables that are thought to be related to physical activity of individuals and populations. Although typically assessed in cross-sectional studies, these correlates of being active or inactive are often referred to as *determinants*. This is because it has been hypothesized that they are causally related to physical activity. Typically, determinants have been classified into within-individual factors (including cognitive and affective variables, such as self-efficacy, self-motivation, perceived benefits), behavioral skills, social and cultural factors, and demographic factors. New variables have been added to this list, broadly grouped as "environmental and policy" variables (Sallis et al., 1998). These may directly influence dimensions of physical activity (Sallis & Owen, 1996) or may be moderators of the relationships between the other classes of variables and physical activity. It is likely that different environmental factors have varying effects on the multiple dimensions of physical activity (Owen et al., 2000). For example, environmental or policy variables might influence the choice of physically active commuting options such as walking or biking. Or, in the future, job-related physical activity may be mandated by occupational health and safety regulations.

The main purposes of this chapter are to present a framework for describing environmental and policy variables and to develop approaches to measuring these variables. Compared to other physical activity measures, or even measures of other categories of determinants, these measures are in an early stage of conceptualization and evaluation. It is hoped that the systematic presentation of the material in this chapter will stimulate further progress in measuring environmental and policy variables related to physical activity. Because environmental and policy variables are believed to be important influences on daily physical activity levels of people in general (Sallis et al., 1998), there is a great need to document and understand these actual associations. Findings of strong and consistent associations can inform the design of population-based interventions and can motivate and guide policy changes that will result in increased activity levels (Sallis & Owen, 1999; Sallis et al., 2000).

Conceptual and Theoretical Foundations

Environmental and policy approaches to public health improvement have a diverse heritage. Part of their origin is in a traditional approach to public health that considers the host, agent, and environment (Brownson, 1998). Various World Health Organization policy initiatives on health promotion focus on the development of supportive environments for health and on the necessity for healthy public policy to support individual behavioral changes (Bauman & Bellew, 1999).

In the behavioral sciences, environmental psychology (Barker, 1968; see Sallis & Owen, 1996) influenced the subsequent development of social ecology frameworks (Stokols, 1996). Ecological models posit that behavior is influenced by intrapersonal, interpersonal, and environmental and policy factors (Sallis & Owen, 2002). This perspective is consistent with multilevel psychological theories, especially social cognitive theory and operant learning theory (Bennett & Murphy, 1997; Sallis et al., 1990).

Specific ecological frameworks to identify environmental and policy influences on physical activity have been proposed. For example, Sallis and Owen (1999) proposed several categories of environmental and policy influences. Booth et al. (2001) proposed more specific models that identified primary settings within which physical activities take place. Those settings include homes, health clubs, other private activity businesses (e.g., dance or martial arts studios), workplaces, shopping malls, parks and recreation centers, senior centers, daycare centers, schools, and neighborhoods. Booth et al. (2001) provided hypotheses for specific influences within each of these settings and identified people or organizations in a position to control policies that affect these settings. Research on environmental determinants is premised on several key concepts that can mediate or moderate the effect of environments of physical activity:

- Multiple levels of influence determine physical activity behavior.
- Behavior-specific models are needed to account for different categories of physical activity.
- Environments may more directly influence physical activity.
- Other levels of influence may affect physical activity (i.e., genetic, biological, intrapersonal, interpersonal, and social factors).

Owen et al. (2000) identified how particular behavioral settings may have an impact on different types of physical activity and on sedentary behaviors. Such specific physical activity–relevant models should be helpful in generating hypotheses to guide the needed research. The challenge is to identify how specific environmental and policy variables might act to influence the multiple dimensions of physical activity, and how to measure these characteristics of environments and policies. In this field that clearly is at an early stage in its development, this theoretical work is a necessary precursor to the development of measures of these variables.

What Are Environmental and Policy Variables?

Our main focus in this chapter is on measures of the physical environment, rather than on the social environment. The physical environment consists of natural and built (or human-made) aspects. The natural environment includes weather and topographical factors that are hypothesized to influence physical activity. The built environment ranges from homes and stores to the design of cities. We recognize that built environment factors are the product of social and economic trends, laws, and regulations. Our focus is on those aspects of environments that might have an effect on physical activity. There are an infinite number of environmental characteristics

potentially related to physical activity, so in the absence of existing data, it is a major challenge to select a few variables to include in any given study. The specific measures described here indicate the variables that have been measured and studied so far.

Policies are rules, laws, or regulations that influence organizational and individual behaviors. Like the physical environment, policies are products of social environments, social norms, and economic conditions. Policies could provide funding for programs, create incentives for individuals to be active at work, or specify requirements for physical education at school. Policies shape environments and are a primary means of creating environmental changes. Policies about sidewalks in new urban developments and the design of stairwells in buildings have obvious relevance to physical activity. Policies can be related to safety, urban planning, transportation, incentives for being active, and even the funding of public education campaigns to encourage physical activity (Dora, 1999; Sallis et al., 1998; Schmid et al., 1995). Measurement tasks are to document the existence, implementation, and adherence to policies. It cannot, however, be assumed that policies will be adhered to once they are in place. Ordinances and policies are often codified but not implemented across all relevant jurisdictions (Moscowitz et al., 2000).

Toward a Framework for Environmental and Policy Measures

Table 15.1 illustrates some of the potential environmental and policy dimensions that may be related to physical activity participation, and for which measurement development is required. These are categorized as settings, sectors, and societal/cultural factors. Although this table is derived from real-world examples, the measurement properties (reliability and validity) of many of the relevant indicators are usually not known.

The settings element identifies environmental indicators within specific locations. Examples are provided, but the range of possible indicators and measures is much larger. Work sites and communities as settings could be assessed for the introduction of exercise facilities or more comprehensive environmental changes that might facilitate physical activity (Linenger et al., 1991). The policy correlates would be incentives or subsidies for those employees participating in activity programs or utilizing built facilities. Similarly, schools, local institutions, or neighborhoods could function as environmental settings to enhance (or reduce barriers to) physical activity participation. Relevant policy correlates would include hours of class time, curricula around health, and required physical education hours per student.

The sectors dimension in table 15.1 illustrates some of the agencies responsible for environmental and policy initiatives relevant to physical activity. Measures include process evaluation of community-based facilities (e.g., number of miles of trails, number of bike lanes, or bike-parking spaces), policy initiatives (e.g., regulations regarding building codes), car usage, public transportation, and the use of public space. These are relatively simple as measurements: do they exist, are statutes enforced, and are regulations disseminated across the relevant systems? More complex is the assessment of policy development, such as the growth and operation of coalitions of agencies to promote and advocate for physical activity. Few research examples exist of the measurement and assessment of these policy processes, usually through qualitative measurements (Hawe & Stickney, 1997). Policy development can

TABLE 15.1 Potentially Measurable Environmental and Policy Factors

	Environmental factors	Policy factors
Settings		
Work sites	Presence of exercise facilities, bike racks, showers	Incentives for usage of facilities
Schools	Sport or physical education facilities; physical education curricula	Physical education policy mandatory min/week; adoption of curricula
Local institutions or local community	Aged care facility programs; miles of bike lanes; safe walking routes to school	Use of institutions for physical activity programs, car-pooling incentives; funding for bike lanes; policy for safe walking routes
Sectors		
Health	Public education (raising awareness about activity-friendly environments)	Inter-sectoral action group formation; effective policy-influencing coalitions; payment for physical activity counseling
Transport	Miles of bike paths, walking trails; presence of bike parking; efficient public transportation system	Subsidized active commuting; street designs for slower traffic; links to clean air policy
Private sector	Provide facilities and programs at work sites	Health insurance incentives for physical activity
Town planning and local government	Acres of park space; presence of sidewalks, street lighting; degree of mixed residential and commercial land use	Building and zoning codes; government funding for nonvehicle transport
Sport and recreation	Provision of recreation facilities (e.g., park, trails, sports fields)	Policies balancing elite sport versus development of facilities for community participation
Societal/cultural		
Media	Reduction in promotion of sedentary entertainment; positive portrayals of physical activity	Restrictions on the promotion of sedentary entertainment
Social norms	Community norms regarding incidental activity, use of parks and trails, use of public transport	Community support for policy changes to support physical activity
Political and economic factors	Extent of political discussion of physical activity policies	Existence of local, state, or national policy to develop low-cost accessible facilities; economic analysis of businesses that require sedentary behavior

also be appraised through investment shifts (e.g., assessing health department funding to support physical activity programs). Similar shifts in resources could be monitored for new recreation initiatives or active commuting promotions. Some policy measures to assess changes in urban design practices could be tracking proposals to increase urban density, ratio of parks to population in new developments, and expenditures to make communities more "walkable" through sidewalk construction, better lighting, or improved pedestrian safety.

The societal/cultural dimension shown in table 15.1 identifies macro-level phenomena that cut across settings and sectors. These are more difficult to measure but are probably related to declines in energy expenditure in developed countries. As indicators of the information environment, print and electronic media could be assessed for messages that promote physical activity and sedentary behavior. Community support for activity-related policies might be possible to monitor, as could the dissemination of initiatives to mandate accessible stairs or improved sidewalks. The discussion of physical activity policies in political discourse could be monitored through content analysis of political speeches.

In dealing with any environmental strategy to promote physical activity, different levels of environmental and policy-related measurement may be required. For example, table 15.2 shows two examples of real-world problems. The first hypothetical problem is low rates of walking in a community. As part of an evaluation of a community-wide intervention, a range of environmental and policy-relevant measurements may be required. These may be the existence of resources to fund community

walking trails (policy measure), the construction of actual trails (environmental measure), the assessment of community attitudes or willingness to use the trails (individual measure), and self-reports or direct observation of trail use (behavioral measure). Direct observation of trail use may be through electronic or infrared counters (Engelhard et al., 2001), which may be more accurate for assessing total trail use but will not distinguish repeat users (and hence will not measure population usage).

Table 15.2 shows another example of these levels of measurement, for a putative physical education policy in schools, and the range of environmental and policy measures required. These examples indicate the diversity of process and outcome measures that may be needed, or should be developed, for the comprehensive evaluation of physical activity interventions with environmental or policy components.

Measures of Environmental and Policy Variables Relevant to Physical Activity

An important challenge is to develop ways of measuring and assessing indicators and variables relevant to environmental and policy approaches to physical activity. This section reviews current efforts to assess physical activity-related policy, and then reviews the range of environmental measurement across different settings. Studies are relatively sparse, but it is important to develop measurements with known and acceptable measurement properties.

TABLE 15.2 Different Levels of Environmental and Policy Measurement Related to Physical Activity

Environmental or policy issue	Policy measure	Environmental measures	Individual measures	Behavioral measures	
				Self-reported measures	Directly observed measures
Developing increased infrastructure for walking in the community	Funding of walking trail development	Miles of walking trails constructed	Surveys of community attitudes to walking trail location, safety	Self-report of trail usage	Direct or indirect observation of trail usage
Promoting or supporting more physical education in schools	Adequacy of policy regarding minimum standard for physical education time/week	Amount of facilities, space, and equipment for use in physical education classes	Awareness of policy by administrators, teachers, and parents	Students' self-reported activity in physical education time/week	Direct observation of physical education classes

Policy Assessment

Existence of policies can be counted, attributes of policies can be coded, and the implementation of policies can be tracked across settings and over time. Policies can be assessed by analyzing written documents, examining documentation of their effects, or by surveying individuals. Policy assessment is a new topic for most physical activity researchers, so examples of measurement strategies from other fields may be useful (Alciati et al., 1998; Brownson, 1998). There are few examples of policy assessments related to physical activity. However, Eyler et al. (1999) conducted policy-relevant research by surveying community leaders to assess their support for various physical activity policies. A mail survey procedure was used to collect the data that have been used to guide policy change efforts.

Baker et al. (2000) reported an effort by the Centers for Disease Control and Prevention to identify promising policy and environmental indicators. Participants rated each indicator on quality, feasibility, and overall usefulness in promoting behavior change. Brownson et al. (1998) surveyed a population sample to assess support for policies related to zoning regulations for bike paths, government funding for exercise facilities, work-site policies, and requirements and budgets for school physical education. Test-retest reliability was assessed, and results indicated substantial variation in reliability coefficients among the questionnaire items. These studies show how data derived from self-reports can be useful for guiding research and practice. Perhaps a next step is to develop quantitative measures of selected specific policies that were reported by Baker et al. (2000).

It is relevant to consider measures of physical activity in specific environments that can be used to assess policy or environmental interventions. Several studies have used stair signage to promote point of choice decisions regarding stair use in malls and train stations (Andersen et al., 1998; Blamey et al., 1995). The measures here have been to count the users of stairs or moving escalators before and after the intervention. Counts of travel mode usage and commuter patterns have been used to evaluate transport-related interventions (Vuori et al., 1994; West Australian Dept. of Transport, 2001). User surveys were the methods employed to assess the impact of trail development in rural Missouri communities (Brownson et al., 2000). Technology can be applied to these purposes by using electronic people counters to assess those who take the stairs or use walking trails.

Environmental Assessment in Health-Related Physical Activity Research

Self-report, direct observation, existing records, and unobtrusive measures have been used to assess environmental variables in physical activity research. This review is organized by the specific environmental settings proposed by Booth et al. (2001) to be most relevant to physical activity.

- *Homes.* A checklist of physical activity equipment and supplies was developed with 15 items, such as the presence of aerobic exercise equipment, toning devices, sports equipment, and a dog (Sallis, Johnson et al., 1997). The scale had a test-retest reliability of R = 0.89. Jakicic et al. (1997) used a similar checklist and found that presence of home exercise equipment was significantly associated with physical activity.

- *Health clubs and other activity-related businesses.* The density of physical activity locations around people's homes was computed by making an inventory of facilities, assembled mainly through telephone books and local publications. In addition to businesses like health clubs, public facilities like parks and school grounds were coded. Those resources were then mapped, along with residence location, and coordinates were entered into a computerized database (Sallis et al., 1990). The objectively assessed density of pay (but not free) facilities around the home was found to be associated with the individual's exercise frequency. This study was carried out using the rather time-consuming and laborious techniques that were available at that time. It is now feasible to carry out similar analyses in a more efficient, precise, and comprehensive manner using geographic information systems software (see the next section). Proximity or convenience of exercise facilities can be assessed by self-report, such as with the 18-item scale developed by Sallis, Johnson et al. (1997). Although the scale had high test-retest reliability (i.e., R = 0.80), it was not associated with reported physical activity. However, an Australian study of older adults did find a relationship between the reported accessibility of local exercise facilities and physical activity (Booth et al., 2000).

- *Workplaces.* An observational instrument was developed specifically to assess environmental variables at work sites that might influence physical activity. The Checklist for Health Promotion Environments at Worksites (CHEW; Oldenburg et al., 2001) assessed the work-site building, the grounds, and the immediate neighborhood. Presence of directly observable facilities, opportunities, information, and barriers was assessed. Items such as bike racks and showers were counted, and characteristics of stairwells were coded. The informational environment was assessed by counting posters and signs mentioning physical activity. Presence of gyms and recreation facilities in the nearby area was assessed. In an initial study in 20 workplaces in Australia, inter-observer agreement was high for most variables with adequate variability.

• *Parks.* The proximity of parks was found to be related to exercise frequency in the mapping study reported earlier (Sallis et al., 1990). A simpler measure of proximity to parks and similar play areas was devised by asking parents to state whether a play area for young children was within a five-minute walk (Sallis et al., 1993). Proximity to play areas was a significant correlate of children's physical activity. In a policy-relevant study, parents reported the importance of 24 characteristics of parks in choosing a park for their children to play in (Sallis, McKenzie et al., 1997). Parents identified safety and availability of amenities such as toilets, drinking water, and lighting as the most important characteristics.

• *Schools.* A recent study used observational methods to evaluate numerous environmental characteristics of schools. These were studied in relation to physical activity during leisure periods, such as after lunch and after school (Sallis et al., 2001). First, all potential activity areas were located. Observers measured size and counted permanent improvements (e.g., markings for games on playgrounds, soccer goals in fields). The type of area was coded: court, field, or indoor. During leisure activity periods, the SOPLAY (Systems for Observing Play and Leisure Activity in Youth) observation system was used to assess student activity levels, as well as the presence of equipment and adult supervision (McKenzie et al., 2000). The environmental variables were reliably coded and accounted for 47 and 63% of the variance in girls' and boys' activity levels, respectively (Sallis et al., 2001). This study shows how measures of environments and physical activity within those environments can be integrated to yield policy-relevant data.

Environmental Assessments in Transportation, Planning, and Urban Design

Outside of the health field, there is extensive literature showing how the design of communities can affect active travel choices, particularly walking and cycling. Such studies have significant implications for understanding the environmental determinants of health-related physical activity. Much of this research uses geographic information systems (GIS), which are software tools that allow points, lines, areas, or sets of these attributes to be related to each other spatially. Each feature in GIS can be linked with a database of characteristics (Croner et al., 1996; Korte, 1997; Richards et al., 1999). A wide variety of natural and built environment variables can be integrated with socioeconomic, crime, traffic, and physical activity data of people in specific residences. This technology greatly expands a researcher's ability to examine associations of physical activity with a wide variety of environmental variables. Because GIS are used by multiple government agencies, as well as researchers in many fields, rich GIS databases are available for analysis, if the data can be located and accessed.

Transportation, planning, and urban design literature has identified a cluster of neighborhood characteristics that are reliably associated with walking or cycling for transportation. This literature has been recently reviewed (Frank & Engelke, 2000; Saelens et al., in press) and the following are three commonly cited characteristics:

• Density of residences is a consistent correlate of nonmotorized transport.
• Mixed land use is characterized by an intermingling of commercial, residential, and recreational buildings. Walking is greater in mixed-use areas, presumably because it is easier to walk than drive to nearby shops.
• High connectivity of streets is exemplified by a grid pattern and by direct routes for walking from place to place. Modern suburbs usually have low connectivity, with long blocks and many cul-de-sacs that lengthen trips, reduce choice of routes, and discourage walking.

Neighborhood attributes are of particular importance because walking and jogging are typically the most common activities (U.S. Department of Health and Human Services, 1996), and these activities are most often done in neighborhoods instead of specific facilities. The neighborhood can be assessed at multiple levels. Microenvironmental variables such as sidewalks, tree cover, and characteristics of nearby parks can be assessed. There is also a substantial literature on urban design and how macroenvironmental characteristics such as density, mixed use, and connectivity are associated with physical activity.

Neighborhood characteristics can be assessed either objectively or by self-report, and there is no clear consensus on which approach is more powerful (Saelens et al., in press). Researchers in the health area have assessed several aspects of land use and urban design using objective and subjective measures. Bauman et al. (1999) found coastal place of residence to be related to physical activity, even after controlling for socioeconomic status. Simple categorization of rural residence was associated with low levels of physical activity among women (Brownson et al., 2000). Perceptions of pleasant environments and convenient access to destinations (which is an index of mixed land used) were associated with walking (Ball et al., 2001). Census data have been used in several countries to examine the relation between neighborhood socioeconomic status and physical activity. Interestingly, residents of poor or deprived neighborhoods had lower levels of physical activity in Sweden (Sundquist et al., 1999) but higher levels of activity were reported in a U.S. study (Ross, 2000).

Several investigators have developed self-reports of specific neighborhood characteristics. A nine-item checklist assessed presence of variables such as sidewalks, heavy traffic, hills, and enjoyable scenery. The scale, plus a five-point rating of perceived safety, had modest test-retest reliability (R = 0.68). In a national study of middle-aged and older women, physical inactivity was associated with absence of enjoyable scenery, lack of hills, and infrequent observation of others exercising in one's neighborhood (King et al., 2000). Brownson et al. (2000) reported the use of a self-report assessment of walking trails. The measure assessed presence of walking trails in the area, trail length, trail surface, distance to trail, use of the trail, and perceived safety when using the trail. Access to indoor exercise facilities was also assessed. A simple self-report of neighborhood safety was positively associated with physical activity levels (MMWR, 1999), showing the utility of well-conceived measures.

Environmental characteristics and physical activity differ greatly across countries, presenting measurement challenges. Rütten et al. (2001) conducted surveys in several European countries and found that perceived environmental variables (e.g., opportunities for activity in the local area, provision of programs by local clubs and agencies, perception of the physical activity friendliness of the physical environment) were related to behavior. Newman and Kenworthy (1991) used mainly key informant interviews to estimate density and commuting to work by walking and cycling. Strong associations between these variables were found in ecological analyses across 32 cities throughout the world.

Summary

Although the development of relevant theory and measurement methods relating to environmental and policy factors that may influence physical activity is in an early stage of development, there is evidence of creativity and progress. Objective and subjective measures of micro- and macro-level environmental variables have demonstrated consistent associations with physical activity. Measures have been developed for several of the key behavioral settings in which physical activity is commonly done. Interest in applying ecological models to physical activity appears to be increasing (Sallis & Owen, 2002). However, there is great promise in expanding research collaborations to include investigators in transportation, planning, and urban design. Researchers in these areas have identified that land use variables are consistently related to walking and cycling for transport. The use of GIS promises to transform research on environmental and policy influences on physical activity by expanding the range of variables that can be examined using objective measures. Despite a promising beginning, there are numerous challenges in this research area.

- The general lack of reliability and validity studies on these new measures and the responsiveness of the measures to changes in the environmental characteristics are not known.
- The relative explanatory power of self-reported perceptions of the environment, versus more objective measures, requires further investigation—as do the relationships between these two domains of measurement.
- There is a need to identify potential influences in a wider variety of settings and to develop more detailed measures of each setting.
- Variables need to be identified that are relevant to all age, gender, cultural, and socioeconomic status groups.
- Environmental measures are needed that can be applied in multiple countries. International studies are particularly needed in this area because each country may have limited variance in environmental characteristics that might underestimate the true associations with physical activity.

- In the area of policy indicators, measurement of policy is often qualitative, but objective documentation of policy variables and the measurement of their implementation are important research tasks. Few examples exist of policy-relevant research related to physical activity, but opinion surveys documenting support for physical activity policies represent initial efforts.

There will likely be significant progress in the conceptualization and measurement of environmental and policy influences on physical activity as more scientific effort is devoted to these issues. Perhaps the greatest challenge is in defining the range of environmental and policy variables that should be assessed. Communities and whole nations can be viewed from multiple levels and perspectives, so it is not obvious which environmental variables have the strongest influence on physical activity.

The next set of challenges is in the area of environmental and policy-related interventions, which will require relevant measures to evaluate their effects. The progress made over the next several years in the conceptualization and measurement of environmental and policy influences will hopefully contribute to innovative interventions. These are desperately required for public health purposes, as levels of physical activity are static or declining in many developed countries (Armstrong et al., 2000; Pratt et al., 1999).

Overall progress in this new and challenging field will depend on whether researchers can generate new hypotheses, develop measures of potential environmental and policy influences, collect data to test associations, refine hypotheses, and persist with this effort as an iterative process. During this process, it is essential to develop and use measures of high-quality and known psychometric characteristics so that the results can be interpreted with confidence. It will be necessary to persist in the face of some of the initially encountered difficulties as new concepts and methods, and approaches from other fields, are tried and perhaps found not to live up to their initial promise.

References

Alciati, M.H., Frosh, M., Green, S.B., Brownson, R.C., Fisher, P.H., Hobart, R., Roman, A., Sciendra, R.C., & Shelton, D.M. (1998). State laws on youth access to tobacco in the United States: Measuring their extensiveness with a new rating system. *Tobacco Control, 7,* 345-352.

Andersen, R.E.., Franckowiak, S.C., Snyder, J., Bartlett, S.J., & Fontaine, K.R. (1998). Can inexpensive signs encourage the use of stairs? Results from a community intervention. *Annals of Internal Medicine, 129,* 363-369.

Armstrong, T., Bauman, A., & Davies, J. (2000). *Physical activity patterns of Australian adults.* AIHW Catalogue CVD 10. Australian Institute of Health and Welfare, Canberra.

Baker, E.A., Brennan, L.K., Brownson, R., & Houseman, R.A. (2000). Measuring the determinants of physical activity in the community: Current and future directions. *Research Quarterly for Exercise and Sport, 71,* 146-158.

Ball, K., Bauman, A., Leslie, E., & Owen, N. (2001). Perceived environmental aesthetics and convenience and company are associated with walking for exercise among Australian adults. *Preventive Medicine, 33,* 434-440.

Barker, R.G. (1968). *Ecological psychology.* Stanford, CA: Stanford University Press.

Bauman, A., & Bellew, B. (1999). Environmental and policy approaches to promoting physical activity. In *Health in the Commonwealth—Sharing solutions 1999/2000* (pp. 38-41). London: Commonwealth Secretariat.

Bauman, A., Smith, B., Stoker, L., Bellew, B., & Booth, M.L. (1999). Geographical influences upon physical activity participation: Evidence of a "coastal effect." *Australian and New Zealand Journal of Public Health, 23,* 322-324.

Bennett, P., & Murphy, S. (1997). *Psychology and health promotion.* Philadelphia: Open University Press, Buckingham.

Blamey, A., Mutrie, N., & Aitchison, T. (1995). Health promotion by encouraged use of stairs. *British Medical Journal, 311,* 289-290.

Booth, M.L., Owen, N., Bauman, A., Clavisi, O., & Leslie, E. (2000). Social-cognitive and perceived environmental influences associated with physical activity in older Australians. *Preventive Medicine, 31,* 15-22.

Booth, S., Sallis, J., Ritenbaugh, C., Hill, J., Birch, L., Frank, L., Glanz, K., Himmelgreen, D., Mudd, M., Popkin, B., Rickard, K., St. Jeor, S., & Hays, N. (2001). Environmental and societal factors affect food and physical activity: Rationale, influences, and leverage points. *Nutrition Reviews, 59,* S21-39.

Brownson, R.C. (1998). Epidemiology and health policy. In R.C. Brownson & D.B. Petitti (Eds.), *Applied epidemiology: Theory to practice* (pp. 349-387). New York: Oxford University Press.

Brownson, R.C., Housemann, R.A., Brown, D.R., Jackson-Thompson, J., King, A.C., Malone, B.R., & Sallis, J.F. (2000). Promoting physical activity in rural communities: Walking trail access, use, and effects. *American Journal of Preventive Medicine, 18,* 235-241.

Brownson, R.C., Schmid, T.L., King, A.C., Eyler, A.A., Pratt, M., Murayi, T., Mayer, J.P., & Brown, D.R. (1998). Support for policy interventions to increase physical activity in rural Missouri. *American Journal of Health Promotion, 12,* 263-266.

Croner, C.M., Sperling, J., & Broome, F.R. (1996). Geographic information systems (GIS): New perspectives in understanding human health and environmental relationships. *Statistics in Medicine, 15,* 1961-1977.

Dora, C. (1999). A different route to health: Implications of transport policies. *British Medical Journal, 318,* 1686-1689.

Engelhard, S., Stubbs, J., Weston, A., Fitzgerald, S., Giles-Corti, B., Milat, A.J., & Honeysett, D. (2001). Methodological considerations when conducting direct observation in an outdoor environment: Our experience in local parks. *Australian Journal of Public Health, 25,* 149-152.

Eyler, A.A., Mayer, J., Rafii, R., Housemann, R., Brownson, R.C., & King, A.C. (1999). Key informant surveys as a tool to implement and evaluate physical activity interventions in the community. *Health Education Research, 14,* 289-298.

Frank, L.D., & Engelke, P. (2000). How land use and transportation systems impact public health: A literature review of the relationship between physical activity and built form. [Online]. Available: **www.cdc.gov.**

Hawe, P., & Stickney, E.K. (1997). Developing the effectiveness of an intersectoral food policy coalition through formative evaluation. *Health Education Research, 12,* 213-225.

Jakicic, J.M., Wing, R.R., Butler, B.A., & Jeffery, R.W. (1997). The relationship between presence of exercise equipment in the home and physical activity level. *American Journal of Health Promotion, 11,* 363-365.

King, A.C., Castro, C., Wilcox, S., Eyler, A.A., Sallis, J.F., & Brownson, R.C. (2000). Personal and environmental factors associated with physical inactivity among different racial-ethnic groups of U.S. middle-aged and older-aged women. *Health Psychology, 19,* 354-364.

Korte, G.B. (1997). *The GIS book: Understanding the value and implementation of geographic information systems.* Sante Fe, NM: Onword Press.

Kriska, A.M., & Caspersen, C.J. (1997). A collection of physical activity questionnaires for health-related research. *Medicine and Science in Sports and Exercise, 29,* S1-S205.

Linenger, J.M., Chesson, C.V., & Nice, D.S. (1991). Physical fitness gains following simple environmental change. *American Journal of Preventive Medicine, 7,* 298-310.

McKenzie, T.L., Marshall, S.J., Sallis, J.F., & Conway, T.L. (2000). Leisure-time physical activity in school environments: An observational study using SOPLAY. *Preventive Medicine, 30,* 70-77.

MMWR. (1999). Neighborhood safety and the prevalence of physical activity—Selected states, 1996. *Morbidity and Mortality Weekly Reports, 47,* 143-146.

Montoye, H., Kemper, H.C.G., Saris, W.H.M., & Washburn, R.A. (1996). *Measuring physical activity and energy expenditure.* Champaign, IL: Human Kinetics.

Moscowitz, J.M., Zihua, L., Hudes, E.S. (2000). The impact of workplace smoking ordinances in California on smoking cessation. *American Journal of Public Health, 90,* 757-761.

Newman, P.W., & Kenworthy, J.R. (1991). Transport and urban form in thirty-two of the world's principal cities. *Transportation Review, 11,* 249-272.

Oldenburg, B., Sallis, J.F., Harris, D., & Owen, N. (2001). *Checklist of Health Promotion Environments at Worksites: Development and measurement attributes.* Manuscript submitted for publication.

Owen, N., Leslie, E., Salmon, J., & Fotheringham, M.J. (2000). Environmental determinants of physical activity and sedentary behavior. *Exercise and Sport Sciences Reviews, 28,* 153-158.

Pratt, M., Macera, C.A., & Blanton, C. (1999). Levels of physical activity and inactivity in children and adults in the United States: Current evidence and research issues. *Medicine and Science in Sports and Exercise, 31,* S526-S533.

Richards, T.B., Croner, C.M., Rushton, G., Brown, C.K., & Fowler, L. (1999). Information technology. Geographic information systems and public health: Mapping the future. *Public Health Reports, 114,* 359-373.

Ross, C.E. (2000). Walking, exercising, and smoking: Does neighborhood matter? *Social Science and Medicine, 51,* 265-274.

Rütten, A., Abel, T., & Kannas, L. (2001). Self-reported physical activity public health, and perceived environment: Results from a comparative European study. *Journal of Epidemiology and Community Health, 55,* 139-146.

Saelens, B.E., Sallis, J.F., & Frank, L.D. (in press). Environmental correlates of walking and cycling: Findings from the transportation and urban design and planning literatures. *Exercise and Sports Sciences Reviews.*

Sallis, J.F., Bauman, A., & Pratt, M. (1998). Environmental and policy: Interventions to promote physical activity. *American Journal of Preventive Medicine, 15,* 379-397.

Sallis, J.F., Conway, T.L., Prochaska, J.J., McKenzie, T.L., Marshall, S.P., & Wildey, M. (2001). Environmental characteristics are associated with youth physical activity at school. *American Journal of Public Health, 91,* 618-620.

Sallis, J.F., Hovell, M.F., Hofstetter, C.R., Elder, J.P., Caspersen, C.J., Hackley, M., & Powell, K.E. (1990). Distance between homes and exercise facilities related to the frequency of exercise among San Diego residents. *Public Health Reports, 105,* 179-185.

Sallis, J.F., Johnson, M.F., Calfas, K.J., Caparosa, S., & Nichols, J. (1997). Assessing perceived physical environment variables that may influence physical activity. *Research Quarterly for Exercise and Sport, 68,* 345-351.

Sallis, J.F., McKenzie, T.L., Elder, J.P., Broyles, S.L., & Nader, P.R. (1997). Factors parents use in selecting playspaces for young children. *Archives of Pediatrics and Adolescent Medicine, 151,* 414-417.

Sallis, J.F., Nader, P.R., Broyles, S.L., Berry, C.C., Elder, J.P., McKenzie, T.L., & Nelson, J.A. (1993). Correlates of physical activity at home in Mexican-American and Anglo-American preschool children. *Health Psychology, 12,* 390-398.

Sallis, J.F., & Owen, N. (1996). Ecological models. In K. Glanz, F.M. Lewis, & B.K. Rimer (Eds.), *Health behavior and health education: Theory, research, and practice* (2nd ed., pp. 403-424). San Francisco: Jossey-Bass.

Sallis, J.F., & Owen, N. (1999). *Physical activity and behavioral medicine.* Thousand Oaks, CA: Sage.

Sallis, J.F., & Owen, N. (2002). Ecological models. In K. Glanz, F.M. Lewis, & B.K. Rimer (Eds.), *Health behavior and health education: Theory, research, and practice* (3rd ed.). San Francisco: Jossey-Bass.

Sallis, J.F., Owen, N., & Fotheringham, M.J. (2000). Behavioral epidemiology: A systematic framework to classify phases of research on health promotion and disease prevention. *Annals of Behavioral Medicine, 22,* 294-298.

Schmid, T.L., Pratt, M., & Howze, E. (1995). Policy as intervention: Environmental and policy approaches to the prevention of cardiovascular disease. *American Journal of Public Health, 85,* 1207-1211.

Stokols, D. (1996). Translating social ecological theory into guidelines for health promotion. *American Journal of Health Promotion, 10,* 282-298.

Sundquist, J., Malmstrom, M., & Johansson, S.E. (1999). Cardiovascular risk factors and the neighbourhood environment: A multilevel analysis. *International Journal of Epidemiology, 28,* 841-845.

U.S. Department of Health and Human Services. (1996). *Physical activity and health: A report of the Surgeon General.* Atlanta, GA: U.S. Department of Health and Human Services, Centers for Disease Control and Prevention.

Vuori, I., Oja, P., & Paronen, O. (1994). Physical activity commuting to work: Testing its potential for exercise promotion. *Medicine and Science in Sports and Exercise, 26,* 844-850.

West Australian Dept. of Transport. (2001). *TravelSmart Homepage*. Retrieved February 4, 2002 from **http://www.travelsmart.transport.wa.gov.au/**.

World Health Organization. (1997). *Obesity: Preventing and managing the global epidemic*. Geneva: World Health Organization.

A

Major Conclusions from the Surgeon General's Report on Physical Activity and Health

1. People of all ages, both male and female, benefit from regular physical activity.

2. Significant health benefits can be obtained by including a moderate amount of physical activity (e.g., 30 minutes of brisk walking or raking leaves, 15 minutes of running, or 45 minutes of playing volleyball) on most, if not all, days of the week. Through a modest increase in daily activity, most Americans can improve their health and quality of life.

3. Additional health benefits can be gained through greater amounts of physical activity. People who can maintain a regular regimen of activity that is of longer duration or of more vigorous intensity are likely to derive greater health benefits.

4. Physical activity reduces the risk of premature mortality in general and of coronary heart disease, hypertension, colon cancer, and diabetes mellitus in particular. Physical activity also improves mental health and is important for the health of muscles, bones, and joints.

5. More than 60 percent of American adults are not regularly physically active. In fact, 25% of all adults are not active at all.

6. Nearly half of American youths 12 to 21 years of age are not vigorously active on a regular basis. Moreover, physical activity declines dramatically during adolescence.

7. Daily enrollment in physical education classes has declined among high school students from 42 percent in 1991 to 25 percent in 1995.

8. Research on understanding and promoting physical activity is at an early stage, but some interventions to promote physical activity through schools, worksites, and health care settings have been evaluated and found to be successful.

B

Healthy People 2010 Objectives Focus Area 22: Physical Activity and Fitness

Overall Goal

Improve health, fitness, and quality of life through daily physical activity.

Objectives

22-1. Reduce the proportion of adults who engage in no leisure-time physical activity.
- Target: 20 percent.
- Baseline: 40 percent of adults aged 18 years and older engaged in no leisure-time physical activity in 1997 (age adjusted to the year 2000 standard population).

22-2. Increase the proportion of adults who engage regularly, preferably daily, in moderate physical activity for at least 30 minutes per day.
- Target: 30 percent.
- Baseline: 15 percent of adults aged 18 years and older engaged in moderate physical activity for at least 30 minutes 5 or more days per week in 1997 (age adjusted to the year 2000 standard population).

22-3. Increase the proportion of adults who engage in vigorous physical activity that promotes the development and maintenance of cardiorespiratory fitness 3 or more days per week for 20 or more minutes per occasion.
- Target: 30 percent.

- Baseline: 23 percent of adults aged 18 years and older engaged in vigorous physical activity 3 or more days per week for 20 or more minutes per occasion in 1997 (age adjusted to the year 2000 standard population).

22-4. Increase the proportion of adults who perform physical activities that enhance and maintain muscular strength and endurance.
- Target: 30 percent.
- Baseline: 18 percent of adults aged 18 years and older performed physical activities that enhance and maintain strength and endurance 2 or more days per week in 1998.

22-5. Increase the proportion of adults who perform physical activities that enhance and maintain flexibility.
- Target: 43 percent.
- Baseline: 30 percent of adults aged 18 years and older did stretching exercises in the past 2 weeks in 1998.

22-6. Increase the proportion of adolescents who engage in moderate physical activity for at least 30 minutes on 5 or more of the previous 7 days.
- Target: 35 percent.
- Baseline: 27 percent of students in grades 9 through 12 engaged in moderate physical activity for at least 30 minutes on 5 or more of the previous 7 days in 1999.

22-7. Increase the proportion of adolescents who engage in vigorous physical activity that promotes cardiorespiratory fitness 3 or more days per week for 20 or more minutes per occasion.

- Target: 85 percent.
- Baseline: 65 percent of students in grades 9 through 12 engaged in vigorous physical activity 3 or more days per week for 20 or more minutes per occasion in 1999.

22-8. Increase the proportion of the nation's public and private schools that require daily physical education for all students.

- Target: Increase in schools requiring daily physical activity for all students (middle school and junior high target = 25%; high school target = 5%).
- Baseline: Percentage of schools requiring daily physical activity for all students in 1994 (middle school = 17%; high school = 2%).

22-9. Increase the proportion of adolescents who participate in daily school physical education.

- Target: 50 percent.
- Baseline: 29 percent of students in grades 9 through 12 participated in daily school physical education in 1999.

22-10. Increase the proportion of adolescents who spend at least 50 percent of school physical education class time being physically active.

- Target: 50 percent.
- Baseline: 38 percent of students in grades 9 through 12 were physically active in physical education class more than 20 minutes 3 to 5 days per week in 1999.

22-11. Increase the proportion of adolescents who view television 2 or fewer hours on a school day.

- Target: 75 percent.

- Baseline: 57 percent of students in grades 9 through 12 viewed television 2 or fewer hours per school day in 1999.

22-12. (Developmental) Increase the proportion of the nation's public and private schools that provide access to their physical activity spaces and facilities for all persons outside of normal school hours (i.e., before and after the school day, on weekends, and during summer and other vacations).

- Target: Not reported.
- Baseline: Not reported.

22-13. Increase the proportion of worksites offering employer-sponsored physical activity and fitness programs.

- Target: 75 percent.
- Baseline: 46 percent of worksites with 50 or more employees offered physical activity or fitness programs at the worksite or through their health plans in 1998–1999.

22-14. Increase the proportion of trips made by walking.

- Target: Increase in trips made by walking (adults: 25% of trips less than 1 mile; children and adolescents [5-15 years old]: 50% of trips to school of less than 1 mile).
- Baseline: Current percent of trips made by walking in 1985 (adults: 17% of trips less than 1 mile; children and adolescents [5-15 years old]: 31% of trips to school of 1 mile or less).

22-15. Increase the proportion of trips made by bicycling.

- Target: Increase in trips made by bicycling (adults: 2% of trips less than 5 miles; children and adolescents [5-15 years old]: 5% of trips to school of less than 5 miles).
- Baseline: Current percent of trips made by bicycling in 1985 (adults: 0.6% of trips less than 5 miles; children and adolescents [5-15 years old]: 2.4% of trips to school of 5 miles or less).

C

Research Recommendations Outlined in the Surgeon General's Report on Physical Activity and Health

Chapter 2: Historical Background—Terminology, Evolution of Recommendations, and Measurement

- Maintain surveillance of physical activity levels in the U.S. population by age, sex, geographic, and socioeconomic measures
- Develop better methods for analysis and quantification of activity. These methods should be applicable to both work and leisure-time measurements and provide direct quantitative estimates of activity.
- Conduct physiologic, biochemical, and genetic research necessary to define the mechanisms by which activity affects CVD including changes in metabolism as well as cardiac and vascular effects. This will provide new insights into cardiovascular biology that may have broader implications than for other clinical outcomes.
- Examine the effects of physical activity and cardiac rehabilitation programs on morbidity and mortality in elderly individuals.
- Carry out controlled, randomized clinical trials among children and adolescents to test the effects of increased physical activity on CVD risk factor levels including obesity. The effects of intensity, frequency, and duration of increased physical activity should be examined

in such studies.

Chapter 3: Physiological Responses and Long-Term Adaptations to Exercise

- Explore individual variations in response to exercise
- Better characterize mechanisms through which the musculoskeletal system responds differentially to endurance and resistance exercise.
- Better characterize the mechanisms through which physical activity reduces the risk of cardiovascular disease, hypertension, and non-insulin dependent diabetes mellitus.
- Determine the minimal and optimal amount of exercise for disease prevention.
- Better characterize beneficial activity profiles for people with disabilities.

Chapter 4: Effects of Physical Activity on Health and Disease

- Delineate the most important features or combinations of features of physical activity (total amount, intensity, duration, frequency, pattern, or type) that confer specific health benefits.
- Determine specific health benefits of physical activity for women, racial and ethnic

minority groups, and people with disabilities.

- Examine the protective effects of physical activity in conjunction with other lifestyle characteristics and disease prevention behaviors.
- Examine the types of physical activity that preserve muscle strength and functional capacity in the elderly.
- Further study the relationship between physical activity in adolescence and early adulthood and the later development of breast cancer.
- Clarify the role of physical activity in preventing or reducing bone loss after menopause.

Chapter 5: Patterns and Trends in Physical Activity
- Develop methods to monitor patterns of regular, moderate physical activity.
- Improve the validity and comparability of self-reported physical activity in national surveys.
- Improve methods for identifying and tracking physical activity patterns among people with disabilities.
- Routinely monitor the prevalence of physical activity among children under age 12.
- Routinely monitor school policy requirements and students' participation in physical education classes in elementary, middle, and high schools.

Chapter 6: Understanding and Promoting Physical Activity
- Assess the determinants of various patterns of physical activity among those who are sedentary, intermittently active, routinely active at work, and regularly active.

- Assess determinants of physical activity for various population subgroups (e.g., by age, sex, race/ethnicity, socioeconomic status, health/disability status, geographic location).
- Examine patterns and determinants of physical activity at various developmental and life transitions, such as from school to work, from one job or city to another, from work to retirement, and from health to chronic disease.
- Evaluate the interactive effects of psychosocial, cultural, environmental, and public policy influences on physical activity.
- Develop and evaluate the effectiveness of interventions that include policy and environmental supports.
- Develop and evaluate interventions designed to promote adoption and maintenance of moderate physical activity that addresses the specific needs and circumstances of population subgroups. Subgroups include racial/ethnic groups, men and women, girls and boys, the elderly, the disabled, the overweight, low-income groups, and persons at life transitions (i.e., adolescence, early adulthood, family formation, and retirement).
- Develop and evaluate the effectiveness of interventions to promote physical activity in combination with healthy dietary practices that can be sustained over time and can be broadly disseminated to reach large segments of the population.

Credits

Figure 3.2—Adapted, by permission, from G.J. Welk, 1999, "The youth physical activity promotion model: A conceptual bridge between theory and practice," *Quest* 51:5-23.

Figure 7.1—Adapted, by permission, from Baranowski and Domel, 1994, "A cognitive model of children's reporting of food intake," *American Journal of Clinical Nutrition* 59:2125-2175. © Am. J. Clin. Nutr. American Society for Clinical Nutrition.

Figure 7.2—Adapted, by permission, from L. Dipietro, C.J. Casperson, A.M. Ostfeld, and E.R. Nadel, 1993, "A survey for assessing physical activity among older adults," *Medicine and Science in Sports and Exercise* 25:628-642.

Figure 7.3—Adapted, by permission, from Willis et al., 1991, "The use of verbal report methods in the development and testing of survey questionnaires," *Applied Cognitive Psychology* 5:251-267. © John Wiley & Sons Limited. Reproduced with permission.

Table 7.2—Adapted, by permission, from C.E. Matthews et al., 2001, "Sources of variance in daily physical activity levels in the seasonal variation of blood cholesterol study," *American Journal of Epidemiology* 153(10):987-995, by permission of Oxford University Press.

Figure 10.3—Reprinted, by permission, from D.R. Bassett et al., 1996, "Accuracy of five electronic pedometers for measuring distance walked," *Medicine and Science in Sports and Exercise* 28:1071-1077.

Figure 10.4—Reprinted, by permission, from D.R. Bassett et al., 1996, "Accuracy of five electronic pedometers for measuring distance walked," *Medicine and Science in Sports and Exercise* 28:1071-1077.

Figure 10.5—Reprinted from M. Sequeira et al., "Physical activity assessment using a pedometer and its comparison with a questionnaire in a large population survey," *American Journal of Epidemiology*, 1995, 142:989-99, by permission of Oxford University Press.

Figure 10.6—Reprinted, by permission, from Y. Hatano, 1997, "Prevalence and use of pedometer," *Research Journal of Walking* 1:45-54.

Table 13.1—Reprinted, by permission, from R.G. Eston, 1998, "Validity of heart rate, pedometry, and accelerometry for prediction of the energy cost of children's activities," *Journal of Applied Physiology* 84(1):362-371.

Figure 13.3—Reprinted, by permission, from M.S. Treuth, 1998, "Energy expenditure in children predicted from heart rate and activity calibrated against respiration calorimetry," *American Journal of Physiology* 275(38):E15.

Figure 13.4, a & b—Reprinted, by permission, from J.K. Moon and N.J. Butte, 1996, "Combined heart rate and activity improve estimates of oxygen consumption and carbon dioxide production rate," *Journal of Applied Physiology* 81(4):1754-1761.

Figure 13.5, a & b—Reprinted, by permission, from J.K. Moon and N.J. Butte, 1996, "Combined heart rate and activity improve estimates of oxygen consumption and carbon dioxide production rate," *Journal of Applied Physiology* 81(4):1754-1761.

Table 13.2—Reprinted, by permission, from M.S. Treuth, 1998, "Energy expenditure in children predicted from heart rate and activity calibrated against respiration calorimetry," *American Journal of Physiology* 275(38):E14.

Figure 14.1—Reprinted with Permission, LifeScan, Inc.

Contributors

David R. Bassett Jr., PhD, Department of Exercise Science and Sports Management, University of Tennessee, Knoxville, Tennessee, USA

Adrian Bauman, MD, PhD, Epidemiology Unit, Liverpool Hospital, Sydney, Australia

Diane J. Catellier, DrPH, Department of Biostatistics, University of North Carolina, Chapel Hill, USA

Darren Dale, PhD, Health and Physical Education Department, Eastern Connecticut State University, Willimantic, Connecticut, USA

Kathleen F. Janz, EdD, Department of Health, Leisure, and Sport Studies, University of Iowa, Iowa City, Iowa, USA

Matthew T. Mahar, EdD, Department of Exercise and Sport Science, East Carolina University, Greenville, North Carolina, USA

Charles E. Matthews, PhD, Norman J. Arnold School of Public Health, University of South Carolina, Columbia, South Carolina, USA

Thomas L. McKenzie, PhD, Department of Exercise and Nutrition Sciences, San Diego State University, San Diego, California, USA

James R. Morrow Jr., PhD, Department of Kinesiology, Health Promotion, and Recreation, University of North Texas, Denton, Texas, USA

Keith E. Muller, PhD, Department of Biostatistics, University of North Carolina, Chapel Hill, USA

Claudio R. Nigg, PhD, Department of Public Health Sciences and Epidemiology, University of Hawaii, Honolulu, Hawaii, USA

Neville Owen, School of Population Health, University of Queensland, Brisbane, Queensland, Australia

David A. Rowe, PhD, Department of Exercise and Sport Science, East Carolina University, Greenville, North Carolina, USA

James F. Sallis, PhD, Department of Psychology, San Diego State University, San Diego, California, USA

Raymond D. Starling, PhD, Pfizer, Inc., New London, Connecticut, USA

Scott J. Strath, PhD, Department of Physical Medicine and Rehabilitation, University of Michigan, Ann Arbor, Michigan, USA

Jerry R. Thomas, EdD, Department of Health and Human Performance, Iowa State University, Ames, Iowa, USA

Katherine T. Thomas, PhD, Department of Health and Human Performance, Iowa State University, Ames, Iowa, USA

Margarita S. Treuth, PhD, Center for Human Nutrition, Johns Hopkins University, Bloomberg School of Public Health, Baltimore, Maryland, USA

Weimo Zhu, PhD, Department of Kinesiology, University of Illinois at Urbana-Champaign, Urbana, Illinois, USA

Index

About the Editor

Gregory J. Welk, PhD, is an assistant professor at Iowa State University in the area of physical activity and health promotion. He is an extensively published author and researcher in the field of physical activity assessments. His work with children and adults in both laboratory and field studies makes him uniquely qualified to address the challenges of collecting, processing, and interpreting data on physical activity.

Dr. Welk previously worked at the Cooper Institute for Aerobics Research, where he studied the broad applications of public health and epidemiological research and gained a deep understanding of the importance of accurate assessments to advance research. Additionally, he is an active member, both regionally and nationally, of the American College of Sports Medicine; the Alliance for Health, Physical Education, Recreation and Dance; and the Society for Public Health Education.

In his free time, Dr. Welk enjoys aerobic activities such as running, biking, cross-country skiing, and swimming, as well as golf and hiking. He makes his home in Ames, Iowa, with his wife, Karen, and their four children.

*You'll find
other outstanding physical
activity assessment resources at*

www.HumanKinetics.com

In the U.S. call

800-747-4457

Australia 08 8277 1555
Canada 800-465-7301
Europe .... +44 (0) 113 255 5665
New Zealand 09-523-3462

HUMAN KINETICS
The Information Leader in Physical Activity
P.O. Box 5076 • Champaign, IL 61825-5076 USA